The Morrow Book of
AMERICAN RESORTS

The Morrow Book of AMERICAN RESORTS

Edited by

THOMAS TRACY

WILLIAM MORROW AND COMPANY, INC.
NEW YORK

Editorial Director: Eunice Riedel
Project Editor and Writer: Thomas Tracy
Staff Writers: Thomas A. Dreyer
Iver J. Kern
Jay B. McInerney
Sara Pyle
Assistant Editor: Wick Sloane
Editorial Assistants: Adair A. Brown
Lenore M. Johnsen
Cynthia M. Winley

Library of Congress Cataloging in Publication Data

Tracy, Thomas.
The Morrow book of American resorts.

1. Resorts—United States—Directories. 2. Recreation—United States—Directories. 3. United States—Description and travel—1960- —Guide-books.
I. Title.
TX907.T588 917.304926 80-23882
ISBN 0-688-03692-2
ISBN 0-688-08692-6 (pbk.)

Printed in the United States of America

First Edition

1 2 3 4 5 6 7 8 9 10

HOW TO USE THIS BOOK

Many early resorts in America were modeled on the European concept of the Grand Hotel—mammoth enclaves, often inspired by Mediterranean villas and palaces, for the rich and socially prominent. Such showplaces still exist, though they are no longer the exclusive precincts of the wealthy, and are included in this book. But with sublime landscape offering unique canyons and deserts, extraordinary shorelines, rain forests, volcanoes and hot springs, barrier beaches, forested mountain ranges, thousands of islands, and vast wilderness, Americans developed far more rustic, action-oriented, and independent modes of vacationing. Whether you want to rough it or be pampered, you'll find an amazing choice of dude ranches, dignified country estates, sleek high-rises, simple and fancy log cabins, modest inns, adobe cottages, primitive fishing camps, plush villas, timbered lodges, resort motels, and entire condominium villages. With such a range of styles and prices, it would be safe to say that almost any person or family could enjoy a vacation at dozens of resorts described in this book. The only problem is picking the right place.

Size: In the descriptions below, we have indicated the number of accommodations each resort offers, so that you can determine whether the establishment is small and intimate or a large-scale enterprise.

If you are looking for privacy, however, resort capacity does not tell the whole story. Some small resorts cluster accommodations close together so that you have less isolation than in a huge complex where small lodging units may be widely scattered on spacious grounds.

If you are less concerned with resort capacity than with room to roam, you will find our descriptions provide helpful details on especially large acreage or beach frontage, or proximity to extensive public lands.

Sports facilities: For your convenience in quickly locating resorts with your favorite activities, at the end of each write-up we have assigned a bold-face letter if the resort is particularly good for tennis (**T**), golf (**G**), fishing (**F**), downhill or cross-country skiing (**S**), horseback riding (**R**), hiking (**H**), or watersports (**W**). We have given these keys only to places that *feature* these sports—not to those that happen to have, say, a tennis court or putting green incidental to other major activities. And the keys are simply for quick reference, and do not represent quality ratings.

Rates: All resorts in this volume have been categorized as **Inexp** (Inexpensive), **Mod** (Moderate), or **Lux** (Luxury) in bold type to help you find your price range quickly. We do not quote specific prices for several reasons: Each season, rates usually change, but the general categories remain the same. All rooms are not always the same rate, and rates can differ depending on whether you are booking in midweek, on a weekend or holiday, on long or short term, or on a package. Finally, many resorts cannot forecast their rates accurately much in advance. Innkeepers will send an updated rate card on request, a service that this book could not hope to duplicate by listing the prices current at the time we went to press.

We have assigned categories based on double occupancy per night in the peak season, and we have taken meal plans into consideration. For establishments that will not rent for just a single night, we divided the cost of the minimum stay by the number of nights.

AP, MAP, or EP? American Plan (AP)–sometimes called Full American Plan–means that all three meals are included in the rates. AP once used to be the rule at resorts and is still found at the more traditional establishments, but it is most frequent at isolated places such as ranches where no alternative food service is available. Most resorts will substitute a box lunch if you wish to be away all day, but few give rebates for meals missed.

Modified American Plan (MAP) means that breakfast and dinner are included in the rates. This plan is often used at places such as ski areas where guests are expected to be away during the day. A few MAP resorts offer no luncheon service; others have box lunch, snack bar, or restaurant options that you pay for separately.

European Plan (EP) means that rates for lodging do not include meals, though a few places offer complimentary coffee or continental breakfast. This plan is found often at motels and where accommodations include housekeeping facilities.

Resorts do not always offer the same plan year round, and many offer a choice of two or three plans.

What is included in the rates? In general, swimming pools, beaches, lawn games, recreation rooms, fishing docks, and other facilities that are relatively easy to maintain are included in the AP, MAP, or EP rates. Frequently, as mentioned in our descriptions, resorts feature a number of extras that are also available at no extra charge–perhaps a boat cruise, saunas, dance bands, sports instruction, etc. While we have not had the space to indicate every free activity at every resort, we have tried to give you an idea of the scope of each place's regular program and facilities.

Where we have not specified that a sport requiring equipment maintenance is free, you can generally count on paying extra. Thus, almost everywhere, unless you are on a package plan that specifically includes them, you will pay extra for greens fees and cart rental for golf, tennis court time, ski lifts, horseback riding and guided rides, motorboats, waterskiing, and other specialized equipment and instruction. In general, the finer these specialized facilities sound, the more you will pay.

There are exceptions. In our write-ups you'll find luxury resorts that include excellent sports facilities at no extra charge—but then the rates reflect this. Quite a number of moderately priced and inexpensive resorts include use of all on-premises facilities in their rates—but then these facilities are unlikely to be of topnotch quality.

Ranches are an exception. Traditionally, these have operated on AP rates that include use of a horse, riding instruction, services of a wrangler, and any other on-premises facilities and activities. Ranches are highly individualistic, however. On some, especially working ranches (where livestock is still raised), you may be offered a couple of guided rides a day plus convivial get-togethers over meals; but otherwise you will be expected to entertain yourself or, if you wish, to join in ranch chores. At the other extreme are ranches that may or may not carry on ranching activities and that schedule a full round of breakfast-to-bedtime activity for adults and children. A few even offer separate rates for riders and nonriders, or even charge hourly for the use of horses and wrangler service. You'll find examples of both extremes, and many in between, among the ranches we describe.

Package Plans: Before booking at regular rates, inquire about packages. Whether they operate on EP, MAP, or AP, almost all of the larger resorts and many smaller ones offer packages at least part of the year. Frequently these represent substantial savings over regular rates. Family packages are common, and are often geared to school holiday periods. Virtually all resorts specializing in golf, tennis, or skiing have package plans that may include court and greens fees, unlimited play, lift tickets, or perhaps instruction. Some resorts also offer special bargains to encourage patronage during slack periods.

Consider the total cost of each package, including transportation to and from the resort. Some packages include airfare, and others you can combine yourself with airline bargain fares to particular destinations. Given gas prices and airline bargains, you can frequently have a cheaper vacation by flying to a far-off resort that offers a good package than by driving to a resort closer to home.

Children: Though almost all the resorts listed in this guide accept children, and almost all can arrange for babysitters, the descriptions below mention those that make a particular effort to encourage family patronage. Some are geared toward families that enjoy privacy, while others specialize in communal singalongs and other group activities. Families who enjoy spending a great deal of time together will find informal cottage colonies, housekeeping condominiums, and hotels where activities are varied enough for the entire family to explore together. On the other hand, parents who want some time to themselves will find establishments that include in their regular rates part-time or full-time supervised programs for children. At other resorts, day camps and other child-care facilities are available by the hour, day, or week at an extra charge. Where a resort is organized around skiing, tennis, or other skill, we have noted those with instruction programs designed for children.

Saving money: As mentioned above, our rate categories are based on high-season prices. If you vacation off-season you will often find startling price differences–rates can drop as much as half. Ski resorts are apt to be especially good buys in summer for those who enjoy mountain vacations. And in weather too cool for swimming, beach resorts are bargains for tennis, golf, riding, and other sports.

Families should note that our rate categories are figured on the basis of *two* people. Very often at EP establishments there is no extra charge for children sharing your room (for this purpose a child may be variously defined as under 12, 14, 16, etc.). Thus a room with two double beds that we judged at the luxury level for two might be cheaper for a family of four than our rating would indicate. Most resorts will also move a crib or cot into your room for a small fee. AP and MAP resorts also usually offer children's rates that could make your total vacation costs lower than our categories indicate. The amount of reduction granted children, and the ages for reductions, vary so enormously that you may have to compare rate sheets from a number of resorts to come up with the best buy for your family.

Many housekeeping facilities rent by the unit rather than by the number of people occupying the premises. Thus a cottage or apartment we have rated as moderately priced for two adults could easily work out to inexpensive when shared by a family or group of friends. Even an expensive-sounding house with several bedrooms, if shared by friends, can work out to be far cheaper per head and more luxurious than standard motel rooms.

Though elaborate resort layouts sound irresistible, consider whether you will really take advantage of the many facilities and organized activities. If you intend to laze, a simpler and cheaper resort might be all you need.

Resorts in this volume include many that are highly specialized and offer topnotch golf, tennis, and other sports. But because greens fees, lift tickets, boat rental, riding fees, and tennis court time can boost your vacation costs considerably over the basic resort rate, it is sensible to analyze your degree of skill and enthusiasm in advance of making vacation bookings. If you are truly dedicated, so that your vacation happiness depends on championship quality, then head only for first-class facilities. Saving money elsewhere will likely not outweigh the opportunity missed for superb sport and learning experience. To get the best buy, comparison shop for packages, unlimited options, off-season specials, and other bargains at the prime sports centers rather than settling for second best.

On the other hand, if you don't take your sports too seriously, you may be wasting money (and may even feel uncomfortable) in highly professional and perhaps competitive surroundings. Greens fees at top courses can be double, triple, and even quadruple what they are elsewhere. Tennis, skiing, and riding costs can also vary enormously from resort to resort. So if your sports ambitions are modest, tailor your vacation choice appropriately.

Energy-saving vacations: The term "destination resort" has been coined to describe the all-inclusive vacation village or full-service resort where all activities and needs are provided on the premises. Whether you drive in or fly in and are picked up at the airport, you will not need a car for the duration of your stay. Facilities are usually concentrated within walking distance, and bicycles are available for any roaming you want to do. If the golf club, ski lift, or any other activity is a bothersome distance from the center, a shuttle bus or other transportation is provided, usually gratis.

However, we have avoided the term destination resort in this book because it applies equally well to many older types of resorts. From a practical point of view, ranches and many traditional hotels (built in the days before auto travel) are also often destination resorts. In fact, if your sole vacation activity is fishing, a waterside primitive fishing camp can be a destination resort for you.

Accommodations: If you cherish any illusions that you know the definition of villa, chalet, cottage, town house, or cabin, such illusions are likely to be quickly dispelled by reading resort literature. In resort terminology, any of these terms may be used to describe an apartment or suite (in extreme cases, only a room) in attached or detached buildings; and these terms offer very little clue to architectural style, degree of simplicity or luxury, or amenities. In the text we have attempted to provide clear descriptions, but after you have narrowed your choices it's a good idea to write for brochures in hopes of seeing photographs of exactly what you'll get.

Housekeeping facilities: If you are renting a cottage, private home, apartment, condominium unit, or other lodgings with kitchen or kitchenette, inquire in advance about exactly what is included—or you may be unpleasantly surprised to find no pots or no linens. Policies also vary widely on whether maid service is provided, and often this is optional at additional cost.

Minimum stays: Resorts operated on EP are least likely to require a minimum stay, while ranches and housekeeping cottages are most likely to require that you register by the week. There are all sorts of variations, from three-day minimums over weekends to variable minimums over holidays and at peak times. Although many establishments have no rigid official policy on length of stay, you may have trouble getting short-term reservations in high season when you are competing against long stayers. Conversely, a resort that has a minimum-stay requirement in high season may be delighted to accept you by the night off-season.

Conventions and groups: Most of the larger resorts need conventions and group bookings to survive, but conventions are usually scheduled off season and frequently have their own dining or even lodging areas to minimize contact with independent guests. If you want to avoid conventions, ask whether any are booked before making reservations.

Reservations: When you make your choice of resort, reserve as far in advance as possible. Many resorts book their choicest rooms as much as a year in advance, and some are fully booked long before their high season opens. However, should you decide on a vacation at the last minute, it's always worth a call to see if there's a cancelled reservation you might fill. In midweek and off-season, even the most popular resorts are apt to have vacancies.

Many resorts require a deposit for reservations, some or all of which may not be refundable if you cancel. Most motels and larger resorts accept credit cards, but ranches and small resorts frequently demand cash or checks. Some places accept checks for a reservations deposit but not on checking out.

Especially at large resorts with varied accommodations and complex rate scales, it's a good idea to ask for written confirmation of your reservation.

Arrival and departure: If you ask in advance, many resorts will arrange for transportation to and from the airport, rail station, or bus stop. This is true even of ranches and other small establishments, where the owners will frequently pick you up themselves. Sometimes this service is complimentary.

Operating season: Dates given in the write-ups below are inclusive, but operating seasons are often subject to slight variations. Many places, such as those catering to a skiing clientele, have flexible dates dependent on the weather. Others vary on the whim of the owners. Operating seasons refer only to lodgings. In most cases the restaurants are open the same season, but occasionally small places curtail hours or close their dining rooms in slack periods. Outdoor pools, golf, tennis, riding, nightclub entertainment, and other facilities may also vary seasonally, and independent of the lodging dates.

Driving direction abbreviations: We have given the state route (SR) numbers, federal routes (US), and interstates (I-), and the number of miles (mi) and direction (N, S, E, W) from prominent landmarks where necessary.

Tipping: The general rule is to tip 15% of the cost of the services you receive, and to distribute this among the employees who have personally assisted you. Minor variations occur, depending on the region of the country and the price category of the establishment. In rural and modest lodgings, it may be customary to tip closer to 10%. In luxury resorts with fine service and many employees to be tipped, you may run closer to 20%. The amount you tip is up to you, but there is no need, even in the most posh places, for your total tips to exceed 20% of your final bill.

Resorts (and independent restaurants) are increasingly replacing tipping with a service charge (usually 8% to 15%) that is added to your final bill. In this case you need not tip anyone except for any employees who have gone out of their way to be helpful.

The only places where tipping is not expected are motels and some small family-owned resorts. Owners are never tipped, and sometimes, to maintain a family feeling, they discourage tipping of their employees. In such a case, if you feel you must show gratitude, a letter of thanks or an inexpensive gift sent after you are home is appropriate.

When do you tip? Employees who perform a *single* service—such as bellboys or hairdressers—are tipped on the spot. Thus, if you are on EP, you would tip at the end of each meal. Employees who perform *repeated* sevices are tipped at the end of the week or at the end of your stay. If you are on AP or MAP, this would include your waiter or waitress and your chambermaid. It is a good idea to place tips for a long stay in labeled envelopes that can be left at the desk in case particular employees are off duty when you check out.

Liquor: Unless we have specified that a resort has a bar, bring your own liquor if you desire it. Some resorts are so isolated that runs to a package store can be inconvenient, and others are in dry areas. In a few parts of the country resorts circumvent liquor prohibitions by calling their bar a "private club" and issuing you free club membership when you register. Another ploy is the "bottle club," where you buy a bottle or bring your own and the bartender mixes your drinks from your own supply.

Hunting and fishing: Write ahead to the resort about seasons, fees, varieties of fish and game, and whether licenses can be obtained on arrival or where you should write in advance for license applications.

Pets: Very few resorts accept pets. A few that do accept pets may charge a discouragingly high daily fee for them.

About this book: Decisions to include or exclude a particular place are based solely on the opinion of the editors. No fees or other remuneration and no free lodgings or meals have been solicited or accepted for inclusion of establishments in this volume.

Descriptions of facilities and conditions are as accurate as we could make them at press time, and we apologize for any inadvertent errors or for any changes that may have taken place between our press time and your arrival.

If you don't see your favorite place in this book or have a disagreement with our description, please let us know by sending your tips and quibbles to us c/o William Morrow & Company, Inc., 105 Madison Avenue, New York, NY 10016.

Other options: In addition to the many places described in this guide, more than 450 country inns, ranches, and other vacation options are described in our companion volume, *The Morrow Book of Havens and Hideaways: A Guide to America's Unique Lodgings,* by Thomas Tracy and James O. Ward.

CONTENTS

HOW TO USE THIS BOOK	5
ALABAMA	17
ALASKA	21
ARIZONA	24
ARKANSAS	40
CALIFORNIA	44
COLORADO	66
CONNECTICUT	85
DELAWARE	87
FLORIDA	88
GEORGIA	112
HAWAII	119
IDAHO	129
ILLINOIS	134
INDIANA	137
KENTUCKY	142
LOUISIANA	150
MAINE	151
MARYLAND	160
MASSACHUSETTS	162
MICHIGAN	176
MINNESOTA	188
MISSISSIPPI	206
MISSOURI	208
MONTANA	212
NEVADA	215
NEW HAMPSHIRE	219
NEW JERSEY	227
NEW MEXICO	228
NEW YORK	235
NORTH CAROLINA	252
OHIO	262

OKLAHOMA 266
OREGON 271
PENNSYLVANIA 281
SOUTH CAROLINA 287
TENNESSEE 294
TEXAS 298
UTAH 308
VERMONT 315
VIRGINIA 329
WASHINGTON 335
WEST VIRGINIA 346
WISCONSIN 353
WYOMING 371
INDEX 379

The Morrow Book of
AMERICAN RESORTS

Alabama

ALEXANDER CITY: Willow Point Villas, *Box 272, Alexander City, AL 35010, (205) 234-5021* ☐ *Open all year* ☐ *7 mi S of Alexander City on SR 63* ☐ These low, modern villas, constructed of wood and stone, are adjacent to the Willow Point Golf Course on Lake Martin and near the clubhouse, pro shop, lounge, marina, tennis courts, and beach cabana. Each villa has four bedrooms, five baths, a leisure room, a fireplace, a bar, and a patio. Entire villas or sections of villas may be rented. ☐ **Mod** ☐ **G, T, W**

DADEVILLE: Still Waters, *1000 Still Waters Dr., Dadeville, AL 36853, (205) 825-7887* ☐ *Open all year* ☐ *SR 49 5 mi S of jct US 280* ☐ Still Waters is a residential resort community with contemporary houses set in wooded lots on the eastern shore of Lake Martin. Guests may rent two-bedroom villas with two baths, a living room, and a complete kitchen. Many of the villas have sun decks and/or fireplaces. Some are on the lake, others on a hill.

The 6500-yard, par-72 West Golf Course at Still Waters was designed by George Cobb, consultant to the Augusta National Golf Club. Some of the holes play along Lake Martin, while others play across and around several interior lakes. The pro shop stocks equipment and sports apparel for tennis as well as golf. Practice facilities include traps, a driving area, and a putting green. There is a greens fee and charge for cart rental.

Dressing rooms are provided at Lake Martin's sandy beach. A marina offers boat sales and rentals, mechanical and fuel service, a launching ramp, and a grocery store. Lake Martin and Still Waters Lake offer year-round fishing for bass, bream, crappie, and catfish.

Other facilities include a restaurant in the Country Club; riding stables and trails (with rental fees for horses); a swimming pool; two lighted, all-weather tennis courts; and hunting for deer, turkey, quail, and squirrel in the 2500-acre game preserve. ☐ **Lux** ☐ **G, R**

DOTHAN: Olympia Spa Golf and Country Club, *P.O. Box 2052, Dothan, AL 36302, (205) 677-3321* ☐ *Open all year* ☐ *5 mi S of Dothan on US 231* ☐ Facilities at Olympia Spa center on a natural mineral hot spring that daily gushes forth 100,000 gallons of 112-degree water from a fissure 3000 feet deep! The waters flow through a large indoor therapeutic pool with 12 whirlpools and through individual whirlpool baths in the men's and ladies' facilities. To complement the baths are an exercise room, steam and sauna rooms, a masseur, and a masseuse. Nearby are tennis courts. The inn has 96 spacious, modern motel-style rooms, and the dining room specialties include breast of capon au vin and Louisiana shrimp Creole.

The 18-hole championship golf course is one of the best in Alabama. Bob Simmons, student and longtime associate of the late Dick Wilson, designed the course for the enjoyment of amateurs and professionals alike. The course takes advantage of the property's gently rolling hills and four lakes. Adjacent is a complete shop with PGA pro, cart and club rentals, a driving range, dressing rooms, and showers. ☐ **Inexp** ☐ **T, G**

EUFAULA: Lakepoint State Park Resort, *P.O. Box 267, Eufaula, AL 36027, (205) 687-8011* ☐ *Open all year* ☐ *5 mi N of Eufaula on US 431* ☐ Lakepoint State Park is a bargian. Rates in the recently constructed lodge and cabins are exceptionally reasonable considering the range of activities available. Besides acres of pine forest, guests have an 18-hole golf course, lighted tennis courts, indoor and outdoor pools, and Lake Eufaula at their disposal. The 45,000-acre lake has a marina, a watersports area, and excellent fishing for bass, large crappie, white bass, and other species.

In the lodge are 89 rooms, many with private balconies overlooking the lake, and six suites consisting of a bedroom, parlor, and a bar. In addition, there are 15 single cabins and 7 duplex cabins. The single cabins are quite roomy; each has 2 bedrooms plus a living room with a fireplace, a kitchen, and a bath. The duplex cabins are immense. Each has twice the accommodations of the single cabins, including 2 baths and 2 kitchens. ☐ **Inexp** ☐ **W, T, G**

GULF SHORES: Gulf State Park Resort, *P.O. Drawer K, Gulf Shores, AL 36542, (205) 968-7531* ☐ *Open all year* ☐ *1½ mi NE of Gulf Shores on SR 182* ☐ This modern two-story motel-style lodge sits right on the beach. All the guest rooms (including 18 suites) have picture windows and balconies overlooking the Gulf of Mexico, paneled walls, carpeting, bright furnishings, telephones, and air conditioning. Some have kitchenettes.

The restaurant, specializing in seafood, has a Gulf view and a bar with dancing and entertainment nightly.

There's a heated outdoor pool, ocean swimming at the doorstep, and an 800-foot-long fishing pier that extends into the Gulf. The lodge also has all-weather tennis courts that can be lit for night play. Inland are two freshwater lakes for swimming, boating, and fishing. Boats may be rented. Just north of the lakes is an 18-hole championship golf course. ☐ **Inexp** ☐ **W, T, G**

GUNTERSVILLE: Lake Guntersville State Park Lodge, *Star Route 52, Guntersville, AL 35976, (205) 582-2061* ☐ *Open all year* ☐ *6 mi NE of Guntersville on SR 227* ☐ Located in the Tennessee Valley, on a bluff overlooking 70-acre

Lake Guntersville, this is an unusually attractive buy for families. Rates are moderate, a bargain for what you get, and weekly rates and golf plans are available.

The two-story lodge offers modern rooms with paneled walls, individual temperature controls, telephones, wall to wall carpeting, picture windows, and balconies or patios overlooking either woodlands or the lake. Along the ridgetop are luxury chalets (these sleep as many as six) designed to blend into the wooded landscape. All chalets have picture windows, wooden decks, two bedrooms, combination living-dining-kitchen area, bath, fireplace, carpeting, color TV, and telephone. Slightly cheaper lakeside cottages without fireplaces, TVs, or telephones sleep up to ten. Cottages and chalets are centrally heated, air-conditioned, and fully furnished with cooking and tableware.

Besides a dining room and coffee shop, the lodge has outdoor poolside buffets, box lunches, and a snack bar.

Facilities include swimming and wading pools, an 18-hole golf course with pro shop and putting green, lighted tennis courts, indoor lounging and game areas, a playground, baby sitting service, bicycle rentals, a lake beach, a fishing dock, boat rental, waterskiing, and hiking. Special activities include hayrides, luaus, and entertainment. ☐ **Inexp** ☐ **W, T, G, H**

POINT CLEAR: Grand Hotel, *Point Clear, AL 36564, (205) 928-9201* ☐ *Open all year* ☐ *US 98 23 mi SE of Mobile* ☐ Point Clear first became a resort in 1840 and it remains one of the South's most grand. The Point Clear peninsula is a 500-acre wooded wedge of land, edged with broad beaches of creamy white sand and surrounded by the turquoise waters of Mobile Bay. The present Grand Hotel, the third on the site, is a rustic-looking but thoroughly luxurious building with three two-story wings radiating from the central lobby. It is set near the tip of the point and surrounded by lush, landscaped grounds with moss-covered oaks. In addition to the suites and doubles in the main building, the more modern Bay House has 50 doubles. Those facing the bay have a balcony or terrace. A wide, covered walkway connects the Bay House to the main lobby. Also available are duplex brick cottages, each suitable for four people.

The spacious main dining room, overlooking the bay, has recessed areas for private entertaining. Gentlemen are requested to wear tie and jacket for dinner. Among the restaurant's featured dishes are prime ribs of beef, red snapper, and flounder stuffed with crab meat.

Surrounding the hotel and among the oaks and pines of the peninsula are three nine-hole championship golf courses (with greens fees). One course is designed by Joe Lee; the other two are by Perry Maxwell. The Lakewood Golf Club has a dining room, ladies' and men's lounges, attended locker rooms, and a pro shop. The resident pro, the putting green, and the practice range should help your game, and the electric carts and expert caddies will definitely make it easier to play.

Even if you don't go near the links, the other activities, especially the water sports, should keep you busy. Point Clear has a 750,000-gallon freshwater pool that is 140 feet in diameter. Mobile Bay is warm and enjoyable for swimming most of the year. A yacht basin provides safe mooring for guests' boats and for the resort's fleet. For an hourly rental fee, you can go waterskiing, sailing in a Sailfish or Rhodes 19 sailboat, or deep-sea fishing in the 53-foot Hatteras yacht, *The Billfisher.* A fishing pier provides relaxing saltwater fishing.

In addition to all this Point Clear offers free supervised children's activities during the summer, skeet shooting, 10 Rubico tennis courts (for a fee) supervised by a pro, and bicycle rentals. ☐ **Mod** ☐ **T, G, W, F**

ROGERSVILLE: Joe Wheeler State Park Lodge, *Drawer K, Rogersville, AL 35652, (205) 247-5461* ☐ *Open all year* ☐ *20 mi E of Florence on SR 72, then 4 mi S to park* ☐ On the south shore of 74-mile-long Wheeler Lake, formed by TVA dams on the Tennessee River, this is a state-run, three-story motor lodge with a nice restaurant and modern, attractive rooms. A comfortable place for families, the lodge has many facilities for children and will provide a list of babysitters. Suites and kitchenettes are available, and the park also offers cabins.

Park facilities include swimming and wading pools, a lake beach, tennis courts, an 18-hole golf course, a marina with boat rentals, fishing, picnicking, and playgrounds. The golf course and marina are on the north side of the river, a short drive across the dam.

Things to see in nearby Florence include the W. C. Handy Home and Museum; the old stagecoach stop, Pope's Tavern; and the Indian Mound Museum. Also on the lake, in Decatur, is the Point Mallard Golf Club in Point Mallard State Park, which is open to the public. ☐ **Inexp** ☐ **W, G**

Alaska

Traditionally, Alaska's tourist season has peaked from mid-June through August, but recently facilities have opened for spring and fall travel, and in some areas winter tourism has been introduced. In summer, when rates are highest and freedom of travel greatest, Alaska's backcountry resorts offer sport fishing, canoeing and rafting, hiking, photographic safaris, and "flightseeing." A few establishments have aircraft for their guests to use; others make fly-out wilderness trips part of their regular program. Resorts remaining open in winter lower rates and offer downhill and cross-country skiing and dogsledding.

The 49th state is dotted with small and modest-size resorts that must be reached by plane. Many are family-run operations whose proprietors take a keen interest in the comfort and needs of each visitor. Often a voluminous correspondence precedes the actual trip. This enables the host and guest to establish a personal relationship well before arrival date. Advance planning is particularly helpful when the resort organizes wilderness expeditions by raft, canoe, or plane.

Although expensive by continental-U.S. standards, Alaska's backcountry camps provide the priceless experience of spending time in one of the few unspoiled wilderness areas remaining on this planet. Food, guides, camping equipment, fishing tackle, and supplementary transportation are frequently part of an all-inclusive package, so that after the initial outlay you often have no further concern about finances.

Space permits listing of only a few examples of Alaska's many unique resorts. For a comprehensive list of equally good choices, request the booklet "The Worlds of Alaska" from the Alaska Division of Tourism, Pouch E, Juneau, AK 99811.

KATMAI NATIONAL MONUMENT: Angler's Paradise Alaska, *c/o Wien Air Alaska, 4100 International Airport Rd., Anchorage, AK 99502, (907) 243-4100; Telex 09025317* ☐ In 1912, a series of severe earthquakes around Mt. Katmai led to the eruption of Novarupta volcano. Volcanic ash covered more than 40 square miles of the Katmai area, and Mt. Katmai (also a volcano) collapsed, creating a chasm about a mile square and a half-mile deep. To preserve this region for scientific study, Katmai National Monument was established.

Greenery is gradually reappearing in the ash-filled valleys, and the mountain ranges and vast forests of this 4200-square-mile preserve are laced with rivers and chains of wilderness lakes. No road access to the monument exists, but Wien Air Alaska maintains daily jet flights from Anchorage to King Salmon Airport. From there float-equipped air taxis will transport you to your destination.

As a concessionaire in the monument, Wien Air Alaska operates three remote camps: Brooks Lodge, Grosvenor Lake, and Kulik Lodge. Notwithstanding the inevitable emphasis on fishing because the fishing *is* excellent, these lodges are family resorts with hiking, float trips, birdwatching, and photography. And, by backcountry standards, the accommodations are very comfortable. Because guests must fly in, and therefore must travel light, the lodges offer package deals including round-trip airfare from Anchorage, accommodation, and special equipment and services.

Brooks River Lodge *Open early June to early Sept* ☐ A summer station for National Park rangers, this lodge on the Brooks River has 16 heated, furnished four-bunk log cabins with electricity, screened windows, hot and cold running water, and an indoor toilet. Some cabins face the river; others nestle in the woods. Family-style meals are served in the log lodge, which has a lounge and a circular fireplace. Activities include canoeing and boating, fishing, photography, and visiting old village sites. Bus and "flightseeing" tours are available to the Valley of the Ten Thousand Smokes, the volcanic region of the monument. ☐ **Inexp** ☐ **F**

Grosvenor Lake ☐ *Open mid-June to Sept 1* ☐ The capacity of the two log cabins here is four people each, making this a very small camp, which is the kind that best coexists with wilderness. The cabins, cookhouse, washhouse, and main lodge are set about the lakeshore among birch trees. Grosvenor Lake has plenty of fish, and wildlife is abundant in the surrounding forest. Charter flights farther into the wilderness can be arranged. ☐ **Lux** ☐ **F**

Kulik Lodge *Open early June to mid-Oct* ☐ Somewhat north of the monument boundaries, this lodge provides many comforts in a wilderness of rugged mountains. The fish-filled, pristine lake is 50 yards away. In the log lodge, the dining room and lounge areas center about a huge stone fireplace. The four-bunk cabins all have toilet and washbasin; some have private showers and others share a nearby shower house.

Although eagles, moose, and bear are the stars of a large cast of wildlife abounding in the immediate area, you may venture farther into the wilderness to fish, photograph, or observe flora and fauna. The lodge arranges frequent flyouts to streams or lakes within a 100-mile radius of the camp. Guests may also join guided trips to smaller camps by jet boat or raft. ☐ **Lux** ☐ **F**

MT. MCKINLEY NATIONAL PARK: Camp Denali, *Mt. McKinley Park, AK 99755, (907) 683-2290 in summer; (907) 683-2302 in winter* ☐ *Open June to Labor Day* ☐ Weather is so unpredictable at Mt. McKinley, 17,000 feet from base to summit, that some park travelers come and go without ever catching a glimpse of the mountain. But from Camp Denali, 90 miles into the park's 3000-square-mile interior and above the timberline, your chances of seeing Mt. McKinley are greatly improved, and the tundra wilderness offers many compensations for unlucky atmospheric conditions.

Denali (the original Tanana Indian word for Mt. McKinley) has room for only 32 guests, who stay in wood-stove-heated cabins or tent cabins scattered along a ridge and around a tundra pond so that each accommodation has a commanding view of the mountain. Showers and toilets are located in separate outbuildings; there is no telephone or bar. Home-cooked meals are served in a dining hall, and the chalet-style log lodge, with its open fireplace and extensive Alaska library, provides a convivial place for guests to get together.

The camp's staff will assist you in planning hikes or fishing trips, in finding the best locations to observe and photograph wildlife, and in interpreting the tundra world. Guides and camping equipment are available.

Three times per season, Camp Denali holds a week-long "Wilderness Workshop," an informal educational experience consisting of field observation, nature hikes, and relaxed evening seminar-type gatherings in the lodge. These workshops, as well as the "Sourdough Vacation," are package plans that include meals, equipment, transportation, and guide. Guests may also choose the two- and three-room housekeeping plans, in which case meals and other amenities are extra. ☐ **Lux** ☐ **F**

PORT ALSWORTH: Van Valin's Island Lodge, *Lake Clark, Port Alsworth, AK 99653, (907) 344-9811* ☐ *Open June–Sept* ☐ *Reached by aircraft* ☐ Running a wilderness camp in Alaska is a labor of love, if Island Lodge is typical. The proprietors live with their two children in the roadless wilderness year-round and do without running water eight months of the year. During this off-season, they carry on a detailed correspondence to help upcoming guests plan every aspect of their trip.

At Lake Clark, you may stay put at the 12-person camp to enjoy the lake and surrounding forest. Or you may prefer to use Island Lodge as a jumping-off point for guided fishing trips via float plan, or for canoe, kayak, and raft expeditions. In this case, weather often plays an important role; most people plan two days in camp, but weather conditions may cause you to eliminate or extend this stay. Each individual trip receives knowledgeable attention from the Van Valins, both through the mail and in person after arrival.

Island Lodge is a rustic log house with a dining area and a comfortable living area. You bunk in a cabin near the lodge, and other than having the luxury of hot showers, living conditions focus on basic comforts. The Van Valins serve up ample family-style meals, and everything, including bread and pastry, is cooked on a woodburning stove. ☐ **Mod** ☐ **F**

Arizona

BULLHEAD CITY: Lake Mohave Resort, *Katherine Landing, Bullhead City, AZ 86430, (602) 754-3245* □ *Open all year* □ *N of jct SRs 68 and 95 at Davis Dam* □ Davis Dam impounds narrow, 28,000-acre Lake Mohave in Lake Mead National Recreation Area. Lake Mohave Resort, run under government supervision, has 50 air-conditioned rooms at Katherine Landing, just above Davis Dam. Some of the rooms have a kitchenette and most have patio or balcony. The resort has a coffee shop, open all day long, bar, and a marina that rents all types of craft, from fishing boats to houseboats. The lake provides good fishing for bass, catfish, blue gill, and crappie. The powerplant at Davis Dam is open for self-guided tours. □ **Inexp** □ **F, W**

CASA GRANDE: Francisco Grande Resort *Open all year* □ *P.O. Box FF, Casa Grande, AZ 85222, (1-602) 836-8711* □ *I-8 W of Casa Grande to exit 172; follow SR 84* □ Jutting up nine stories above manicured lawns dotted with palm trees is the semicircular, earth-colored tower of the Francisco Grande Resort. The tower's 64 modern air-conditioned rooms and suites (and four deluxe penthouses with sunken baths) have balconies overlooking the Grande's PGA-rated 18-hole golf course, which has putting green, driving range, and pro shop. More modest accommodations, for one or two persons, are available at the motor inn next to the main structure.

Facilities include tennis on Laykold courts, with access to a ball machine at extra cost, outdoor pool, wading pool, and poolside service. Trail rides into the desert include transportation to and from the stables. During the Christmas season, counselors supervise children. Dancing and entertainment can be found in the cocktail lounges. The restaurant features prime rib and cuts of lamb. □ **Mod to Lux** □ **G, R, T**

LAKE HAVASU CITY: Nautical Inn, *Box 1885, Pittsburg Pt., Lake Havasu City, AZ 86403, (602) 855-2141* □ *Open all year* □ *On McCulloch Blvd.* □ Long, twisting Lake Havasu was formed on the Colorado River by Parker Dam. In 1968 the eminently gray and famous London Bridge was purchased, disassembled stone by stone, and reconstructed here. It now connects Lake Havasu City to an island that has a marina, golf and tennis facilities, a public beach, and even an airport. Adjoining the bridge is an "English Village" with the City of London Arms restaurant. The island and much of the lakefront for some 40 miles south of the city make up Lake Havasu State Park.

At Pittsburg Point on the island, this motel shares all the facilities of the park. Golfers especially benefit because they have the park's 27 holes and the motel's 9 holes at their disposal. In addition, the motel offers a dock, lighted tennis (and pro), coffee shop, lounge, and frequent nighttime entertainment. All of the guest rooms have air conditioning and color TV, and some have patio, balcony, or kitchenette. **Inexp to Mod** ☐ **T, G, F, W**

MCNARY: Hon-Dah Motel, *Box 597, McNary, AZ 85930, (602) 336-4311* ☐ *Open all year* ☐ *At McNary Junction, jct SRs 260 and 73* ☐ On the Fort Apache Reservation, this Indian-run lodge offers 10 motel rooms and 15 well-spaced log cabins. Rooms have TV and telephones, and cabins have kitchenettes. Also in the complex are a coffee shop, bar, package store, and grocery.

This area is great for outdoor types. The North Fork of the White River and Hawley Lake both have good fishing for rainbow and brook trout. Horseback riding and fishing and hunting guides are available. About 20 miles east is Sunrise Ski Area, also run by the White Mountain Apache Tribe. The ski slope, served by three chairlifts, has a 1400-foot vertical drop and a heavy annual snowfall with excellent powder. At the base is a chalet with snack bar and rentals. Accommodations are available at Sunrise Lodge (334-2144), three miles farther east on SR 73, and plans call for more facilities to be built.

The first week in September the tribe holds its annual fair and rodeo, but less formal dances and rodeos are scheduled throughout the year ☐ **Inexp** ☐ **S, F, R**

MESA: Saguaro Lake Ranch Resort, *13020 Bush Highway, Mesa, AZ 85205, (602) 985-1330* ☐ *Open all year* ☐ *From Scottsdale, take SR 87 25 mi NE to Saguaro Lake turnoff, then 6.5 mi SE to Ranch* ☐ In one of Arizona's most magnificent scenic areas, this 25-acre ranch, which claims to be the oldest guest ranch in the state, is set right on the Salt River near Saguaro Lake in Tonto National Forest. It has a spectacular view of the Bulldog and Superstition mountains, made famous in the writings of Zane Grey. Guests stay in a few rooms in the main lodge or in a string of frame cottages with stucco walls and shaded, stone porches. The cottages, some with fireplaces, air conditioning, or kitchenettes, are rustically furnished in early Arizona style, and are set amid lush riverside vegetation and towered over by stark, sheer cliffs. The ranch restaurant specializes in steak fries and luncheon buffets. Beer and wine is available, but bring your own hard liquor, for which the ranch provides set ups.

The ranch has 35 gentle and spirited Western ponies, and more than 100 miles of riding trails. Pack trips can be arranged from Nov. 1 to May 1. In summer, the ranch provides canoeing, rafting, and tubing on the river, as well as shuttle service for float trips. A heated pool, sunning area, and outdoor games are provided in a lawn area bordered with desert blooms. Trout and bass are abundant in Saguaro Lake (½ mile away) where sailing, waterskiing, and boat rentals are available. You can also swim in the lake or river. Your friendly hosts will tell you where to birdwatch or observe animals; will find you babysitters; and will arrange for you to play golf or tennis nearby. The AP rates include your horse. Breakfast and moonlight rides, chuckwagon picnics, and cookouts are regularly scheduled. ☐ **Mod** ☐ **R, F, W**

NOGALES: Rio Rico Resort, *Box 2050, Nogales, AZ 85621, (602) 281-1901* ☐ *Open all year* ☐ *Near jct of Calabasas Interchange and I-19* ☐ Here, set on a beautiful mesa in the Santa Cruz Mountains, clusters of white two- and three-story motel-style units surround an olympic-size pool, wading pool, and lawn game area. To one side are the restaurant (seafood and Mexican dishes are specialties), coffee shop, and bar, as well as lighted Laykold tennis courts. Below the mesa is the 18-hole championship golf course, designed by Robert Trent Jones, with clubhouse rentals and other services. From the back tees, the par-72 course measures 7095 yards. The resort can arrange for riding, skeet and trapshooting, babysitters, and tours to points of interest–including Pena Blanca Lake (12 miles) with fishing, swimming, and boat rentals.

Accommodations consist of double rooms (some sleep three), suites of double bedrooms plus living room with a couch that opens into a double bed, and apartments with two bedrooms plus a kitchen and living room combination. All guest rooms have either an outside patio or balcony, cable TV, and climate control. EP, AP, or MAP, and package plans are available. ☐ **Mod** ☐ **G, T**

PHOENIX-SCOTTSDALE: Nowhere else in America is the concentration of first-class resorts greater. The Valley of the Sun, as the city fathers call the area, has more than 1000 tennis courts, 68 golf courses, and more swimming pools per capita than anywhere else on the planet. Though it may be hotter than 100 degrees in summer, the humidity is low.

Phoenix is the more urban of the two cities; Scottsdale is its suburb. Sophisticated and cultured Phoenix boasts an astounding number of fine museums for a city its size. The Arizona Museum, the Heard Museum, and the Pueblo Grande Museum have collections of Indian artifacts; the Phoenix Art Museum has an impressive collection of Western and Mexican art; and the Arizona Mineral Museum has mining memorabilia and mineral displays. The city also offers the Phoenix Zoo and the Tropic Gardens Zoo, and several gardens, including the Desert Botanical Gardens and the Japanese Flower Gardens. Encanto Park, in the center of the city, has golf, tennis, lawn sports, and picnic areas, and Squaw Peak Park has horseback riding and hiking trails.

Scottsdale features several family entertainment centers, such as Olde MacDonald's Farm children's zoo and Rawhide Old West theme park. But it has significant cultural attractions, such as the Scottsdale Center for the Arts, Frank Lloyd Wright's Taliesin West, and the Paolo Soleri Cosanti Foundation.

Arizona Biltmore, *Box 2290, Phoenix, AZ 85002, (602) 955-6600* □ *Open mid-Sept to Mid-May* □ *10 mi NE of downtown Phoenix, at 2400 E. Missouri Ave.* □ Among the palatial resorts clustered at the foot of Camelback Mountain in the northeastern section of Phoenix, the Biltmore is King of the Hill. Frank Lloyd Wright had much to do with its conception: though he wasn't the official architect, he devised the technique of using concrete blocks (some of them patterned) as the basic building material, and he provided the inspiration for the fundamental design–a four-story sprawl of concrete cubist masses, a blend of Indian and Art Deco styles that imitates the monoliths and mesas of the Southwest.

The tone of the resort, shut off from the surrounding world by lawns, is reserved and dignified: the Biltmore is one of the very finest resorts in the country, and it knows it. But there are splashy touches in the decor: ceilings of pure gold leaf, an octagonal ballroom, and an enormous tiled swimming pool in the grand style. Beautifully planted flower gardens weave in and out of the building's enclosed spaces, and the exposed concrete block walls pleasantly blur the distinction between indoors and outdoors.

There are about 300 rooms and suites in the hotel itself and 14 garden cottages. Rooms are spacious, and the appointments as luxurious and tasteful as you'd expect from a spot that's attracted (and continues to attract) the wealthiest and most famous of guests. Many of the rooms have private balconies or patios, with views of the grounds, gardens, and mountains.

A sports director heads up recreational facilities, which include two excellent golf courses–the flat Links (par 71, 6000 yards) and the relatively hilly Centre course (par 72, 6500 yards)–plus putting green and driving range. There are 19 tennis courts with a full-time pro. The 80-stall stable provides mounts for guests who wish to explore the 26 miles of bridle trails. And near the 95-foot-long pool, with wading pool and poolside service, are complete exercise facilities, with saunas, steam baths, and therapy pool. Indoor and outdoor game areas, a recreation room, and bicycle rentals are also available. Planned activities are numerous, and the hotel offers some supervised play and has a sitter list.

Less athletic types can stroll the spectacular gardens and orange groves, shop in the hotel's arcade, or drink in the attractive bar. The Biltmore's main dining room is deservedly famous; also available are an informal Mexican restaurant, outdoor buffets and barbecues, and snacks and box lunches. □ **Lux** □ **G, T, R**

Carefree Inn, *Carefree, AZ 85331, (602) 488-3551* □ *Open all year* □ *On Mule Train Rd.* □ Much of this resort's attraction is its secluded location at the edge of Tonto National Forest, where it is aloof from the hustle and bustle of Phoenix.

The attractive 200-room, two-story hotel, built around one of the two heated pools and a magnificent garden patio, has a fine mountainside view. Rooms and suites are luxurious and spacious, and the usual amenities, such as color TV, are augmented with extras including private lanais or balconies, refrigerators, and electric blankets. Food is excellent in the two restaurants, and there is a comfortable bar. Carefree boasts a half-dozen tennis courts, some with lights, lots of riding horses (lessons available), bicycles, and indoor and outdoor game areas. Regularly scheduled activities include dancing and live entertainment. Golf totals 36 holes at the difficult Desert Forest course and the slightly easier Boulders course, plus pro, putting green, and driving range. Services include babysitters, supervised play, box lunches when you want to explore the surrounding forest, and a 4200-foot airstrip. ☐ **Lux** ☐ **T, G, R**

Del Webb's La Posada, *4949 East Lincoln Dr., Scottsdale, AZ 85253, (603) 952-0420* ☐ *Open all year* ☐ *Near jct of Lincoln Dr. and Tatum Blvd.* ☐ La Posada is a 30-acre resort with one-story buildings designed in Spanish mission style with archways, courtyards, tile roofs, and adobe walls. Some of the areas are exceptionally beautiful and well-crafted. Every room has a patio, dressing room with vanity and roomy wardrobe, and the usual amenities. The most luxurious accommodations are the Casita Suites. Each has a parlor and a bedroom, and a wetbar with refrigerator. A second bedroom can be added on if desired. A butler's pantry and butler service are provided by special arrangement. Ten Cabana suites are also available. These rooms, each with wetbar and refrigerator, have modern versions of the Murphy bed, which double as wall decoration when raised. The parlor-sleeping area can be joined with one or more additional bedrooms to create a suite.

Two Spanish restaurants and a French restaurant, plus many shops and services, are on the grounds. Although guests here may use all the facilities, including the golf course, within walking distance at Mountain Shadows (see below), La Posada has extensive sports facilities on the premises–including six tennis courts, two racquetball/handball courts, two saunas, massage and exercise rooms, two therapy pools, putting green, and large heated swimming pool. ☐ **Lux** ☐ **T, G**

Del Webb's Mountain Shadows, *5641 East Lincoln Dr., Scottsdale, AZ 85253, (603) 948-7111* ☐ *Open all year* ☐ Set on 90 acres at the foot of Camelback Mountain, this luxury resort consists of low buildings built around two freeform, landscaped swimming pools. The main pool, twice olympic size, is set in a sunken garden terrace with patios for dining, dancing, and relaxing and is almost completely ringed by lanai rooms. Surrounding this ring is a second group of guest rooms overlooking lawns and gardens and available at a more modest cost. To one side of the pool is the elegant continental dining room, which serves three meals daily, including buffets and Sunday brunch. In the evening, there is music and dancing. Adjoining is an intimate bar. Another complex of guest rooms surrounds the second smaller pool at the golf course. These units face either the pool or the fairways. Every spacious room has a private entrance–and lanai, balcony, or patio–separate dressing room with vanity, a color TV, radio and hi-fi music, and individual temperature control; some units have cocktail bars and refrigerators. Rooms can be combined into two-room suites, and a few three-room suites are available.

The golf course here is an executive par-56, 18-holer, and the clubhouse provides pros, rentals, cars, caddies, and other services. The hotel can also arrange for you to play several 72-par championship courses nearby. Near the clubhouse are putting greens and six lighted tennis courts, for which there is no fee. Riding can be arranged nearby. Other facilities here include supervised play and babysitters, lawn games, recreation room, a social director to organize special events, car rental, notary public, and other services. Noted for good service, this resort will pamper you. □ **Lux** □ **G, T**

Hermosa Inn, *5532 North Palo Cristi Rd., Paradise Valley, AZ 85253, (602) 955-8660* □ *Open early Sept–May* □ *Off Lincoln Dr. on 36th St.* □ The oldest part of this inn-resort is the lodge, built in 1935 by Lou Megargee, noted cowboy painter. Constructed of adobe bricks and timbers gleaned from an old railroad trestle, the hacienda-style lodge, with its fanciful turrets, beehive fireplaces, and secret passageways, is just as much a work of the imagination as one of Megargee's paintings. Perhaps the most impressive remnants of Megargee's construction methods are the large beams and wooden ceilings in some of the guest rooms.

Set on seven nicely landscaped acres, the hermosa offers accommodations that range from basic double rooms to whole villas with living room, kitchen, two baths, and two bedrooms. All rooms have private patios with a view of Camelback Mountain or Squaw Peak. Rates are EP, but there is a restaurant and cocktail lounge. The heated pool, playground, and five tennis courts provide ample recreation. Golf and riding can also be arranged and a list of babysitters provided. □ **Mod** □ **T**

The Inn at McCormick Ranch, *7401 North Scottsdale Rd., Scottsdale, AZ 85253, (602) 948-5050* □ *Open all year* □ *Near Jct with Indian Bend Rd.* □ This luxury lakeside resort is buffered from the outside world by the 4200-acre McCormick Ranch. The angular, three-story main structure, suggesting the architecture of Indian pueblos, is on the grassy shore of Camelback Lake. Guest rooms and suites are attractively furnished in contemporary Southwestern style and have huge picture-window views of the lake or the beautiful McDowell Mountains that change color as the sun shifts. Rooms have large private balconies or patios, and suites have refrigerators. Also fronting the lake, and built in harmonious design, are two- or three-bedroom condominiums that may be rented by the day, week, or month. These are super-luxurious–with kitchens, wetbars, fireplaces, huge expanses of glass to take in the views–and are designed for privacy.

The Four Peaks restaurant, overlooking the lake, presents fine food elegantly and with deft service. The Superstition Lounge has nightly dancing and entertainment.

Across the lake are two 18-hole championship golf courses, with putting greens, driving range, and pro shop services. A free ferryboat will take you back and forth across the lake. Also on the grounds are lighted tennis courts, also with pro. Riding and instruction are available nearby. The lake has a fishing dock with sailboats and paddleboats. Behind the lodge is an L-shaped pool with poolside service, and the health club offers saunas and whirlpools. The hotel has bicycles on which you may explore the ranch. There are also the usual lawn games and lounges for card playing or relaxing. A social director arranges cookouts and other events, and you can take tours of wilderness and desert areas. Babysitters are available. The inn, which offers EP, AP, MAP, and package plans, prides itself on friendly service that encourages many return visits. ☐ **Lux** ☐ **G, T, R, W**

John Gardiner's Tennis Ranch, *5700 East McDonald Dr., Scottsdale, AZ 85253, (602) 948-2100* ☐ *Open Oct–May* ☐ *near jct with 59th Pl.* ☐ With 24 hard outdoor courts, professionally staffed clinics, ample facilities, and the latest in teaching equipment, this is believed by many fans to be the best tennis resort in the world. The ranch is set on the side of Camelback Mountain, giving the casas and casitas superb views of Paradise Valley stretching out below. At the edge of the complex of courts is the clubhouse with spacious dining room, oversize pool, lounge, and pro shop.

Although some people do come here to play a little tennis and relax around the pool, most join the intensive clinics to improve their game. The ball machines and instant-replay television systems are used to help both novice and advanced players. Clinic rates include lodging, all meals, saunas, two massages, whirlpool, and 22 hours of tennis instruction.

Guest accommodations are in casitas–two-bedroom houses with living room, kitchenette, fireplace, two baths, and private balcony–or in casas–deluxe accommodations, each with private tennis court and pool. A single bedroom or a bedroom plus living room may be rented in a casita. ☐ **Lux** ☐ **T**

Marriott's Camelback Inn, *5402 E. Lincoln Dr., Scottsdale, AZ 85252, (602) 948-1700* ☐ *Open all year* ☐ *Near jct 44th St.* ☐ This 120-acre resort is nestled against Mummy Mountain, not Camelback Mountain, which it overlooks. It includes within its luxurious provinces 413 rooms and bi-level suites in two-story white buildings with red-tile roofs. These are surrounded with gardens and the fairways dotted with palms and reflecting pools. Each unit has an outside entry, patio, nice extras such as refrigerators, and a view of the mountains. In addition to the Chaparral restaurant, you can dine Western-style in the Navajo Room, or enjoy a poolside buffet or a starlit steak fry on Mummy Mountain. Lounges include one at poolside and La Cantina for dancing and entertainment. You can settle in here on AP, MAP, EP, or at family rates.

The beautiful golf courses add up to 36 holes, with putting greens, driving range, pro, and clubhouse services. There are lighted tennis courts, also with a pro. Horseback riding and instruction are available. Also here are two large swimming pools, whirlpools, bicycle rental, and the usual lawn games. Shops include beauty salon, barbershop, gift boutique, and coin laundry. Parents will find babysitters, and during certain holidays an all-day supervised play program is operated.

During winter months a social director supervises activities. Extra services here include a complimentary bus to a shopping mall, theater tickets, tours to local points of interest by Jeep or limousine, and even air tours to places as far as the Grand Canyon. ☐ **Lux** ☐ **G, T, R**

The Pointe, *7677 N 16th St., Phoenix, AZ 85020, (602) 997-2626* ☐ *Open all year* ☐ *Near Squaw Peak Park* ☐ This huge, still-developing resort, convention center, and residential community sits on 2700 acres of desert foothills in the heart of the Phoenix-Scottsdale metropolis. Each of the two identical lodging wings, designed in a Spanish style, rises four stories above three interior courtyards. Room balconies overlook these courts, creating an intimate community atmosphere. In the two largest courts are greenery, stone fountains, walkways, and swimming pools that are heated and have whirlpool areas, underwater music, and swim-up bars. The four smaller rear courtyards have sunken gardens with Mexican-tile walkways and putting greens.

All 582 guest units are suites with refrigerators and wetbars, which will be stocked at your request. In-room steam baths and executive suites with conference areas are available. If you want more space, the Pointe offers villas with a den, wetbar, dining area, and bath on the lower level, and one or two bedrooms, bath, and sundeck on the upper level.

The central dining facility is the Pointe of View, a building tiered up the side of a hill to take advantage of the fine view of the Phoenix skyline. Here are airy dining rooms, cocktail areas, and a discotheque. Near the riding and racquet club is the Hole in the Wall, an informal Western-style restaurant featuring mesquite-broiled steak and barbecue ribs. There is also a cookout area. Additional restaurants and bars may have opened by your arrival in an area with shops designed to resemble a Mexican market.

The riding and racquet club has eight lighted tennis courts, pro shop, swimming pool, racquetball/handball courts, and a stable with 100 horses. The riding, jogging, and hiking trails meander through the 2700 acres of the mountain preserve. Guests enjoy privileges at the 18-hole course at nearby Orange Tree Country Club. Complimentary transportation is provided. ☐ **Lux** ☐ **T, R**

Ramada Valley Ho, *6850 Main St., Scottsdale, AZ 85252, (602) 945-6321* ☐ *Open all year* ☐ *In downtown Scottsdale* ☐ For guests who prefer to be in urban Scottsdale and still have immediate access to some sports facilities, the 289-room Ramada Valley Ho, with restaurant, lounge, and the usual services, might be a solid choice. The hotel is built around a verdant central courtyard with broad lawns, date palms, and shrubs. Most of the rooms have balconies overlooking the courtyard and one of the three pools. There are also three tennis courts and a whirpool bath at the hotel. Included in the rates are complimentary transportation and charge privileges for golf at the Scottsdale Country Club. □ **Lux** □ **G**

The Registry Resort, *1717 North Scottsdale Rd., Scottsdale, AZ 85253, (602) 991-3800* □ *Open all year* □ *Corner of Scottsdale and Indian Bean Roads* □ Composed of 39 one- and two-story units on a 45-acre site, the Registry is built around a core consisting of a million-dollar kitchen surrounded by a huge convention ballroom. Connected to this building is a L-shaped wing containing 120 spacious guest rooms and suites, many with a wet bar, refrigerator, balcony or patio, and mountain view. Behind the main building are 203 villa rooms ranging in size from doubles to two-level suites. Lodgings are unusually attractive and comfortable.

The Terrace Buffet, with flower beds of geraniums, begonias, and mums, faces the central court with its olympic-size pool. Here guests can enjoy breakfast, lunch, or the Sunday brunch. The informal Brioche, with wood paneling, wicker furniture, and potted plants, is for homemade breakfast pastries or light lunches and dinners. The formal and richly decorated La Champagne has only one seating for dinner that might include roast quail in champagne, or abalone topped with lobster mousse and lobster champagne sauce. The Phoenician Room is a three-level Las Vegas-type dinner theater showcasing big-name stars. The Kachina Lounge, decorated with kachina dolls, Navajo rugs, and Papago baskets, is a special pride.

Recreation facilities include three pools, 21 lighted tennis courts, bicycles, lawn games, and a health club. Privileges are extended to two adjacent golf courses with pro services and putting greens. A social director sets up horseback riding, sightseeing tours, and other activities on or off the premises. EP, AP, and MAP are available at this superior resort. □ **Lux** □ **R, G**

Rio Verde Resort and Ranch, *25800 Four Peaks Blvd., Rio Verde, AZ 85255, (602) 991-3350* □ *Open Nov–June* □ *from Scottsdale take Scottsdale Rd. N to Pinnacle Peak Rd., then E past Reata Pass to resort* □ The guest villas here are set around the Rio Verde Country Club in Tonto National Forest, a full-recreation club offering horseback riding, fishing, tennis, and swimming as well as golf. The two- and three bedroom villas are done in classic Southwestern style with white-stucco exteriors and red-tile roofs. All have all the usual amenities, plus washer and drier.

The club's 18-hole championship course is laid out through forests of palo verde, mesquite, and saguaro. The par-71 course, which measures 6458 yards from the back tees, has been the site of major tournaments, including the Arizona Open. A new course is to be constructed soon and there are plans for a third course. In a mountain valley, the resort enjoys a cooler average temperature than the Phoenix-Scottsdale area and a more relaxed pace. Opportunities for sailing or power boating are available a few minutes away □ **Lux** □ **G**

Scottsdale Country Club, *7702 East Shea Blvd., Scottsdale, AZ 85260, (602) 948-6000* ☐ *Open all year* ☐ *From Scottsdale ctr, take Scottsdale Rd. 5 mi N to Shea Blvd.* ☐ Here 67 one-story motel-style units have picture windows and private patios overlooking the fairways and lakes of the well-kept, 6136-yard, par-70 golf course. The lodging units, furnished with a ranch flavor, have color TV and climate control; some have kitchenettes. The Road Runner, a cheery and informal restaurant, features Western-style lunches and dinners, and there is a cocktail lounge. On the 180-acre premises are a full-time pro, a driving range, a pro shop, an outdoor pool, and two tennis courts. Babysitters and reciprocal agreements with nearby resorts are among the services. ☐ **Lux** ☐ **G**

The Wigwam, *Litchfield Park, AZ 85340, (602) 935-3811* ☐ *Open mid-Sept to early June* ☐ *Near jct of Litchfield Rd. and West Indian School Rd.* ☐ One of the finest and loveliest resorts in the nation, the Wigwam, operated on American Plan, offers outstanding facilities on 70 acres beautifully landscaped with ornamental trees and shrubs. The property is buffered by an additional 14,000 acres controlled by the Wigwam estate. Originally developed in 1918 for executives of the Goodyear Tire and Rubber Company, the resort is built entirely in Territorial style. The Main Lodge houses the public rooms, expertly decorated to avoid an institutional look—the elegant terrace dining room with dinner dancing, the cozy cocktail lounge and bar, and tasteful lounges for TV, games, movies, or relaxing before a roaring fire. Just outside is the freeform swimming pool, where buffet lunches are served outdoor. The schedule of activities often includes desert breakfast rides or steak broils.

More than 400 guests can be lodged in two-story Territorial buildings set in clusters among manicured lawns, tall palms, and well-kept gardens of colorful desert blooms. The twin-bedded rooms and one- to three-bedroom suites with separate parlors all have an entrance from a private patio or terrace, climate control, and refrigerator; many have dressing rooms or fireplaces. Executive suites and four-bedroom casas are also available.

The Goodyear Golf and Country Club has three 18-hole courses—two by Robert Trent Jones. The Gold, a 7200-yard, par-72, is notoriously clever and said to be one of America's best. All the courses are impeccably maintained and have snacks and beverage rest areas. The clubhouse, with tartan rugs, offers lunch and dinner in comfortable rooms, cocktail lounge, and locker rooms. Adjacent are golf and tennis pro shops. The Tennis Centre has eight Plexipave courts (six lighted), a lounge, pros, practice lanes, pitching machines, and other teaching aids.

Stable horses may be rented for rides or all-day trips, and stagecoach, hay rack, and breakfast rides are offered. Lawn games, trap shooting and skeet and indoor game areas are available. Provision is made for children with playgrounds, supervised activities, and special rates and family packages. The Wigwam prides itself on fine service and offers many extra touches and conveniences from morning papers to limousine service. ☐ **Lux** ☐ **T, G, R**

SEDONA: Surrounded by Coconino National Forest, this town thrives as unofficial headquarters for Oak Creek Canyon, which extends 16 miles north of the city and is one of the most beautiful areas of Arizona. This is spectacular landscape, known locally as Red Rock Country, where many Westerns have been filmed. Near the river, 1200 feet down, is lush greenery that provides breathtaking contrast with the jagged rock formations that change from shades of pink and lavender to deep oranges and reds, depending on your viewpoint and the angle of the sun. The Forest Service maintains a dozen recreation areas in the canyon, and the Sedona Chamber of Commerce provides maps and information on the area. Jeep tours of the backcountry, trail rides with outdoor cowboy cookouts, sightseeing plane trips, and guided natural history hikes are offered in town. Sedona also has a Robert Trent Jones golf course open to the public, public tennis courts, and a Mexican market. Oak Creek is regularly stocked with trout, and the canyon is a favorite with birdwatchers.

Best Western Poco Diablo Resort, □ *P.O. Box 1709, Sedona, AZ 86336, (602) 282-7333* □ *Open Apr–Nov* □ *2 mi S of Sedona on SR 179* □ On five acres at the foot of towering red cliffs, this EP complex has a central building with dining room and cocktail lounge with dancing and entertainment. Flanking this are two-story lodging wings with 22 singles and 34 doubles. The 33 villa suites feature fireplaces, wetbar, hot plates, and refrigerators, and some have whirlpools (honeymoon villas have two-person sunken whirlpools). Some units have private patio or balcony, and all have good views; villas overlook the creek or the fairways.

Included in the rates are two heated outdoor pools with poolside service and whirlpools, four tennis courts (two lighted), and two sports courts. A pro shop has rentals and lessons. The par-three, nine-hole golf course, with duck ponds, waterfalls, and weeping willows, has a moderate greens fee. The creek has fishing and swimming, the grounds are good for birdwatching and have nature trails, and the backcountry Jeep tours will stop here to pick you up. Poco Diablo will also arrange for babysitters or for riding (two miles away). □ **Mod** □ **G, T, F**

Rondee Best Western, *P. O. Box 1021, Sedona, AZ 86336, (602) 283-7121* □ *Open all year* □ *I-17 N from Phoenix, then N on SR 179* □ For a spectacular view of Oak Creek's Canyon, Rondee can't be beat. Each of its 40 motel rooms has a balcony on which to muse over this beautiful vista. The motel has a good restaurant with candlelight dining and Greek specialties, a bar, a pool, and a whirlpool. Two trails lead down into the canyon, where you may fish. The motel maintains bird feeders and has a bird list. A good base for the Red Rocks area. □ **Mod** □ **F**

SPRINGERVILLE: Amberian Point Lodge, *P.O. Box 135, Greer, AZ 85927, (602) 735-7503* □ *US 60 NE from Phoenix to Show Low, then SR 260 SE to Greer turnoff 373, 1 mi S of post office* □ Here, at the end of the paved road, in

Apache-Sitgreaves National Forest, is a small hideaway for those who like wilderness. On ten acres of hilly pine and aspen forest, with streams and meadows, at the headwaters of the Little Colorado River, is a lodge offering EP accommodations; a 7 am to 10 pm restaurant featuring Mexican dishes, country gravy with biscuits, and steak; and a fireside bar said to be the best stocked in the White Mountains. TV and telephone are availabile in the lodge, which has nine double rooms with paneled walls. A few paneled contemporary cabins have exceptional views; some are on a stream, or have fireplaces, porches, or kitchenettes. One, with two bedrooms and cathedral ceiling, sleeps six. All units have maid service and heat.

Although this lodge has no organized sports facilities, it is surrounded by Forest Service recreation sites and is a good base for hiking, viewing wildlife, participating in Forest Service interpretive programs, and backpacking or riding into beautiful Mt. Baldy Wilderness. For an extra fee, you can choose among 15 horses (some gentle enough for beginners) for trail rides. Streams and lakes provide swimming and fishing, and you can rent boats nearby. In winter, you can go snow tubing or cross-country skiing, or find downhill skiing at the fully equipped Sunrise ski area. For self-sufficient outdoors people, this unpretentious lodge provides more than enough excitement. ☐ **Inexp** ☐ **F, H, W, R, S**

TUCSON: Arizona Inn, *2200 East Elm, Tucson, AZ 85719, (602) 325-1541* ☐ *Open all year* ☐ *Off Cambell Ave., between Grant Rd. and Speedway Blvd.* ☐ Once the wintering place of the Rockefellers, and still a haven for celebrities, the compact Arizona Inn combines Southwestern architecture with genteel elegance. The inn is still owned by the family that built it half a century ago and is restrained in its good taste. The low, pink adobe buildings, attached or closely spaced, are towered over by trees, and are contained within 14 acres of neatly clipped lawns and shrubs, flower gardens, and brick walkways. Both lodge rooms and individual cottages, including the spacious Rockefeller House, are available. Rooms and suites are unusually warm and comfortable in traditional decor.

The public rooms are subdued yet grand. The dining room, white with a high ceiling, is especially nice, and the food is first-class continental. It serves three meals, but there are also luncheon buffets by the pool. Also here are a library, bars, and entertainment.

The Arizona Inn is not so much a sports resort as it is a place to relax and be pampered in tasteful surroundings. However, it does have a large pool and wading pool bordered by a shady arbor and several lighted Har-Tru tennis courts, croquet and other lawn games, putting green, and playground. Golf and horseback riding will be arranged nearby, and the emphasis is on catering to your every whim. ☐ **Lux**

Hacienda del Sol, *Hacienda del Sol Rd., Tucson, AZ 85718, (602) 299-1501* ☐ *Nov 1 to May 1* ☐ *Take Campbell Ave. N to River Rd.* ☐ Surrounded by a plain of scrub and cactus and, in the distance, the barren peaks of the Santa Catalina Mountains, Hacienda del Sol is a virtual oasis in the desert. Tall date palms and orange and grapefruit trees rise above the Spanish mission wall and entrance gate, a sign that the grounds within are well watered. The low, contemporary pueblo units cluster around verdant lawns, fountains, or flower or cactus gardens. Rooms here typically have viga ceilings and Spanish-style furnishings. Lodge rooms and private casitas are available. The poolside patio with putting green and the lounge and dining areas have sweeping vistas of the desert plain and mountains. A short walk from the lodge area are the tennis courts and riding stables, and golf privileges are available. ☐ **Mod** ☐ **R, T**

The Lodge on the Desert, *306 North Alvernon Way, P.O. Box 42500, Tucson, AZ 85733, (602) 325-3366* ☐ *Open all year* ☐ *Near corner of N Alvernon Way and Broadway* ☐ This small, informal complex, designed in Spanish-colonial style, consists of eight low lodges on landscaped grounds. These adobe-walled lodges provide 3 singles, 28 doubles, and 4 suites; some units have kitchenettes or balconies, and about half have fireplaces; all units have tiled baths, large closets, and climate control. Family rates and EP are available all year, and AP in winter, when breakfast is served in your room. The dining rooms and cocktail lounges have Mexican-tile floors, beamed ceilings, and fireplaces, and there is patio dining. The menu is continental, and jacket and tie are required in winter. Barbecues, buffets, and box lunches are offered.

Around the pool are lawns with lounging chairs. A playground and babysitters are provided. Within a mile are 15 municipal tennis courts and the 36-hole Tucson Municipal Golf Gourse. The lodge can tell you about other golf courses and about riding stables. ☐ **Inexp to Mod**

Rancho del Rio, *2800 N. Sabino Canyon Rd., Tucson, AZ 85715, (602) 298-2351* ☐ *Open all year* ☐ *10 mi NE of Tucson* ☐ This four-acre ranch is a charming small resort motel (EP) offering tennis, riding, swimming, and a superb gourmet restaurant that is perhaps the finest in Tucson. The guest rooms have many first-class features, including walk-in closets, remote control color TV, and large picutre windows. With only eight doubles and three suites here guests have ample use of the sports facilities. Besides the heated swimming pool, fed by a waterfall from an old aqueduct, the ranch has a whirlpool and three Plexipave tennis courts with pro. Del Rio stables, on the property, offers horseback riding daily on the hundred acres surrounding the ranch. Trails lead to the nearby Santa Catalina Mountains, Sabino Canyon, and Seven Falls.

The Tack Room restaurant's specialties include fine Provimi-fed veal served with Pfifferlinge mushrooms and rack of lamb bouquetière, carved at the table and served with a bouquet of vegetables. The room's Southwestern decor—beamed ceilings, arched windows, and wrought iron chandeliers—plus its superb food and wine list, should please even the fussiest epicure. ☐ **Mod** ☐ **R, T**

Tanque Verde Ranch, *Route 8, Box 66, Tucson, AZ 85710, (602) 296-6275* □ *Open all year* □ *Speedway Blvd. E to end* □ A historic ranch with resort facilities, the Tanque Verde has the best of both worlds. Guests may spend any part of their days horseback riding, playing tennis on one of the four lighted courts, or simply lounging around the spa with its colorful Mexican motifs, indoor and outdoor pools, saunas, whirpool, exercise room, and lounge.

Although most of the ranch appears to be of recent vintage, old charm prevails in the main building, constructed before the turn of the century. One interesting remnant of the early days is the old-fashioned corral of mesquite logs "woven" between upright posts.

The guest rooms in the lodges and cottages are a pleasant blend of contemporary furnishings and rustic brick walls and beamed ceilings. Some units have patios and fireplaces, and all have climate control. The bright and spacious dining room, overlooking the pool, serves award-winning meals, and there is a bar. The AP rates include riding, cookouts, educational programs on local flora and fauna, entertainment, and other activities.

The riding trails at the ranch stretch out across almost one million acres of the Coronado National Forest and the Saguaro National Monument. This area is a birder's paradise, with an extraordinary variety banded on the ranch. For children, there are supervised activities and playgrounds as well as separate dining facilities during certain seasons. □ **Mod** □ **R, T**

Westward Look, *245 Ina Rd., Tucson, AZ 85704, (602) 297-1151* □ *Open all year* □ *From downtown, take US 89 N to Ina Rd.* □ Set on 67 acres in the foothills of the Santa Catalina range, this resort offers 92 double rooms and one suite in a cluster of interconnected one- and two-story modern lodges. Casita units, near the tennis courts, are cheapest; Posadas are near the pools and have patios and dressing rooms; Fiestas have sitting rooms, dressing rooms, wetbars and refrigerators, and at-door parking. EP, AP, and MAP are available. The Gold Room restaurant has good continental and American cuisine, candlelight, and a patio, and the bar has dancing and entertainment.

This resort is noted for fine tennis facilities and has eight Laykold courts (three lighted), pro shop with rental equipment, and a nationally ranked pro. A brick terrace with poolside services edges the pool and hydrotherapy pool just outside the Gold Room. Also on the grounds are a putting green and playground. The resort can arrange for babysitters, golf on nearby courses, riding and instruction. □ **Mod** □ **T**

White Stallion Ranch, *Route 9, Box 567, Tucson, AZ 85704, (602) 297-0252* □ *Open Oct 1 to early May* □ *On Twin Peaks Rd., 17 mi NW of Tucson* □ This 4000-acre working ranch takes only 47 guests and offers a somewhat sophisticated, yet totally Western vacation, with rates including room, all meals, horseback riding, two Laykold tennis courts, redwood hot tubs, the heated swimming pool, indoor and outdoor game areas, library, hayrides, barbecues, shopping trips, and all ranch activities. Quarter horses and long-horned cattle are still raised here, and the wranglers hold informal rodeos twice each week. The ranch's isolated spread covers desert at the foot of the rugged Tucson mountains, with good bird and animal watching and rock-hounding. Guests are housed in a string of white one-story cottages among cactus gardens and some tall saguaros outlined against the scrub. These units are spacious, with paneled walls, wood ceilings, wall to wall carpeting, and ranch-style furnishings. Family-style meals are served at long tables in the mission-style dining room. Steak barbecues are held on the patio on Saturday evenings, and complete dinners are often cooked on the woodburning oven outdoors. The saloon is Old West style, and there is a nice covered lounging deck. ☐ **Mod** ☐ **R, T**

TUMACACORI: Rancho Santa Cruz, *P.O. Box 8, Tumacacori, AZ 85604, (602) 398-2261* ☐ *Open mid-Nov. to early May* ☐ *US 89 and I 19 48 mi S of Tucson* ☐ Located along the old pioneer route between Tucson and the Mexican border is the early 19th-century Mission site of Tumacacori and the historic Rancho Santa Cruz. It's a rustic pueblo-style ranch house with 11 double rooms and three larger apartments. There is a heated outdoor pool. Its popularity is less the result of its physical comforts (which are considerable) than the genuine Western hospitality of the friendly proprietors. Guests are said to return again and again due to the ranch's authentic old charm and the beauty of the nearby Coronado National Forest's rugged terrain.

Among the attractions are the trail rides that are included in the AP rates. All-day rides through thc cactus and timber-studded mountains give you a good opportunity to imagine the lifestyle of the early nomadic Apache. Side trips are made to the old Mission, a massive building steeped in history, and to the local artists' colony at Tubac. Also in Tubac is the Tubac Country Club, where golf and tennis are available to ranch guests. Meals featuring Western specialties are served in the ranch's private restaurant. Overall, Rancho Santa Cruz should be considered if you're in search of a taste of the old Southwest. ☐ **Mod to Lux** ☐ **R, G, T**

WICKENBURG: Rancho de los Caballeros, *Box 1148, Wickenburg, AZ 85358, (602) 684-5484* ☐ *Open Oct.-May* ☐ *2 mi W of Wickenburg on US 60, then 2 mi S on Vulture Mine Rd.* ☐ This exceptionally beautiful ranch offers 14 modern bungalows set around a large, rambling main building. The grounds are expertly landscaped: stone walls, rail fences, or desert flowers border the walkways, and along the verdant lawns are rock gardens bristling with large cacti. The low bungalows and main building resemble traditional pueblo buildings. Rooms have viga ceilings and Indian rugs. Some of the guest rooms have the carved and colorfully painted Mexican-style woodwork and furniture that highlight public rooms in the main building. Many of the bungalows have private terraces or glass-enclosed sun porches. Accommodations vary from luxurious Sunset Rooms to less expensive family rooms. The main building has several dining areas, and the high-ceilinged living room has stone floors and a large open hearth with immense copper hood.

Probably the most popular activity is horseback riding. Surrounding the ranch are 20,000 acres of foothills and desert to explore on the all-day, morning, or afternoon rides. There's a small fee for the rides which wranglers supervise. Additional facilities include a heated swimming pool, four tennis courts, skeet shooting (fee), and a beautiful, old-fashioned pool table. Guests have privileges at the Wickenburg Country Club and a complimentary children's counselor will take the kids swimming, riding, and hiking. The ranch has a private airstrip. The AP rates include all meals, cookouts, sightseeing trips, and other special activities. □ **Mod** □ **R, G**

Wickenburg Tennis and Guest Ranch, *P.O. Box P, Wickenburg, AZ 85358, (602) 684-7811* □ *Open all year* □ *Entrance to ranch is off SR 89, N of Wickenburg* □ This is a small place–only 10 rooms in the wood-beamed main lodge and another 41 in the casitas–but its personalized, top-quality accommodations and tennis instruction have given it nationwide acclaim. The adobe casitas are the standouts here. Two-story, with fireplace on the rooftop patio as well as in the downstairs living room, the casitas are done in traditional Southwestern style, with handcrafted furnishings and antiques among the cowboy-rustic decor.

The accent here is on ranch activities and tennis. Guests get around on horseback, and moonlight cookouts with steak, beans, and guitar-strumming cowboys are regularly scheduled. The 11 tennis courts are serviced by a pro shop and a teaching staff equipped with the most modern instructional aides. After and between sets guests can take craft courses in leather working or pottery. A naturalist conducts tours into the resort's desert wildlife refuge to identify the flora and to discover reclusive reptiles and birds. Other facilities include restaurant and bar, swimming pool, lawn games, playground, some supervised play (babysitters are available), a library, and occasional entertainment. Golf on a nine-hole course can be arranged nearby. □ **Lux** □ **T, R**

Arkansas

ARKADELPHIA: DeGray State Park Lodge and Convention Center, *Box 375, Arkadelphia, AR 71923, (501) 865-4591* ☐ *Open all year* ☐ *from I-30 in Arkadelphia, take SR 7 12 mi N to park* ☐ This 96-room lodge on an island in Lake DeGray is constructed of redwood and stone and offers striking views of the broad lake and the surrounding forests.

Guest rooms are done in contemporary style, as is the spacious dining room with raftered ceiling, stone fireplace, and large windows overlooking the woods.

The 915-acre park was Arkansas' first state resort park, constructed at a cost of $6.5 million. On the park grounds are a 7200-yard championship golf course, two tennis courts, sand beaches, and a marina where you may rent party boats and fishing boats. The 13,400-acre lake has abundant bass, crappie, blue gill, catfish, northern pike, and walleye, as well as opportunities for sailing and scuba diving. The woods surrounding the lake, laced with hiking trails, are inhabited by deer, beaver, wild turkey, and quail. ☐ **Inexp** ☐ **W, F**

FAIRFIELD BAY: Indian Rock Resort, *P.O. Box 3008, Fairfield Bay, AR 72088, (501) 884-3333* ☐ *Open Mar.-Nov.* ☐ *1 mi S of SR 16, 4 mi E of Shirley* ☐ Indian Rock is a 12,000-acre resort on the hilly shores of Greers Ferry Lake. Among the heavy stands of pine, oak, maple, and black gum are villas, chalets, condominium suites, and guest houses. Most luxurious are the villas, which have one- and two-bedroom suites, fully equipped kitchens, cathedral ceilings and fireplaces, and maid service. The most elegant dining is at the Racquet Club; the Wild Boar restaurant is for family dining.

The sports facilities here are extensive. With 10 lighted tennis courts and the latest in instructional equipment, Indian Rock is one of the top tennis resorts in the country. The rugged terrain makes the 18-hole golf course both challenging and beautiful. Four lakes and numerous streams lie along and between the fairways. From the men's tees, the course measures 5678 yards, but from the back tees the distance stretches to 6528 yards.

On the 40,000-acre lake is a full-service marina with rentals for fishing, (bass, trout, walleye, and catfish), scuba diving, waterskiing, and sailing. For trail rides, 30 horses are available. The resort has three pools (one indoor). ☐ **Mod to Lux** ☐ **T, G, F, W**

HEBER SPRINGS: Red Apple Inn, *Eden Isle, Heber Springs, AR 72543, (501) 376-8818* ☐ *Open all year* ☐ *From US 67 in Searcy, take SR 16 to Heber Springs* ☐ This quietly elegant resort is set on 600 lakeside acres and made intimate and handsome with stone patios, fountains, and landscaped courtyards with rock gardens. The inn, of weathered wood and stone, is built into wooded hills, giving most rooms a fine view of the lake. In contrast to the modern yet rustic lines of the lodge architecture, the rooms are furnished in Mediterranean style. In addition to the 56 doubles, two luxury suites and four fireplace rooms are available.

The restaurant, serving formal continental cuisine as well as hearty Southern fare, is known for its homemade breads and pastries, prime rib of beef, cheese souffle, and quiche. The Red Apple has one of the largest marinas on the lake, providing covered dock space and facilities for fishing, sailing, and waterskiing. The inn also has a pool, three lighted tennis courts, and a 9-hole, par-36 golf course with challenging hills and water hazards. ☐ **Mod to Lux** ☐ **F, W**

HOT SPRINGS: At the center of town is the 4000-acre Hot Springs National Park which administers the springs themselves and distributes the waters. The 47 thermal springs were known to the Indians, who showed them to Hernando de Soto when the explorer passed through Arkansas more than 400 years ago. Except for the display springs near the end of the Grand Promenade–a pleasant brick walkway–the springs have been enclosed and are not open to the public. The heart of the town is small and quaint–a string of old-fashioned bathhouses, a few grand old hotels. One of the town's biggest attractions is Oaklawn Race Track, which hosts the Thoroughbred Race Meet from February to April. Farther south is Lake Hamilton, which, together with lakes Catherine and Ouachita, forms a 100-mile-long impoundment on the Ouachita River. State parks on Lake Ouachita, a state park on Lake Catherine, and numerous motels and cottage colonies on Lake Hamilton provide boating and watersports facilities. Hot Springs Golf and Country Club (Malvern Road) has tennis courts as well as 45 holes of golf, and gives guests at some neighboring hotels sports privileges (see below).

Arlington Hotel, *Central Ave, Hot Springs, AR 71901, (501) 623-7771* ☐ *Open all year* ☐ *On SR 7 (Central Ave.) at Fountain St.* ☐ This is one of those grand, old-fashioned, monolithic resort hotels in the center of town. It has more than 500 rooms, some with refrigerators, all with color TV. There are two pools, several dining, drinking, and dancing areas, some shops, and spa facilities on the premises. Golf and tennis privileges are available at Hot Springs Golf and Country Club. ☐ **Inexp to Mod** ☐ **T, G**

Majestic Hotel, *Central Ave, Hot Springs, AR 71901, (501) 623-5511* ☐ *Open all year* ☐ *On SR 7 (Central Ave.)* ☐ This is the other grand, old-fashioned, monolithic resort hotel in the center of town. It has about 400 rooms, some with refrigerators or kitchens. The large circular pool is surrounded by a spacious courtyard and has bar service. Guests at this hotel also have privileges at Hot Springs Golf and Country Club, and the hotel has its own fishing and boating lodge on Lake Hamilton. ☐ **Inexp to Mod** ☐ **T, G**

LAKEVIEW: Gaston's White River Resort, *Lakeview, AR 72642, (501) 431-5202* ☐ *Open all year* ☐ *from US 62 in Mountain Home, take SR 178 E to Lakeview, then 2 mi S to resort* ☐ Gaston's is composed of over 40 cottages strung along the shore of the White River. (For flyers, a 2400-foot private airstrip adjoins.) The frame cottages come standard or deluxe, some with kitchens, some with fireplaces, some with patio. Most have walnut paneling and air conditioning. A four- and a ten-bedroom cottage are available.

Gaston's has a tennis court and a swimming pool, but the main activity here is float fishing. Below the Bull Shoal dam, trout fishing in the White River is excellent. On a float trip, guests glide downriver in a 20-foot "John" boat supervised by an experienced guide. It is perhaps the most relaxing way to fish, hitting a new spot with every cast, while high limestone cliffs and forested hills drift by. Everything is provided, from soft drinks and lunch to fish cleaning and packing.

The restaurant at Gaston's, overlooking the river, has a complete menu, specializing in steaks and prime rib as well as catfish and trout. ☐ **Inexp to Mod** ☐ **F**

MOUNTAIN HOME: Blue Paradise Resort, *Route 6, Box 379, Mountain Home, AR 72653, (501) 492-5113* ☐ *Open all year* ☐ *take US 62 7 mi E of Mountain Home to turnoff for resort* ☐ Blue Paradise is a family and fishing resort set on 34 acres. The one- and two-bedroom cottages, deluxe or standard, are simply furnished in contemporary style. All have air conditioning, TV, and equipped kitchens; some have fireplaces. For relief on hot days there's a small, L-shaped pool.

Lake Norfolk teems with largemouth and smallmouth bass, crappie, and catfish. Guests may rent aluminum fishing boats, bass boats, ski boats, pontoon boats, and motors at the covered dock. Use of a 14-foot aluminum boat is included in the daily rate. ☐ **Inexp** ☐ **F**

SILOAM SPRINGS: Dawn Hill Golf and Racquet Club, *P.O. Box 296, Siloam Springs, AR 72761, (501) 524-9321* ☐ *Open all year* ☐ *2 mi N of Siloam Springs on SR 43 take paved County E to resort* ☐ Dawn Hill is a resort community in a lovely area of the Ozarks. The townhouses, as they are called, are handsome one- and two-story houses with wood siding and stone chimneys. All are furnished in contemporary style and have spiral staircase, loft bedrooms, and fully equipped kitchens with dishwasher and garbage disposal.

Suites, consisting of a bedroom with king-size bed and a living room with a queen-size fold-out couch, are also available.

The community is focused on the golf course and tennis courts. The par-72 course measures 6890 yards from the back tees. Flint Creek runs alongside some of the fairways to provide scenic beauty as well as some fish. The four Plexipave tennis courts are lit for night play and the large outdoor pool is just a short walk away. Also here are a disco room and restaurant. □ **Mod** □ **G, T**

California

AVALON: Las Casitas, *P.O. Box 615, Avalon, CA 90704 (213) 510-0226 or (213) 388-1151 and Pavilion Lodge, P.O. Box 278, Avalon, CA, (213) 510-0465 or (213) 388-1151* ☐ *Open all year* ☐ *On Santa Catalina Island; boat service from San Pedro, Long Beach and Newport Beach, air service from Long Beach, San Pedro and Orange County* ☐ Three blocks from the beach and center of town, Las Casitas is a cluster of 34 tile-roofed housekeeping bungalows in a landscaped grove. These stucco bungalows have wall to wall carpeting, color TV, equipped kitchens, maid service, and private patios. On the grounds are three Laykold tennis courts, a swimming pool, and recreation room. (The same management offers more conventional motel units at nearby Pavilion Lodge). Across the street from Las Casitas is the Catalina Visitors Country Club with a 9-hole golf course, pro-shop with rentals, and dining and cocktail lounge.

Catalina Island, 22 miles long, has been a resort since the 19th century. You may tour the massive stucco casino, which still stands at the end of Avalon's harborside promenade as a symbol of the island's heyday. Today the island is noted for big-game fishing (marlin, tuna, swordfish) and marina services abound; there is also fine swimming and scuba diving in the clear water. Stables rent horses for riding into the hills above Avalon. Popular attractions include the botanical gardens at the Wrigley Memorial and the herd of buffalo at the Conservancy. The island has tram service, boat trips and bus tours, or you may rent a bicycle or a miniature car. ☐ **Mod** ☐ **F, W, R, G**

BAKERSFIELD: Rio Bravo Tennis Ranch, *11200 Lake Ming Rd., Bakersfield, CA 93306, (805) 366-3251* ☐ *Open all year* ☐ *On Alfred Harrell Highway, off SR 178, near Hart Park* ☐ Here low, shingled resort-modern lodges cluster around the tennis courts and pool surrounded by 16,000 acres of gently rolling hills. A one-minute drive from the lodge are the county-owned 18-hole Kern River Golf Course and 110-acre Ming Lake, and the Kern River. Those who bring their own boats and equipment can waterski on the lake, and whitewater raft trips down the river are available.

Most people come here for the tennis facilities—19 all-weather courts (15 lighted), pro shop, and professional staff with modern instructional equipment–but the ranch also has a stable, two swimming pools, and outdoor whirlpools. Every room here has a private patio or balcony and contemporary furnishings. The deluxe suites have fireplace and wetbar. The dining room, sleek and modern, serves continental food, and there is an informal bar. ☐ **Mod to Lux** ☐ **T, R, G, W**

BASS LAKE: Ducey's Bass Lake Lodge and Pine Chalets, *P.O. Drawer A, Bass Lake, CA 93604, (209) 642-3131* ☐ *Open all year* ☐ *From SR 41 2 mi N of Oakhurst, take SR 222 to SR 232* ☐ At the edge of Sierra National Forest, Bass Lake is surrounded by mountains thickly forested with pines and has abundant trout, bass, kokanee, crappie, and blue gill. With 42 miles of shoreline, it is roomy enough for waterskiing and sailing, and this resort offers all types of boat rental as well as a private sand beach.

Ducey's Bass Lake Lodge and Pines Village, both EP, are about a mile apart on the north shore of the lake. The lodge has 11 all-season cabins that sleep five and have rustic furnishings, private baths, and kitchens; 19 summer cabins that have no heat and must share a central bath facility, and motel units.

The Pine Chalets are newer, fully carpeted, and paneled with knotty pine. They offer full kitchens, baths, sleeping lofts for children, fireplaces, balconies, private sundecks, and barbecues.

The Pines Family Restaurant has a full menu selection for breakfast, lunch, and dinner. Less than a mile away, Ducey's Restaurant serves dinners with entrees such as lobster tail, filet mignon, and prime rib. ☐ **Inexp to Mod** ☐ **W**

BIG SUR: This is more of an unpretentious artist's colony than a mirrors-and-chrome resort area. The mountains here drop so steeply to the sea that no spot is level enough for a town center. Strung along the narrow, twisting main road, at the few spots flat enough for roadside parking, are the post office and grocery stores. From the road, dirt lanes marked by small, unobtrusive signs wind up or down the mountains to the inns and private homes, most secluded from view. The popular, much-photographed Pfeiffer Beach is reached by one of these, Sycamore Canyon Road, a bumpy 3-mile drive to a parking area and then a walk. The magnificent shore has beautiful sandy stretches for swimming, wave-formed rock arches, surf caves, and tidal pools alive with birds and marine flora and fauna. Sea otters often sport off-shore. Big Sur disdains sports facilities and night life. It is for those who want to relax and enjoy the spectacular natural sights. Since facilities are inconveniently laid out, you will need a car.

The inns are mostly of wood and stone construction with spacious rooms. For variety, most guests sample the cuisine of the other lodges or eat in (groceries specialize in prepared food and have excellent selections of wine; hard liquor is not available). One stop on almost everyone's tour is Nepenthe, a glass-enclosed restaurant with outdoor dining area set at the edge of a cliff. The food is good, the prices moderate, and the view mesmerizing.

Big Sur Lodge, *Big Sur, CA 93920, (408) 667-2171* ☐ *Open all year* ☐ *On SR 1, 26 mi S of Carmel* ☐ At this comfortable lodge in Pfeiffer Big Sur State Park guests have all the park facilities at their disposal. At the entrance is the main lodge with office, restaurant (open Easter to mid-October), and gift shop. The restaurant, good for families, features a salad bar and homemade soups and has a terrace dining area with a view of the river.

Up the mountainside in an area of redwoods are clusters of shingle-roofed cottages of various sizes. Some are sleeping cottages and others have kitchens. The largest have living room with fireplace, two bedrooms, kitchen, and porch designed for privacy. In the center of the cottage complex is a small pool and snack bar (both open in summer only).

The park has miles of hiking trails, a lovely river stocked with trout, picnic areas, and naturalist programs. One trail has a series of physical fitness exercise stations constructed along its length. The park has lots of wildlife (thus no pets are allowed), good birdwatching, and groves of redwoods. ☐ **Inexp to Mod** ☐ **H**

Deetjen's Big Sur Inn, *Castro Canyon, Big Sir, CA 93920, (408) 667-2377* ☐ *Open all year* ☐ *Off SR 1* ☐ This cluster of redwood-timbered cottages and lodges is for those who want to quietly savor fine food, magnificent scenery. and splendid isolation. Units include duplexes and triplexes with one or two bedrooms and shared baths; they are appointed with handmade furniture. The most rudimentary and imaginative lodgings are in the hollowed trunks of redwoods, perfect for low-budget backpackers and other sleeping-bag bearing guests.

The restaurant's menu runs to vegetarian and seafood dishes. Sometimes lunch and dinner are served before a roaring fire on the hearth to the accompaniment of live chamber music. There are no recreational facilities here, though a trail does lead through the woods to a stream in the canyon. ☐ **Mod**

Ventana, *Big Sur, CA 93920, (408) 667-2331* ☐ *Open all year* ☐ *2½ mi S of Pfeiffer Big Sur State Park on SR 1* ☐ This very chic lodge is set high in the Santa Lucia Mountains, providing most rooms with picture-postcard views. The ranch-modern cedar and stone lodge contains 24 guest units, tastefully furnished with such niceties as handmade Nova Scotia quilts and hand-painted, carved headboards. All have terraces, and about half come with Franklin fireplace or wetbar (sink plus refrigerator set in a dining alcove). Guest quarters 9 and 23 are townhouse suites, with a sitting room, wetbar, and fireplace downstairs and a bedroom and bath upstairs. A complimentary continental breakfast is served each morning in the lobby or in your room, if you prefer. Adjoining the lodge is a spa with saunas, Japanese hot baths, and a 90-foot swimming pool.

The excellent continental restaurant, open for 3 meals 7 days a week, has an outdoor dining terrace with a 50-mile view of the Sur coast. Adjacent is a store selling tasteful merchandise, from Pendleton shirts to epicurean kitchen gadgets.

Although the lodge caters to young, moneyed couples, it discourages them from bringing children since there is little to keep a child entertained here. ☐ **Lux**

CARLSBAD: La Costa Hotel & Spa, *Costa Del Mar Road, Carlsbad, CA 92008, (714) 438-9111* ☐ *Open all year* ☐ *From 1-5, take La Costa Rd. exit E to El Camino Real* ☐ This plush, sybaritic resort has a spa run on American Plan: exercise rooms, whirlpools and saunas, masseurs and masseuses, and a dining room with low-cal dishes. The men's spa area is equipped with exercise machines and muscle building and toning areas; the women's area, mirrored and carpeted, emphasizes beauty and figure improvement, and has a beauty and skin-care salon. Private whirlpool baths and larger Roman baths in glass-enclosed atriums are in each area. All types of classes, from karate to isometrics and many services, from pedicures to herbal wraps, are available. The spa dining room manages to make halibut in wine sauce, lamb curry, and chocolate mousse in non-fattening portions. For a small fee, guests not enrolled in the spa plan may use the spa facilities, but with so many other things to do you may not have time for the spa.

La Costa's 27-hole golf course, designed by Dick Wilson, is the site of the PGA Tournament of Champions. The 17th hole is sometimes referred to as "The Monster" by the pros. It's 573 yards into the prevailing wind—with traps on the left and water on the right. Tennis fanatics are satisfied by 25 hard courts, videotape system, automatic ball service, individual instruction, and 4000-seat exhibition court. Out past the tennis courts are the stables, where groups leave daily on rides over 21 miles of trails on resort mounts. You may have your own horse boarded and groomed here.

The 318 EP accommodations vary from single rooms in the main hotel to private villas with two or three bedrooms, full kitchens, dining area, and living room with fireplace. Equally varied are the restaurants. You may choose between the spa dining room, a seafood restaurant, a steakhouse, an Italian restaurant, a continental restaurant or the clubhouse lounge. ☐ **Lux** ☐ **G, T, R**

CARMEL: Quail Lodge, *8205 Valley Greens Dr., Carmel, CA 93923, (408) 624-1581* ☐ *Open all year* ☐ *From SR 1 S of Carmel take Carmel Valley Rd. to the resort* ☐ On the Carmel River, plush Quail Lodge has a lovely setting on 245 acres containing ten lakes. Guests are housed in exceptionally handsome cottages with wooden beams, glass walls, and sloping roofs with overhangs. Each cottage consists of four spacious bedrooms and an extra-large bedroom/sitting room that has a fireplace and a bar. Each of these fine rooms has color-coordinated furnishing, a private patio, a private entrance, and a private bath. This plan enables the lodge to offer what it calls "personalized" accommodations—you may rent an entire cottage, one or more bedrooms plus the sitting room, or simply a bedroom. Many rooms have lake or fairway views. You can check in here on EP, with golf and tennis extra, or on sports-inclusive plans.

The resort offers good, year-round golf at the Carmel Valley Golf and Country Club, which has a shop, swimming pool, and tennis courts, plus cocktail lounge and breakfast and lunch facilities.

Quail Lodge has its own pool and an excellent restaurant, the Covey, offering a continental menu and local specialties such as abalone. Jackets are required. There is also a bar. ☐ **Lux** ☐ **G, T**

CECILVILLE: Lor-O Ranch, *P.O. Cecilville, CA 96018, Nov.-Apr. (213) 353-7236; May-Oct. (916) Sawyers Bar 4681 Toll Sta* ☐ *Open May-Oct.* ☐ *From SR 3 in Callahan, turn SE toward Cecilville; at East Fork Campground go S 6 miles along South Fork to the ranch* ☐ When this ranch was established in the 1880s, it first raised apples, potatoes, and hay; then it turned to beef production. Although it's a guest ranch now, the Lor-O still raises its own vegetables, beef, chickens, and eggs. Even the wheat for the buckwheat is ground on the premises.

An informal family ranch, the Lor-O offers American Plan cottages overlooking a pond and the magnificent Trinity Alps beyond. The deluxe cabins have a private bath and will sleep five. The standard cabins, sleeping four or more, share a central bathhouse. The rates include weekly steak and fish fries, campfire breakfasts, picnics, as well as all other meals, open bar, and trail rides. In the weekly rates an overnight pack trip to the upper mountain is also included.

Around the ranch you can swim in the pool, pan for gold, hunt wild flowers, pitch horsehoes, play Ping-Pong, or fish in the pond or river. Farther afield fishermen will find rainbows in the forks of the Salmon River from May until June. The ranch can arrange hunting pack-trips for deer, bear, mountain lion or bobcat. Your friendly hosts will allow the kids to tag along for milking and other chores, teach them skills such as riding, provide babysitters, and supervise the playground. ☐ **Mod** ☐ **F, R**

CORONADO: Hotel del Coronado, *1500 Orange Ave., Coronado, CA 92118, (714) 435-6611* ☐ *Open all year* ☐ *On SR 75, 4 mi SW of I-5* ☐ This 19th-century oceanside giant is architecturally one of the grandest of America's grand hotels. It is an unlikely but wonderfully successful combination of Spanish and High Victorian styles: a red roof and white walls, and a tropical enclosed garden lined with pillared balconies, suggest missionary influence, but up close the hotel is intricately modeled with an exuberant profusion of wooden turrets, cupolas, and gingerbread decoration.

Of all the showplace public areas, the Crown Room dining hall is still the most impressive. This enormous long hemi-cylinder is enclosed by a curved, coffered ceiling of sugar pine and damask-covered walls; crown-shaped chandeliers hang down the center of this dramatic arch. But then, most of the interior is spectacular: dark-wood columns and polished carved woodwork everywhere–balconies, stairways, beams, and mantels–set off by richly colored carpeting. The place was built in 1887 to attract the wealthy to the as-yet-undeveloped city of San Diego, and the hotel is now a National Historical Landmark.

Located on the peninsula (almost inland) of Coronado, the del Coronado faces a wide sand beach. Beside this are seven lighted tennis courts, the heated swimming and wading pools, encircled by red-roofed cabanas, and a health spa with full exercise facilities. On Glorietta Bay is the private yacht harbor and marina, and across it, the Coronado Golf Club.

Guest rooms in the hotel and in the newer oceanside towers vary in degree of luxury; the long list of options includes ocean and bay views and elaborate suites. In addition to the newer, more intimate Prince of Wales Grille, named after the hotel's most celebrated guest, who first caught a glimpse of a certain Mrs. Simpson (nee Wallis Warfield) at a grand ball here in 1920. There's still dancing, and cocktails in the Ocean View Room and Casino Lounge, and at the handsome lobby bar as well. □ **Mod to Lux** □ **T, W**

FALLBROOK: Pala Mesa Golf & Tennis Resort, *2001 South Highway 395, Fallbrook, CA 92028, (714) 728-5881* □ *Open all year* □ *15 mi N of Escondido on I-15* □ This resort has modern motor-court-type lodgings overlooking the tenth fairway of its golf course. The spacious rooms have wall-to-wall carpeting and beamed ceilings. Some have kitchenettes, and most have private balconies.

Set on gently rolling, wooded terrain, the 18-hole, par-72 course measures 6213 yards with a rating of 69.6. For practice, there's a putting green and driving range, and the pro shop has a complete line of equipment for sale and rental. Another pro shop, for tennis, is next to the four concrete and two grass courts. The small heated pool, beautifully landscaped, is near the lodgings.

Overlooking the first fairway are the restaurant and lounge. On Friday nights the dining room features a buffet with a choice of 30 different salads and entrees. You may stay here on EP or on golf or tennis packages.

At nearby Swan's Lake, the Pala Mesa maintains a health spa with whirlpool, sauna, and exercise room. □ **Lux** □ **G, T**

IONE: Camanche Northshore Resort, *Rural Rte. 1, Ione, CA 95640, (209) 763-5121* □ *Open all year* □ *25 mi N of Stockton on SR 88, take Liberty Rd. W to resort* □ This large resort on Camanche Lake, (stocked with kokanee, blue gill, and bass), was designed to provide a moderately priced vacation for families. The resort has motel rooms as well as 14 mobile-home housekeeping units converted to resemble cottages by the addition of porches and shrubs. Each of these "Dualwide" homes can be divided into single or double rooms or a suite. Also here are a restaurant, saloon, and general store. The resort takes advantage of the 7700-acre lake by providing every type of boat service and rental imaginable, from the usual rentals to river canoe float trips to sailing lessons. Moreover, there are free tennis courts and a stable offering everything from a barbecue dinner ride to an amateur rodeo. Also available is a hunting preserve, first aid station, and child-care station. Comanche offers a variety of package vacation options that can include unlimited golf or horseback riding, scenic flights over the Mother Lode country, hunting, fishing, skiing, or kayaking. For the unfussy family looking for activity, these packages offer it at rock-bottom prices ☐ **Inexp to Mod** ☐ **F, W, F**

JUNE LAKE: Although somewhat eclipsed by its mighty neighbor 17 miles to the south (see Mammoth Lakes), June Mountain is a good-sized ski area in its own right. It boasts five double chairs, one T-bar, and a 2560-foot vertical drop. Extensive snowmaking equipment services the runs beneath chairs 2 and 5. The Alpenhouse Restaurant is at the base of the hill. Other facilities include a ski shop, repair shop, cafeteria, and ski lockers.

In summer, June Lake is attractive not only for its lake but because the town is midway between the events at Mammoth Lakes and the scenic beauty of Yosemite and Mono Lake.

June Lodge, *P. O. Box 265, June Lake, CA 93529, (714) 648-7713* ☐ *Open all year* ☐ *On SR 185, 2 mi E of I-395* ☐ June Lodge opened in 1928 as the home of the June Lake Rod and Gun Club and soon became a popular retreat for movie stars. Charlie Chaplin, Humphrey Bogart, and Clark Gable were guests, and several movies were filmed here. As you might expect of a former hunting lodge, the rustic interior design includes trophy heads and horns hung from the exposed beams and trusses. All rooms open onto an interior balcony overlooking the lobby and the large four-sided fireplace. Rooms have such modern conveniences as color TV and telephone. The June Room features a large dance floor, live entertainment, cocktails, and dining in a Country Western setting. The Veranda Bar overlooks June and Gull lakes.

The lakes are stocked with rainbow, brook, and German brown trout, and the dining room will prepare and serve any fish you might catch. The ski area is only three-quarters of a mile away. ☐ **Mod** ☐ **S, F**

KLAMATH RIVER: Beaver Creek Lodge, *Klamath River, CA 96050, (916) 465-2246* ☐ *Open all year* ☐ *From Yreka N on SR 263 8 mi, then W on SR 96 14.3 mi until entrance* ☐ In this secluded region of Klamath National Forest, the main attractions are freshwater fishing and hunting. For seekers of steelhead trout, deer, or bear, Beaver Creek's basic but pleasant two- to six-person housekeeping cottages make a good home base. The cabins are spaced along the banks of the still-untamed Klamath River. While there is no restaurant, (the closest is four miles) provisions for your cottage kitchen are available about a mile away at the general store. September through March is the steelhead migration season on the Klamath, and the lodge arranges guided fishing and float trips and freezer service. Fishing lures, licenses, and flies can be purchased at the general store. Nearby is a golf course and in the winter, the Mt. Ashland ski center is less than an hour's drive. ☐ **Inexp** ☐ **F, W**

LAKE TAHOE: It's an odd combination: You can sit on the shore of this immense, clear lake, looking at the reflection in the water of the tall pines that surround it, and the image is undisturbed enough for you to forget which is rightside up; yet on the Nevada side, the forest abruptly gives way to glittery hotels and casinos. You could go from an afternoon hike in the mountains to a tour of Vikingsholm Castle to a lake cruise and then to a casino, having experienced as wide a sampling of the kinds of things people do on vacation as you could manage anywhere in the United States.

A dozen public beaches, many marinas with boat rental and fishing services, boat trips, tennis, riding stables, and golf courses make Lake Tahoe one of the most *complete* vacation areas anywhere. The U.S. Forest Service Visitor Center (on SR 89 NW of Camp Richardson) offers many interpretive programs, sports, and the unique underwater creek view where you can see the life cycle of trout. Surrounding national forests–Tahoe, Toiyabe, and Eldorado–are excellent for every level of woodland adventure, from drive-through to backpack.

Lake Tahoe owes much of its wintertime popularity to excellent ski areas. The most renowned include Squaw Valley, (see below), Alpine Meadows, Powder Bowl, and Sierra Ski Ranch, and on the California-Nevada border, Heavenly Valley (vertical drop 3600 feet, 15 double chairs, one T-bar, two pomas, and four Mitey-Mites). The highway circling the lake–SRs 89 and 28 and I-50–passes an endless parade of lodgings, ranging from cabins to condominium complexes.

Charmey Chalet, *P.O. Box 109, Tahoe Vista, CA 95732, (916) 546-2529* ☐ *Open all year* ☐ *¼ mi W of Tahoe Vista on SR 28, at 6549 North Lake Blvd.* ☐ The Charmey, just off the north shore of the lake, combines the comfort of a modern motel with the informality of a cabin in the woods. The 22 units are housed in two strings of attractive wooden cottages with sloping shake roofs and board-and-batten siding; some face the terraced garden, others pine woods on the lake, or the Sierras. Two two-bedroom suites have separate living rooms (one has a fireplace), and many doubles have kitchenettes. All units have telephones and cable color TV. On the grounds are a heated outdoor pool and a hot tub, and across the way is a private beach. There's no restaurant, which is not a problem given the location, but a continental breakfast is served in the lounge during summer. ☐ **Mod**

Donner Ski Ranch, *P.O. Box 66, Norden, CA 95724, (916) 426-3578* ☐ *Open mid-Dec. to mid-Apr.* ☐ *At the Soda Springs-Norden exit off I-80, take Old Highway 40 3½ mi E to area* ☐ This modest-size ski area was one of California's first. Three double chairs and two pomas take skiers up 825 feet to the top of Signal Hill, overlooking Donner Lake and Donner Pass. Twenty-one runs, up to three-quarters of a mile long, lace the 300 skiable acres. At the base are ski shop, cafeteria, lounge, and lodge. Rentals, lessons, and lockers are available.

The base lodge offers 2 double rooms, 2 single rooms, and dorm space for 40. Skiers who rent dorm rooms can bring their sleeping bags or rent linens for the twin beds and bunks. □ **Inexp** □ **S**

Squaw Valley Lodge, *P.O. Box 2393, Olympic Valley, CA 95730, (916) 583-0121* □ *Open all year* □ *On SR 89, 10 mi S of I-80 at Truckee* □ This ski resort sits at the base of the slopes and lifts of Squaw Valley, site of the 1960 Winter Olmpics. Twenty-five major lifts, from chairlifts to a huge cable car, carry skiers from the base at 6200 feet to runs as high as 8900 feet. The slopes, on the broad flanks of five mountains, are so well-developed that guests don't have to ski to the base to enjoy a meal. The Gold Coast Restaurant is at the end of an 18-minute gondola ride, and the High Camp Restaurant is at the end of the cable car ride, at 8200 feet.

The lodge has all types of accommodations, from modern suites to rustic rooms with shared baths. The suites have a spacious living room and loft. Some have kitchenettes. The standard hotel rooms come in two sizes; the smaller rooms, the "bungalows," have telephones but not the color TVs found in the deluxe rooms. The lodge is surrounded by numerous shops and restaurants, most housed in the Mall or Olympic House.

Squaw Valley is a year-round sports center. A short walk away from the lodge is Blyth ice skating area, where Carol Heise won her gold medal in 1960. The valley is also the home of the Vaughan Hitchcock Wrestling Camp, the Squaw Valley Gymnastics Camp, and the Phidippides staff of world-class marathoners. The Squaw Valley Tennis Club has six championship courts available to guests at the lodge who have package plans during the summer. □ **Mod to Lux** □ **S, T**

Sugar Bowl, *P.O. Box 5, Norden, CA 95724, (916) 426-3651* □ *Open late-Nov. to mid-Apr.* □ *From I-80 at Soda Springs-Norden exit, take Old Highway 40 E to area* □ When this ski area opened in 1939, it featured the first chairlift in California. Since then it has grown to encompass two peaks served by eight double chairlifts and one gondola. Fifty percent of the slopes are of expert difficulty. The remainder are about equally divided between novice and intermediate levels. The vertical drop is 1500 feet, and the longest run is three miles. Night skiing is offered on Tuesdays and Saturdays.

Base facilities include a ski shop, a ski school, a cafeteria, a dining room, and three bars. No cars are allowed in the base area; skiers are transported between the lodge and indoor parking garage by an enclosed tramway.

The lodge has 100 units to rent, from standard doubles with balconies to share-bath bunkrooms. Its Scarlet Pimpernell is an elegant dining room. On Wednesdays from January to Easter the lodge holds a Tyrolean night featuring a special buffet and music. □ **Mod to Lux** □ **S**

Tahoe Marina Inn, *Box 871, South Lake Tahoe, CA 95705* □ *Open all year* □ *1 block W of US, 1 mi SW of town* □ This motel is right on the sandy beaches of the lake. The lodge offers 49 guest rooms plus 14 condominium units with kitchenettes. All rooms have color TV, telephone, private patio or balcony, and fine views of the lake and surrounding mountains. Some have a fireplace. Also on the motel's two acres are coffee shop and heated pool. Boat rentals are nearby. ☐ **Mod to Lux** ☐ **W**

The Waystation, *P.O. Box 1267, South Lake Tahoe, CA 95705, (916) 541-6220* ☐ *Open all year* ☐ *US 50 at Ski Run Blvd.* ☐ This modern resort hotel on the south shore has 400 feet of private beach. The three-story building, separated from the broad, sandy beach by a narrow strip of grass, has balconies with sweeping views of the lake. Rooms are wood-paneled and designer decorated in bright colors. The hotel also has five "Lakeside" suites, each with kitchenette with wetbar. The Waystation's restaurant, the Dinner House, features veal Oscar, beef Wellington, prime rib, and coq au vin among its dishes.

Boat and ski rentals are available on the property, plus a heated pool, and a complete ski shop. Free transportation is provided to Heavenly Valley ski area, the casinos, a nearby riding stable, and the airport. ☐ **Mod to Lux** ☐ **W, S**

MAMMOTH LAKES: The name of the area is a bit inappropriate. Although the lakes do attract summer visitors, Mammoth Mountain is the magnet that draws the crowds. It is one of the 3 largest ski areas in the country; even on those days when 14,000 skiers dot the slopes no one feels hemmed in. If you threw a blanket over a group of six understanding people, you would have, on a small scale, a fair model of the complex configuration of peaks, valleys, slopes, ridges, and bowls that make up Mammoth. Currently the mountain is strung with 19 chairlifts, two gondolas, two T-bars, and two poma lifts, but it could easily support twice as many. The scope of Mammoth's success as a ski area can be gauged by its plans for the future: Before long it may be connected to June Mountain Ski Area—17 miles north—by ski lifts spanning the San Joaquin Ridge. (See June Lake.)

The base facilities can only be described as complete. The town, which begins where the lifts end, has cafeterias, lounges, restaurants, ski shops and schools, and virtually unlimited accommodations. Most of the lodges and resort condominiums list with one of three major reservations services (all in Mammoth Lakes, CA 93546): Mammoth Lakes Reservation Bureau, P.O. Box 8, (714) 934-2528; Mammoth Reservation Service, P.O. Box 277, (714) 934-2522; and Chateau Reservations, P.O. Box A-2, (714) 934-2600.

Extensive warm-weather activities include an ambitious music and dance festival that packs classical, folk, jazz, or Country Western concerts into the last week of July and the first two weeks of August. Many other events—motorcycle races, square dances, antique shows, arts and crafts exhibits—are scheduled throughout the summer. The Sierra Nevada Inn (see below) runs a tennis camp from June to September.

Crowley Lake, the largest of the area's lakes, has waterskiing and Hobie Cat rentals, as well as good fishing for trout and perch. Quite a few of the lakes are stocked, and the High Sierra lakes above 9000 feet harbor the prized golden trout, the state fish.

The entire area is in Inyo National Forest, just east of Yosemite. Among the area attractions are Devil's Postpile National Monument and Hot Creek, a thermal stream. The Visitor Center (on SR 203) runs naturalist programs and other activities including guided tours on skis in winter. Private outfitters conduct pack trips into the wilderness for fishing and sightseeing, and one outfit will even fly skiers to remote peaks by helicopter. For more information, contact the Chamber of Commerce, P.O. Box 123, Mammoth Lakes, CA 93546.

Alpenhof Lodge, *Box 1157, Mammoth Lakes, CA 93546, (714) 934-8558* ☐ *Open all year* ☐ *On Minaret Rd. off SR 203* ☐ Completely European, from its clock tower to the flower boxes outside its windows, the Alpenhof has good on-premises facilities: recreation room, sauna, whirlpool, and 40-foot-long heated pool. All rooms have color TV; some have fireplaces or kitchenettes. The Clock Tower restaurant has an outdoor dining terrace and the bar has disco music and decor. A shuttle bus connects the lodge to Warming Hut #2 and chairlifts 7, 8, and 16. ☐ **Mod** ☐ **S**

Sierra Nevada Inn, *P.O. Box 918, Mammoth Lakes, CA 93546, (714) 934-2515* ☐ *Open all year* ☐ *43 mi N of Bishop off of I-395, then W on SR 203 3 mi to Old Mammoth Rd.* ☐ Practically in the shadow of Mammoth Mountain, the buildings of the Sierra Nevada Inn, with their steeply pitched shingled roofs, are fine examples of modern alpine architecture. Housed within are 132 comfortable double rooms with two queen-sized beds, color TV, and carpeting. Also, 24 more luxurious rooms sport full baths and kitchenettes; some have free-standing, hooded metal fireplaces. For groups and families, two-room suites are capable of holding 10 persons, and a two-level, chalet-style suite can sleep 14 persons. All accommodations have contemporary furnishings and paneled and stucco interior walls.

Summer activities include soaking and swimming in the pool and whirlpool, tennis on three courts, and good fishing and hiking among the scenic alpine forests. For dining and entertainment, the Sierra's Rafters restaurant and the nearby Mill City Mining Co. offer a good selection of steaks and salads. ☐ **Lux** ☐ **S, T, H, F**

MILL CREEK: Childs Meadows Resort, *Rte. 5, Box 3000, Mill Creek, CA 96061, (916) 595-4411* ☐ *Open all year* ☐ *On SRs 36E and 89, 49 mi E of Red Bluff* Only nine miles from the southwest entrance to Lassen National Park, this resort sits on a meadow surrounded by forested slopes, and offers 40 conventional motel rooms with tiled showers; many have kitchenettes, these have maid service but no kitchen utensils; units can be interconnected to sleep up to eight people. In addition, two-bedroom bungalows have a living room with a fold-out bed, kitchen (no utensils), full bath, screened sleeping porch, and front porch with a meadow view. These, rented by the week and without maid service, sleep up to eight. The restaurant has a good selection of California wines. The coffee shop is open in the summer for breakfast and lunch, and during ski season the dining room is open weekends and holidays. Also here are a seasonal bar, general store, coin laundry, recreation room, and picnic areas with grills.

Childs Meadows has a tennis court, a swimming pool, a shuffleboard court, and a string of horses rentable by the hour (instruction available), or for pack trips. The resort's trailhead provides links to the Lassen Park and Pacific Crest trail systems. For trout fishing, Mill Creek is one-quarter of a mile away. Other streams and lake recreation sites are in surrounding Lassen National Forest. In winter, the resort's touring center offers cross-country ski instruction, rentals, and a network of prepared trails. For downhill skiers, Lassen Park Ski Area is 15 miles north and is usually open weekends and holidays. Lassen Park Road and Visitor Center are open, weather permitting, from about mid-June to mid-October. □ **Inexp to Mod** □ **R, W, S, F**

MONTEREY PENINSULA: This broad, rock-sheathed peninsula is known for stunning scenery, birds and marine mammals along its coast, great golf, quaint fishing fleets, and opulent mansions. It is the home of Cannery Row and Pebble Beach, both the ramshackle and the rich (with the emphasis on the latter). A favorite way to see a large chunk of the peninsula is to take the 17-Mile Drive, a toll road that overlooks the lush golf courses, movie-star mansions, and the craggy coastline.

Along the north shore are miles of public beach and the town of Monterey. For years the Spanish and Mexican capital of California, Monterey has several interesting adobes. Many of the historic homes can be seen by taking the "Path of History Walking Tour," easily made by following the line painted down the streets. The Royal Presidio Chapel here and the Mission San Carlos Borromeo del Rio Carmelo in Carmel are especially nice examples of the area's missionary architecture.

Just south of the peninsula is Point Lobos State Reserve, with bonsai-like cypresses on the point and magnificent sea lions and innumerable birds on the offshore rocks.

Asilomar Conference Center, *Box 537, Pacific Grove, CA 93950* □ *Open all year* □ *On the W end of SR 68* □ This resort is set on the Monterey Peninsula within the grounds of the beautiful Asilomar State Beach. Asilomar was established as a conference center in 1913. The original buildings and grounds were planned by the distinguished California architect Julia Morgan. The hotel, a dozen small lodges, and other buildings are of wood and stone, to follow Mrs. Morgan's dictates that they blend unobtrusively into their setting. Some of the 314 rooms have fireplaces and some share a bath. Many have private balconies and superb views of the water.

From the beach, good for fishing and beachcombing but unsafe for bathing, sea otters and seals can often be seen, birdwatching is good, and butterflies and whales pass by during their migration period. Much of the park's 100 acres are forested with pine and cypress, and guests often spot deer and raccoon. Pacific Grove has an 18-hole municipal golf course, a museum of natural history, and a light station with spectacular coastal views.

On the park grounds are a heated swimming pool, tennis courts, indoor and outdoor game areas, and a cafeteria. American Plan is available. ☐ **Inexp** ☐ **F**

Del Monte Hyatt House, *1 Old Golf Course Rd., Monterey, CA 93940, (408) 372-7171* ☐ *Open all year* ☐ *Near jct SRs 1 and 68, between Sloat Ave. and Aquajito Rd. off Mark Thomas Dr.* ☐ The low-slung architecture and intelligent landscaping make this hotel seem rural and isolated, althouth it is only minutes away from the heart of Monterey. Most of the rooms face the golf course, reputedly the oldest course west of the Mississippi. Set amid live oak and Monterey jack pine, the 18-hole course provides a good test, especially on the 7th and 16th holes. A full line of golf equipment is available at the pro shop, and a bar and grill is adjacent. The lodge also has six tennis courts (two lighted), three swimming pools, and a whirlpool.

Each of the 400 rooms and suites has colorful contemporary furnishings, color TV, and telephone. Some have a fireplace or a kitchenette. The most elaborate lodging is the President's House; four bedrooms with private terraces, five baths, den, kitchen, living room, and pool.

The menu at the Pirate's Cove restaurant boasts such California delicacies as abalone and bay shrimp, as well as continental and game dishes, steakhouse fare, and overstuffed sandwiches. ☐ **Lux** ☐ **G, T**

The Lodge at Pebble Beach, *Pebble Beach, CA 95953, (408) 624-3811* ☐ *Open all Year* ☐ *13 mi S of Monterey on the 17-Mile Dr.* ☐ Staying at this elegant lodge (which offers EP and MAP) is expensive, but splurging will put you on a dazzlingly beautiful cove of Carmel Bay. The setting is so spectacular, with pine forests, groves of Monterey cypress, rocky outcroppings, and sandy beaches, that the lodge was originally established to service sightseers who made the 17 mile, round trip from Monterey just to gaze. Nothing about this resort is splashy—the emphasis is on supreme and understated elegance and superb service. You are housed in a series of two-story white villas that surround the main lodge and look out onto fairways and the bay. The 133 rooms and suites have large expanses of glass to make the most of the views; many units have the balconies, fireplaces, bathroom telephones, or other extras you'd expect in a luxury resort.

The most formal restaurant is the well-regarded Club XIX, where reservations and jackets are mandatory. Dining is by candlelight and the menu offers French dishes such as filet of beef en croute as well as local specialties such as abalone. Other dining and drinking options include the Cypress Room and Tap Room at the lodge, and in the shopping complex, the informal Mulligan restaurant or picnic supplies from the Company Store.

Golf is naturally a big draw here. The lush and famed Pebble Hill Golf Links (par 72, 6800 yards) is generally rated one of the top ten in the nation and extends right along the water's edge. The Pebble Beach Pro Shop with rental and other services, can also arrange for you to play Spyglass Hill (par 72, 6800 yards, designed by Robert Trent Jones) or the 9-hole, par-3 Peter Hay course.

Tennis facilities include 13 Laykold courts, full-time pro, and pro shop. Pebble Beach stables host many events, including polo matches, and here you rent horses for 34 miles of bridle trails. The lodge also offers swimming and wading pools with poolside service, saunas, a private beach with lifeguard, and a chic little shopping center. Fishing is permitted only at Fanshell Beach, a crescent-shaped white-sand cove.

The entire Pebble Beach property is a nature preserve, where deer graze on the golf courses and in the forests, and where many of the plant forms are unique–Huckleberry Hill is among the areas of unusual botanical interest. Birdwatching is very rewarding, not only on the shore and in the bays, but also in the groves. On the offshore rocks are herds of noisy sea lions, and in the bays leopard and harbor seals as well as sea otter can be viewed from close quarters. Along the 17-Mile Drive are parking and picnic areas at vantage areas for birds and sea mammals, and there are also walking trails along the shore and in the groves. ☐ **Lux** ☐ **G, T, R**

MORRO BAY: Best Western Golden Tee, *19 Country Club Rd., Morro Bay, CA 93442, (805) 772-7313* ☐ *Open all year* ☐ *1 mi S of Morro Bay* ☐ What this motel has going for it is a magnificent view of Morro Rock and the bay, and a handy location in Morro Bay State Park. The two-story buildings are stepped up the steep slope from the dock, giving about half the units fine views of the bay, its bird life, and sea otters from the picture windows and private balconies; avoid units at the back that do not share this view. The dining room (fish specialties) and bar (entertainment) also overlook the bay. There is a pool.

The lodge is in Morro Bay State Park, which has excellent birdwatching, and the heron rookery borders one side of the lodge. A pleasant 18-hole municipal golf course with clubhouse spreads out over the hill behind the lodge. Also in the park are a museum of local natural history that offers naturalist programs and a deck for viewing sea mammals, swimming beach, fishing, boat rental, picnic grounds, and clam digging. Two attractive state beaches on the ocean are a short drive away, and the town of Morro Bay offers an aquarium and harbor cruises. ☐ **Mod** ☐ **G, W**

NAPA: Silverado, *1600 Atlas Peak Rd., Napa, CA 94558, (707) 255-2970* ☐ *Open all year* ☐ *Off SR 121, 4 mi N of Napa* ☐ This luxury 1200 acre condominium resort is set around two challenging, par-72 Robert Trent Jones golf courses. The focus is the 1870 Silverado mansion, standing in a grove of beautiful old magnolias and oaks. The mansion, a blend of Italian and French architecture, connects with the clubhouse and contains several dining areas. The Silverado Room, the main dining room, has two walls of large windows overlooking the golf courses. Next to the clubhouse are Plexipave tennis courts; some are lit for night play. Pros, rentals, instruction, and other services are available for both golf and tennis.

Each condominium–from studios to three-bedroom suites–has its own entrance, private patio or balcony, woodburning fireplace, air conditioning, color TV, wetbar, and fully equipped kitchen. Rented on European or American Plans, these units cluster around the five swimming pools or look out on the fairways.

Wine tours are popular here, because some 40 vineyards are within a half-hour drive. Silverado also provides complimentary transportation to a nearby ranch for horseback riding. ☐ **Lux** ☐ **T, G**

OJAI: Ojai Valley Inn and Country Club, *Ojai, CA 93023, (805) 646-5511* ☐ *Open all year* ☐ *14 mi NE of Ventura on SR 33* ☐ The Ojai Valley takes its name from the Indian word for "nest"–an evocative term since the valley is surrounded by the Sierra Madres, which protect it from the Los Angeles smog. This resort's main buildings, an appealing blend of Spanish and rustic architectural elements, are built into the hillside at several levels, giving guests superb views of the valley. Guest cottages are staggered along the fairways. The resort operates on the American Plan, with extra fees for tennis, golf, and riding, and offers several formal and informal restaurants and cocktail lounges, as well as buffets and poolside and terrace service.

The difficult 18-hole, par-70 course, which virtually encircles the resort, is laid out on naturally rolling terrain. A pro shop has carts and caddies, and the 18-hole putting green is free. The eight hard-surface courts for tennis also have pro service. The resort maintains stables and hundreds of miles of riding trails. Breakfast rides and barbecues are featured regularly. Also on the 220 acres are an outdoor pool, croquet grounds and children's playground. Babysitters are available. ☐ **Lux** ☐ **T, G, R**

PALM SPRINGS: Shielded from the desert winds by the San Jacinto Peak, this oasis was popular long before the Hollywood celebrities came here to golf and tan. The Cahuilla Indians, who still own much of this very valuable land, were enjoying the mineral springs when Spanish explorers first passed through. Now the entire area is a resort where everyone, it seems, has a swimming pool, the modern version of the mineral spring.

At last count, Palm Springs was adorned with 32 golf courses, most of them private or open only to hotel guests. The municipal course (1885 Golf Club Dr.), however, is a fine test open to the public. Other good public 18-hole courses are the Westward Ho (3 mi W of Indio on Jefferson St.), the Palm Desert (77-035 New York), and the Desert Air (42-300 Bob Hope Dr.). Tennis courts abound, and several stables conduct horseback rides.

The Palm Springs Aerial Tramway is a spectacular 2½-mile ride to the top of San Jacinto (8500 feet). At the upper station is an alpine chalet, a restaurant, and winter sports rentals. Seven miles south of the city is Palm Canyon, lined with 3000 Washington palms, some over 2000 years old. Farther south is Salton Sea, which provides fishing and boating facilities.

The Canyon Hotel, ☐ *2850 South Palm Canyon Dr., Palm Springs, CA 92262, (714) 323-5656* ☐ *Open Sept–June* ☐ *off SR 111* ☐ With its own 18-hole golf course, the Canyon is a rarity among hotels in Palm Springs. The white, ultramodern hotel is done with a hint of Moorish architecture. The 250 rooms are decorated in a colorful contemporary style and have picture windows and balconies. Each of the 200 suites has a mini-kitchen and wetbar, and all units have air conditioning and color TV.

In addition to the three heated pools, you may use three whirlpools, and a health spa with sauna. The 6400-yard golf course, set amid fruit trees, is challenging. The 10 hard-surface tennis courts are well laid out, with bordering flower beds; three are lit for night play. Both a golf and a tennis pro, and a golf shop are on hand. Sports lessons, babysitting, and horseback riding can be arranged.

The Canyon has several dining and drinking areas. Perry's serves continental cuisine in a formal ambience. Raffles lounge has a Sydney Greenstreet atmosphere with rattan furniture and overhead fans, and the Greenhouse lounge is planted with ferns and real palm trees. ☐ **Lux** ☐ **T, G**

La Quinta, *La Quinta, CA 92253, (714) 564-4111* ☐ *Open late Sept to mid May* ☐ *7 mi E of Palm Desert on SR 111 take Washington ST. to Lago La Quinta* ☐ In the style of the Southwest's traditional luxury resorts, La Quinta is composed of a number of adobe bungalows set around a grass courtyard studded with tall date palms. The two-, three-, and four-room bungalows are furnished in classic Spanish Mediterranean style, and many have tiled private patios with views of the exceptionally well-manicured grounds and formal gardens. The dining room, also in Spanish style, is an elegant room with pillars and chandeliers. There are also patio lunches, steak fries, and dancing in the cocktail lounge. Rates are AP during high season and holidays.

The heated pool is surrounded by a tile terrace and six tennis courts, supervised by a pro. Also on the 800-acre property is a stable of horses for the miles of trails surrounding the hotel, and for children, a playground with a wading pool. ☐ **Lux** ☐ **R**

Marriott's Rancho Las Palmas Resort, *41000 Bob Hope Dr., Rancho Mirage, CA 92270, (714) 568-2727* ☐ *Open all year* ☐ *5 mi S of jct I-10* ☐ Marriott bought the Rancho Las Palmas Country Club and added a resort on the 26 palm-dotted acres. The country club alone is quite impressive: a 27-hole Ted Robinson golf course with clubhouse services and pros, 25 tennis courts (eight lighted) with their own clubhouse, observation tower, pro shop, and snack bar and two freeform pools with adjacent whirlpools. But the haciendastyle resort was created on an incredibly lavish scale. The two-story lodges are rough-textured stucco and red tile roofs to evoke a Mediterranean atmosphere. Each of the 348 soundproofed double rooms and suites has color TV, climate control, direct-dial telephone, and outside entry with patio or balcony; most overlook the golf course.

Public rooms are an especially nice Spanish mission style with beautiful tile floors enhanced with pots of flowers, viga ceilings, wrought-iron chandeliers, arches decorated with flower motifs, Mexican chairs, and wooden balconies and shutters. Fountains, small lakes, swimming pools, and water courses scattered throughout the grounds reflect sky and mountains. The resort's array of dining facilities includes the Fountain Terrace for family dining in a fountain garden setting and the Sunrise Terrace with outdoor buffet and a mountain view. The elegant Miguel's is a cozy lounge for entertainment and dancing. Cabrillo in early California style, and the exclusive Cuarto del Vino for private parties are open for dinner only. Also on the grounds are game rooms, a gift shop, liquor shop, beauty and barber shops, secretarial services, public notary, airline desk, and car rentals. Hotel services include arranging tours to sightseeing targets such as the Living Desert Museum and courtesy airport limousine. AP, MAP, EP, and packages are available. ☐ **Mod to Lux** ☐ **T, G**

Ocotillo Lodge, *1111 East Palm Canyon Dr., Palm Springs, CA 92262, (714) 327-1141* ☐ *Open all year* ☐ *Near the corner of E. Palm Canyon Dr. and Palm Canyon Dr.* ☐ This elegant lodge offers suites, bungalows, and villas set amid tall palms and flower-bordered pathways. The centerpiece of the lodge is the large fan-shaped pool, overlooked by the chic Ocotillo Room restaurant and Vista lounge. Also adjacent to the pool is the Polynesian-style Sunshine Bar, which features luau parties throughout the winter season. The third bar, JB's, has rustic wood paneling, Tiffany-style lamps, and an intimate atmosphere.

Each of the designer-decorated guest units has a sitting room, bedroom, color TV, kitchen (wetbars only in suites), and private patio or terrace. Some rooms have a fireplace.

The lodge has a tennis court, and a riding stable and par-31 executive course are nearby. Service is very good, and the lodge arranges for babysitters, guided trips, and other extras. The lodge, owned by Dr. Jerry Buss, who also owns the L.A. Lakers, is the home of the Lakers during their fall training. ☐ **Lux**

White Sun Guest Ranch, *P.O. Box 1000, Rancho Mirage, CA 92270, (714) 346-6177* ☐ *Open Oct 1 to June 1* ☐ *65 mi SW of San Bernardino via I-10 and SR 111 to Rancho Mirage and Magnesis Falls Dr.* ☐ Here's a small (50 acre) friendly oasis resort landscaped with lawns and palms. Low row bungalows with white-gravel roofs and overhanging eaves house 29 doubles, 8 singles, and 24 luxurious suites. Most units face the swimming pools or lawns and the best have fireplaces and patios. All are air conditioned and have color TV.

Here there are four swimming pools–the nicest is a small oval-shaped one in a tree-shaded cloister. Flanking the lobby is a pitch-and-putt course. Tennis players will find five lighted courts. In addition, a combination basketball-volleyball court is next to the small children's playground. Riding horses can be rented nearby. Rates are on American Plan, but steak fries are extra. ☐ **Mod** ☐ **T**

RANCHO SANTE FE: The Inn at Rancho Santa Fe, *P.O. Box 869, Rancho Sante Fe, CA 92067, (714) 756-1131* ☐ *Open all year* ☐ *Take Lomas Santa Fe exit off I-5; follow drive 5 mi E to resort* ☐ On this pretty 21 acre tract, the Santa Fe Railroad planted thousands of eucalyptus trees under the misguided impression that the wood could be used for railroad ties. When the wood proved unsuitable, the railroad converted the land to citrus cultivation and residential development. The Guest House for prospective land buyers, built in 1923 of adobe in early Californian style, now forms part of the main building. Over the years, guest cottages of similar design have been added and the grounds have been beautifully landscaped to create a charming and peaceful retreat. Accommodations in the main building and cottages vary from double rooms to three-room suites; virtually all have lanais or private patios or terraces, many have kitchenettes, and some have fireplaces. Also available are private two-room cottages with studio couches, kitchen, fireplace, two baths, and lanai, and three-bedroom homes.

All units are rented EP, with room and maid services. The main building has a comfortable lounge, and a sunny garden room and cocktail lounge. Dress code is observed evenings. On summer weekends, there is dancing on the Patio Terrace.

Outside the main building is a large pool with a terrace for luncheon and beverage service. Nearby are courts for games such as shuffleboard or badminton, and a large bowling green. Also on the grounds are three tennis courts, where instruction is available. In summer, the inn maintains a day-use beach cottage with dressing rooms at Del Mar. The inn has membership in several private golf courses, including an 18-holer just outside the property and the nearby 27-hole championship course at Whispering Palms Country Club. The office will arrange guest cards and starting times. The office will also arrange for babysitters, horseback riding, or car rental and help you plan tours of attractions in and around San Diego. ☐ **Mod to Lux** ☐ **T, G**

SAN DIEGO: Rancho Bernardo Inn, *17550 Bernardo Oaks Dr., San Diego, CA 92128, (714) 487-1611 or toll-free (800) 421-0000* ☐ *Open all year* ☐ *20 mi N of San Diego on I-15* ☐ This superior contemporary motel offers 150 spacious rooms and suites housed in a cluster of low-lying Spanish-mission-style buildings that match homes of the community of Rancho Bernardo. Most of these luxurious lodgings, with all modern amenities including wetbars in the suites, have private patios or balconies looking out on the pool or beautifully landscaped grounds. Two excellent restaurants, nightly dinner dancing and entertainment and two cocktail lounges are available, and dress code is observed. The two swimming pools have poolside service.

In addition to the West Golf Course, a beautiful par-71, 18-hole tour of hills, lakes, and waterfalls the ranch has a 27-hole executive course. Tennis facilities include 16 courts, some with lights, and a lavish clubhouse, and the Tennis College, a well-equipped instructional center specializing in two-day crash courses. The ranch also has lawn games and can arrange for horseback riding or bicycle rental. ☐ **Lux** ☐ **G, T, R**

SANTA BARBARA: Marriott's Santa Barbara Biltmore, *Montecito, Santa Barbara, CA 93108, (805) 969-2261* ☐ *Open all year* ☐ *5 mi S of Santa Barbara on US 101* ☐ The beautiful setting–ocean, mountains, sparkling white mission-style homes, a climate perfect for the lush growth–brought out the best in architect Reginald Johnson, who was commissioned to design this hotel in 1926. He created a perfect compromise between residential scale and the demands of a large resort. The building is a masterpiece of understatement–a low, rambling, tile-roofed hacienda, with graceful archways and corridors leading to hidden garden courts.

Each of the 178 rooms or 11 cottages are each decorated differently, but most units have refrigerators, and some have fireplaces, private patios, kitchenettes, or even private whirlpools. La Marina, the intimate Spanish-style dining room with exposed beams serves haute cuisine with the Pacific Ocean as the backdrop. Two cocktail lounges have nightly entertainment.

On the 21-acre grounds is a pool, or you may swim for a fee at the private ocean beach or olympic-size pool at the cabana club across the street. Bicycles are free. The hotel will make arrangements for golf, tennis, or fishing, and horseback riding nearby. ☐ **Lux** ☐ **W**

Circle Bar B Ranch, *1800 Refugio Canyon Rd., Goleta, CA 93017, (805) 969-1113* ☐ *Open all year* ☐ *Off US 101/SR 1, 20 mi N of Santa Barbara* ☐ Set in a wooded canyon 3½ miles from the Pacific, this informal family ranch offers 10 double rooms in the lodge and cabins. The dining room, serving family-style meals on MAP, has a rock wall and fireplace on one side and a window wall overlooking the pool and patio on the other.

Like the small heated swimming pool, the cabins are shaded by olive trees. The nicest of the cabins is the Honeymoon Cottage, with fireplace and king-size bed. Guests spend much of their time riding horseback, taking the ranch's trails along the ocean or through the mountain meadows. ☐ **Mod** ☐ **R**

San Ysidro Ranch, *900 San Ysidro La., Montecito, CA 93108, (805) 969-5046* ☐ *Open all year* ☐ *Take San Ysidro Rd. exit off US 101; then W 2 mi to San Ysidro Lane; follow San Ysidro Lane to ranch* ☐ Don't be misled by the name: This is not a run-of-the-mill dude ranch. With consummate taste and care, owners Jim and Susie Lavenson have restored to elegance this hillside bungalow colony that is one of the oldest resorts in California. San Ysidro was originally a cattle ranch for the Santa Barbara mission. Spanish missionaries erected a way station here in 1786, and rebuilt it after an earthquake in 1812. The oldest building still standing is an 1825 adobe cottage, now a museum and historical landmark. The ranch opened as a resort in 1893, and was later owned by Ronald Colman, who used it as a haven for his Hollywood friends. The ranch has been a hideaway of an extraordinary list of celebrities from Winston Churchill to John Galsworthy, and John and Jacqueline Kennedy honeymooned here.

The 540 acres are a lush collage of terraced gardens, lawns, forest, and rocky hillside tucked into the foothills of the Santa Inex Mountains which overlook the Pacific Ocean. Each of the 38 clapboard or stone bungalows, sited for privacy, offers its own version of elegant seclusion. The lodgings–from double bedrooms to a four-bedroom cottage with private whirlpool baths–are individually and beautifully decorated. Many units have extras such as fireplaces, kitchenettes, sundecks, or ocean views. Some occupants are tempted to hide-out and rely on room service, but the ranch has handsome public areas. The cozy lounge offers free coffee, newspapers, an "honor bar" where you mix your own drinks, and the ranch's only TV set. The charming dining room, with exposed-beam ceiling, arched doors, and a sunny porch, was converted from a citrus packing house dating from the ranch's days as a farm. The fine French cuisine might include prawns or veal as entrees. Men must wear jackets to dinner. In addition, Sunday brunch, box lunches, or weekend poolside barbecues are available. The bar, in an 1850 wine cellar, has dancing and entertainment.

The landscaped pool, three tennis courts (fee), and game areas, and children's play area are set off to one side of the cottage complex on a rise offering an ocean view. At the other end are the stables, where you may board your horse or rent one for guided trail rides, (instruction is available). There are also many hiking trails into the mountains. For golfers, play privileges are available at Montecito Country Club. Deep-sea fishing is out of Santa Barbara harbor. If you have children, the ranch will try to get baby sitters. ☐ **Lux** ☐ **R, T**

SOLVANG: The Alisal, *P.O. Box 26, Solvang, CA 93463, (805) 688-6411* ☐ *Open all year* ☐ *From Solvang, turn SW on Alisal Rd., then 3 mi to main gate of the Alisal* ☐ Combined within 10,000 acres of oak- and sycamore-studded terrain are the excitement of a working ranch and the amenities of a luxury golfing retreat. This small ranch prides itself on pampering guests with a large staff. Accommodations, which vary from single-occupancy chambers to large five- to seven-person bungalows, are scattered in old ranch-style buildings around the grassy and wooded grounds. To preserve ever precious quiet, the units are without telephone, TV, or radio. Most, however, have porches, and all are attractive and come with fireplaces.

From the string of 65 horses, the head wrangler will choose a mount to suit your ability. Guided morning or afternoon rides are regularly scheduled, and you may arrange for additional riding or for instruction. Children must be seven years old to ride. An 18-hole, par-70 golf course is serviced by a pro shop and a clubhouse with snack bar and cocktail lounge. The tennis buff will find seven courts and an instructor. The MAP rates cover hearty breakfasts and dinners of typical western dishes in the ranch dining room. The rustic cocktail lounge has entertainment; the library has a TV and fireplace. More elegant dining can be had for additional cost at the Sycamore Room. Lunches are available at the Sycamore Room or the Clubhouse. MAP includes all facilities except golf and horseback riding.

Also on the grounds are a large freeform pool, seasonal sailboat rental at the 90-acre private lake with fishing dock, recreation room, archery and outdoor games. Hayrides or other events may be scheduled. The ranch will help you with child care or with excursions to the Danish-style town of Solvang, nearby missions, and other sightseeing. ☐ **Lux** ☐ **R, G, T, W**

TRABUCO CANYON: Coto de Casa and The Vic Braden Tennis College, *22000 Plano Trabuco Canyon, P.O. Box 438, Trabuco Canyon, CA 92678, (714) 581-2900* ☐ *Open all year* ☐ *14 mi E of I-5 at the El Toro Rd exit S of Irvine* ☐ The focus of this 5000-acre resort community is the tennis complex, created by Vic Braden, one of the game's best known instructors. One of Braden's innovations was to set training lanes side-by-side, radiating out from a central core in which the ball machines are concentrated, thus fitting a number of lanes in a small space. Instructors walk along the periphery of the circle, keeping tabs on the students' progress. Facilities like this, together with the four video rooms for replaying shots, the three-story observation deck, and the 16 courts, are all the most professional player will ever need.

The rest of 5000-acre Coto de Casa resort includes a stable with 30 miles of equestrian trails, a hunting preserve for game birds, swimming pool, and a clubhouse with complete spa facilities and bowling, basketball, handball, racquetball, and game room area. The gourmet dining room has cocktail lounge and a patio. The ultramodern guest lodges come in a variety of styles. ☐ **Lux** ☐ **T, R**

TRINIDAD: Bishop Pine Lodge, *900 Patrick's Point Dr., Trinidad, CA 95570 (707) 677-3314* ☐ *Open all year* ☐ *23 mi N of Eureka on US 101 to Trinidad exit, then N on Patrick's Point Dr.* ☐ Just a few miles inland from the rugged seacoast park at Patrick's Point are the 13 shingled cabins of the Bishop Pine Lodge, clustered around a garden under azalea and redwood trees, these one- to four-person cabins, aside from being rather close together, are charming. An effort has been made to avoid plastics and distractions such as TV and telephones, and they are furnished in a colorful early American style with replicas of old oil-lanterns, wood-framed beds, and walls of knotty pine and redwood; all have small porch areas for sitting and enjoying the garden. The lodge does not have a restaurant, but 10 units have kitchenettes, and you can cook in the outdoor barbecue pits. There is a cafe just a few hundred yards from the entrance.

Activities in the area include ocean and surf fishing for salmon and steelhead, and clamming, rock hounding, and sunbathing, and swimming at the nearby beaches. Boat rental and fishing services are plentiful along the coast. Redwood National Park, one of the best places to view those towering giants, is about 20 miles north. It is also a good place for bird watchers and to see a herd of elk. ☐ **Inexp** ☐ **F**

TRINITY CENTER: Coffee Creek Ranch, *P.O. Star Route 2, Box 4940, Trinity Center CA 96091, (916) 266-3343* ☐ *Open late Mar to late Oct and mid-Nov to early Mar* ☐ *N or SR 3 from Trinity Center of Coffee Creek Rd.* ☐ Here's an informal resort for outdoor families in the beautiful Shasta-Cascade mountains. Cozy, wood-paneled, carpeted accommodation is in two rooms with private bath in the main lodge; homey one-bedroom cabins in an apple orchard and grove, each with potbelly stove; and two-bedroom cabins with heat and fireplace. All-you-can-eat meals featuring Black Angus beef and home-baked bread are served in the cheery dining room with wide-beam ceiling and wagon-wheel lanterns, (country-music dancing is also held here), and steak fries and barbecues are also scheduled. There is no bar, but set ups are provided.

Rates include lodging, all meals, maid service, use of the heated swimming pool, bow and arrows for the archery range, guns for the rifle range and trapshooting, canoes for the pond, fishing, hayrides, badminton and other games, gold panning and children's games, and daily supervised activity for kids aged two to nine. Horseback riding plus pack trips and hunting parties are available at extra charges. In winter the daily rates include cross-country skiing, sledding, ice skating on the pond, and shuttle service to ski areas; rental equipment can be reserved nearby. ☐ **Mod** ☐ **R, W, S**

Colorado

ASPEN-SNOWMASS: A booming silver-mining camp in the 1880s, Aspen is booming again as a vacation spot and an alpine Eden. Few towns anywhere can boast a more beautiful setting than this valley beneath the steep rock faces of 13,000-foot mountain peaks. The town is a charmingly eclectic blend of Gay Nineties frontier town, alpine village, and chic resort elements. In front of the many bars and restaurants in town, pickup trucks and off-road vehicles share the curb with Rolls-Royces. On the ski slopes above town you will see some of the most serious ski bums in the world as well as a bumper crop of bunnies showing off their new clothes.

Skiing is the main industry here, and skiers have a choice of four mountains. The closest is a few minutes' walk from any part of town. Aspen Highlands has the highest vertical drop in Colorado (3800 feet) and is designed especially for the advanced skier, while Buttermilk is primarily for beginners and intermediate skiers. Snowmass and Aspen, the other two areas, both have vertical drops greater than 3000 feet. A six-day ticket good at any of the four areas is available. All have ski school, rental, and lodge facilities.

Besides a reputation for world-class skiing, Aspen has been developing an international reputation for its cultural activities. Beginning in June the Aspen Music Festival Orchestra performs for nine weeks. For lovers of dance, Salt Lake City's Ballet West is based in Aspen every summer. And many of the lectures at the renowned Aspen Institute for Humanistic Studies are open to the public.

Hikers, climbers, hang gliders, rockhounds, and fly-fishermen find Aspen a convenient base from which to assay the Colorado Wilderness. And for those with a taste for civilized comforts, the town offers more than 80 restaurants and nightspots.

Lodging rates drop considerably in summer (occasionally to half the winter rates). Aspen offers more than a dozen motels, dorm rooms, and alpine-style inns such as the Bavarian Inn—801 W Bleeker, (303) 925-7391—and the Pomegranate/Plum Tree Inn—Box 1368, (303) 925-2700—at moderate rates. Outside of town are year-round guest ranches: Aspen's T Lazy Seven Guest Ranch—Box 240, (303) 925-7254—is three miles out and has two-room, ranch-style, kitchenette apartments, a heated pool, sauna, fishing, riding, sleigh rides, snowmobile rentals, and pack trips. Elk Mountain Lodge & Cabins—P.O. Box

116, (303) 925-7369–is ten miles south near the restored Ashcroft ghost town and ski-touring center. It has nicely spaced rustic and modern cottages that sleep one to six people, hiking, mountain climbing, riding, pack trips, hunting, Jeep tours, and trout fishing.

Most accommodations, however, are condominums that range from studios to medium-size apartments to luxury four-bedroom homes. An example is the in-town The Gant–P.O. Box K-3, (303) 925-5000–with one- to four-bedroom units with balconies, fireplaces, and kitchens, two swimming pools, outdoor hot tubs, five tennis courts (clinics, video analysis, ball machines), Jeep tours, and other activities, and luxury prices. Aspen Chateaux Condominium Rentals–P.O. Box 4949, 720 East Hyman, (303) 925-1400–offers moderate to luxury units in a number of condominum developments.

For a full list of lodgings ranging from the spartan to the regal, write: Aspen Reservations, Inc., Box 4546, Aspen, CO 81611.

Snowmass, in a glorious, mountain-ringed setting five miles from Aspen, is a separate enclave of lodges and condominums that share recreational facilities and amenities. At the heart of the village is a chic mall with handsome gourmet restaurants, outdoor cafes, bars, boutiques, and shops for sporting goods rentals and gear. Here there are modern inns offering attractive double rooms and suites at moderate to luxury prices. Stretching to either side along the valley are hundreds of condominium units of every conceivable size–priced moderately in summer and in the luxury range in winter. Most units have fireplaces, picture-window and balcony views, and plush extras.

At the Mall is the Recreation Office, which oversees the 17 swimming pools, whirlpools, saunas, indoor and outdoor game areas, summer and winter children's programs, and the teen center. Winter activities include cross-country skiing, ice skating, snowcat and snowshoe tours, dog-sled rides, NASTAR, and sleigh-ride barbecues.

Snowmass Country Club, with pro shop rentals and services, has 11 Laykold tennis courts and an 18-hole, 7071-yard championship golf course with tiered and undulating greens. The 8000-foot altitude here gives sports an extra zing. Snowmass Stables provides riding, instruction, and overnight pack trips. Trout fishermen will find excellent sport in the Roaring Fork and Frying Pan rivers. Also available are guided Jeep tours, whitewater rafting on the Colorado or Roaring Fork, guided and self-guided hiking, a Parcours (jogging and fitness course), archery, sailing, craft workshops, and art fairs. A highlight are tours of the summer kennels of the famed Krabloonik dog-sled teams. For information, contact Snowmass, Box 5566, Snowmass Village, CO 81615, (303) 923-2000.

BAYFIELD: Wilderness Trails Ranch, *Rte. 1, Box B, Bayfield, CO 81122 (303) 884-2581 (summer), 247-0722 (winter)* ☐ *Open Jun 1 to early Sept* ☐ *28 mi NE of Bayfield via Vallecito Lake Rd.* ☐ Wilderness Trails is an AP family ranch secluded in a lush valley within San Juan National Forest. The main ranch house, an attractive long building with green trim, houses a guest lounge with a stone fireplace, the trading post (which sells basic supplies and Indian jewelry and rugs), and a rustic dining room which opens onto a long balcony. Surrounding the ranch house are the one- to five-bedroom log cabins with private bath and private porch. Room service is available.

Daily trail rides tailored to individual ability explore the surrounding national forest. Horseback riding instruction is available throughout the week. A weekly pack trip is offered, and guests are encouraged to take part in the weekly rodeo. Vallecito Lake, Colorado's third largest, is just two miles away, and the ranch rents boats. Near the ranch house is a stocked trout pond, and many nearby mountain lakes and streams for dedicated fishermen. Trapshooting, volleyballs, target shooting, and mountain climbing are the other main activities here.

Children have their own mealtimes and their own activities, which include instruction in riding, fishing, roping, Western lore, and square dancing. □ **Mod □ R, F, W**

BRECKENRIDGE: Fortunately, in turning this historic mining town into a ski resort, the Breckenridge citizenry preserved the old false-front stores and Victorian-era log and frame homes. Thus, while thousands of visitors pass through annually, Breckenridge retains a small-town feel. Many of the historic buildings have been converted to charming boutiques and restaurants—you can savor aged Colorado beef in a former bordello or miners' boardinghouse. And there's plenty of nightlife, from quiet cocktails to discos.

Dillon Reservoir is only nine miles north, and in summer Breckenridge offers tennis, golf, horseback riding, hiking, canoeing, fishing, alpine slide, whitewater rafting, and Jeep trips to ghost towns in the surrounding mountains.

In winter, the focus is on skiing in the Ten Mile Range that looms above the town. With a 2213-foot vertical drop, this is a full-service ski area with ten double chairs, a triple chair, Poma lifts, and T-bars servicing more than 65 runs. The ski school includes a nursery and children's classes. Also here are rentals, patrol, cafeteria, and bar. Cross-country skiing is also popular here, and experienced guides lead tours into the backcountry. Most lodgings here are available on package plans, and the Ski the Summit plan includes lift tickets interchangeable with those of nearby Copper Mountain, Keystone Mountain, and Arapahoe Basin. Downhill conditions are best December through April.

Accommodations include cheap dorms, moderately priced motels and lodges, and moderate to luxury studios and multiple-bedroom units in lodges, condominium apartments, or private homes. For a complete list, write to: Breckenridge Resort Association, P.O. Box 1909, Breckenridge, CO 80424.

COLORADO SPRINGS: The Broadmoor, *Colorado Springs, CO 80901, (303) 634-7711* ☐ *Open all year* ☐ *From Colorado Springs take US 85 5 mi S, then SR 122 1 mi W* ☐ After being banished from the Antlers Hotel for allegedly riding his horse into the bar, multimillionaire Spencer Penrose decided to build his own luxury hotel and playground. The Broadmoor opened in June 1918 with a cast of the titled and the monied and has been one of the most popular luxury resorts in the country ever since. Few hotels boast a more spectacular backdrop: bold Rocky Mountain peaks–including Pikes Peak–rise dramatically beyond the manicured lawns and gardens, lake and fairways. Somehow the nine-story Italian palazzo with its pink-stucco facades and red roofs seems perfectly at home in this setting.

Marble staircases, ceilings with intricate tile and plasterwork, alcove and ceiling frescoes in Wedgwood blue and white, and crystal chandeliers are the kind of details that Penrose insisted upon and that make the Broadmoor one of the grandest hotels in the country. Guest rooms and suites are large and opulently furnished in a variety of styles. The lodgings at the new Broadmoor West are the most modern and expensive. Altogether the resort has eight cocktail lounges (horses are not permitted in any of them) and almost as many restaurants. The Penrose Room, an elegant Edwardian-style restaurant decorated with British antiques and artwork, offers continental cuisine with vintage wines, some of which date back to Penrose's legendary stock hoarded during Prohibition. The adjacent cocktail lounge enjoys a spectacular view of the Rocky Mountains and the golf courses to the west, and the Great Plains to the east. There's dinner dancing in the Tavern and in the Penrose Room. The Terrace Lounge features top names at the piano, and Spec's Spot has dancing and live entertainment. The Golden Bee, modeled after an old English pub, features honky-tonk and ragtime music.

The three 18-hole golf courses have hosted a large number of amateur and pro tournaments over the years. The East course was designed by Donald Ross, and Robert Trent Jones laid out the West course. The third course was designed by the Arnold Palmer organization. Within the main clubhouse are a game room, whirlpool baths, squash courts, a cocktail lounge, a dining room, and tennis and golf shops. Spread out around the grounds are 16 tennis courts, two of them indoors for winter play. The tennis program includes frequent clinics, round robins, and tournaments. Ski Broadmoor, five minutes away by car, has chairlift, ski school, and rental equipment. Adjacent to the ski area is the Broadmoor shooting grounds with facilities for trap, skeet, and rifle shooting. The Broadmoor has a year-round skating rink and arena for recreational skating, figure skating, curling, and hockey. Also on the 500-acre grounds are three swimming pools with poolside service, saunas, beauty and barber shops, liquor store, movie theater, bicycle rental, a social director, and other services. In summer, a children's program and a children's dining room are operated, and babysitters are available all year. You will probably want to explore the mountains by cable car, car, or cog railway, and nearby attractions include the U.S. Air Force Academy and the Cheyenne Mountain Zoo. ☐ **Lux** ☐ **G, T, S**

COPPER MOUNTAIN: Copper Mountain, *P.O. Box 1, Copper Mountain, CO 80443, (303) 668-2882* ☐ *Open all year* ☐ *70 mi W of Denver off I-70* ☐ Copper Mounain is a still-developing, year-round resort village offering a dozen modern lodge rooms and 18 suites, plus more than 200 condominium units in large lodges spread across the valley at the base of the ski lifts. The condominiums vary from studios (all with kitchen, some with fireplace) to one- to three-bedroom apartments with living room, kitchen, and fireplace. Most of the lodges have a laundry, some have saunas, and two have indoor swimming pools. Private homes in the valley are also available. At the western end of the complex is a separate resort for Club Med members. At the village center and at the lifts are a number of restaurants (Italian, German, Mexican, and other cuisines), some with disco dancing or entertainment. Although everything is within walking distance, a shuttle service connects village areas and the lifts.

The Copper Mountain ski area has snowmaking, with skiing as early as mid-November and as late as mid-April. The base elevation is 9600 feet, and the vertical drop is 2213 feet. Serving the 46 trails are ten double chairlifts and two surface lifts. In addition to the adult ski school, NASTAR, a Junior Ranch (ages four to six), and a Senior Ranch (ages seven to twelve), a program is offered for nonskiing toddlers. More than 30 km of cross-country trails are maintained, and guided backcountry tours are offered. Also here are a rental shop and ice skating on the lighted pond. Winter plans include options such as Ski the Summit (lift tickets interchangeable at Arapahoe Basin, Breckenridge, and Keystone) or a day of skiing at Vail, and there are special rates for children and first-time skiers.

In summer, a chairlift operates to the mountain summit for a barbecue lunch. Nearby is a nature center with marked trails among the Ponderosa pine and spruce. The Golf and Racquet Club offers six Laykold courts, pro for lessons and clinics, ball machines, video replay, and a nine-hole golf course winding around the village. The stables offer hourly rental or trail rides. Two ponds in the village are West Lake, stocked for children, and East Lake, with small rental boats. More serious fishermen can try the nearby lakes and streams, and Lake Dillon is ten miles away. Fly-fishing instruction is available. Also in the village are many game courts, bicycle rental, and a large playground that was designed by children. The Copper Mountain Information Center will arrange whitewater rafting on the Colorado River and other off-premises trips. Summer also brings educational and cultural events, including concerts, plays, films, and seminars. ☐ **Mod to Lux** ☐ **S, T, F, R**

CRESTONE: Baca Grande Inn and Resort,*P.O. Box 126, Crestone, CO 81131* ☐ *Open all year* ☐ *8 mi E of SR 17 or SR 62* ☐ Set in a broad valley of rolling grassland against a background of the Sangre de Cristo peaks, Baca Grande is an easygoing resort encompassing a working ranch and condominium complex. The inn here is a low-slung, contemporary adaptation of pueblo architecture. The 40 EP rooms are large and airy, equipped with TV, telephone, radio, and private bath, and furnished in Spanish style. The cocktail lounge and the restaurant, with a Mexican-American menu, are the communal centers. The inn looks out on a 14-acre pond stocked with trout. Sailboats, rowboats, and pedal boats are available for rent. Lodge guests also have the use of a 9-hole, 2100-yard golf course, a tennis court, and a heated outdoor swimming pool.

The 150,000-acre ranch stretches up into the hills. Guests can watch the cowboys at work or take to the horses themselves. The Baca Grande Stable offers riding lessons and steady or spirited mounts on which to explore the nearby mountains. □ **Inexp** □ **R, F**

DURANGO: Colorado Trails, *Box 848, Durango, CO 81301, (303) 247-5055* □ *Open June 1 to early Sept* □ *12 mi from Durango on Florida Rd.* □ Colorado Trails is a family-oriented guest ranch on 500 acres of green valley beneath timbered mountains. The lodge is a timber and stone ranch building with a long porch and high-ceilinged, post-and-beam interior. Some of the older cabins are of similar contruction while others, such as the A-frame alpine cabins, are contemporary in design, with window-walls-and modern furnishings. The ultramodern X-wing cabins have two sets of two connecting bedrooms, one with queen-size and one with twin beds. All units have private bath. Meals are served in the handsome dining room, which has wide-plank spruce floors and ceilings and a large stone fireplace.

Riding instruction is provided from the first day and the trail rides are divided according to ability into small, guided groups. Day trips, overnight pack trips, and breakfast rides are part of the program. Once a week the ranch holds an informal rodeo in which guests compete. The ranch supplies the rifles, shotguns, and bows for its archery, riflery, and trapshooting range. Guests also have a heated outdoor pool and two Laykold tennis courts. Other facilities include a children's game room, a whirlpool bath, a snack bar, and trading post store. Evenings are busy with barbecues, square dancing, hayrides, and staff talent shows. Children and teenagers have their own programs. The AP rates include horses, all on-premises facilities, and all sports equipment–everything except overnight pack trips. Nearby Durango offers river trips, jeep tours, the Silverton narrow-gauge railway, and other activities □ **Mod to Lux** □ **R, F, T**

Tall Timber, *SSR Box 70, Durango, CO 81301, (303) 247-1412-unit 13* □ *Open early May to early Mar* □ *Take the Silverton from Durango* □ The Silverton Railroad, a narrow-gauge railway, winds through dense forest and clings precariously to the jagged cliffs above the canyon del Rio de las Animas Perdidas–the River of Lost Souls. Between 1890 and 1907 the railroad hauled more than $300 million in ore out of these mountains and carried passengers like Doc Holiday and Wyatt Earp. More recently it has appeared in the films "How the West Was Won" and "Butch Cassidy and the Sundance Kid." It is also the only way to Tall Timber Lodge in the middle of San Juan National Forest.

Tall Timber is a genuine retreat. There are no TVs or telephones here. The railroad and an emergency radio phone provide the only links with the world outside. People who can't stand to hear themselves think should consider another vacation. On the other hand, it's not quite a Zen monastery. The accommodations are luxurious and there is plenty to do. The obvious pastime here is hiking the valleys and mountain trails of this spectacular wilderness area. Both stream and lake fishing are nearby. For hiking-sore muscles there's a Finnish sauna and hot spa, as well as a pool, tennis court, and putting green. Riding can be arranged at extra cost.

Tall Timber guests stay in two-story chalets, rough-timbered outside, cozy and luxurious inside. In the two-bedroom Timber Hearths and one-bedroom Timber Suites an inside balcony from the second-floor master bedroom overlooks the massive stone fireplace. In the peak summer season chalets rent by the week; shorter stays are possible at other times. Rates are AP, and the dining room features international cuisine and fine wines, but informal attire. You needn't change out of your hiking boots to enjoy a drink at the bar or a fine meal. This combination of roughing it and luxury is what Tall Timber is all about. □ **Mod to Lux** □ **H, F**

Tamarron, *P.O. Box 3131, Durango, CO 81301, (303) 247-8801* □ *Open all year* □ *18 mi N of Durango on US 550* □ This modern, 630-acre resort, surrounded by the forested hills of the San Juan National Forest, seems designed for an active crowd with a well-developed taste for the good life. Year-round recreation in modern, luxurious surroundings is the theme here. Rooms in the modern glass and timber lodge are large and opulently furnished in contemporary Spanish motif. "Standard" rooms with two queen-sized beds, full bath, kitchenette, and vanity would qualify as deluxe at many other establishments. The deluxe rooms are luxury suites with a sleeping loft. Also luxurious are the one- to three-bedroom town-house units with full bath, kitchen, bar, and living room. Many have fireplaces and balconies.

Excellent skiing is just a few miles away at Purgatory, which has a base elevation of almost 9000 feet and four double chairlifts, school, rentals, nursery, restaurant, bar, and cafeteria. Cross-country skiing, sleigh rides, and snowcat expeditions are among the other winter activities. Tennis here is a year-round sport. The four Laykold courts and the two platform tennis courts are indoors. The 7015-yard golf course is open spring through fall. Horseback riding is another three-season sport. The stable boasts 20 mounts and a program of instruction, breakfast rides, and trail rides. Fishing is in a private lake. After a day of sports you can relax in a whirlpool, steam or sauna in the men's and women's health clubs and follow it up with a professional massage. A complete spa program which includes diet, exercise, and skin analysis is also offered.

You may book here on EP, with sports extra, or on sports package plans. Dining choices include a plush, velvety dining room with continental cuisine and the San Juan Dinner Club with a different ethnic meal nightly. Tamarron is a full-service resort with nightly entertainment, children's programs, barber and beauty shops, liquor store, delicatessen, and many other extras. ☐ **Lux** ☐ **G, T, R, S**

ESTES PARK AND ROCKY MOUNTAIN NATIONAL PARK: Straddling the continental divide, Rocky Mountain National Park encloses in its 262,000 acres some of the most spectacular scenery of the entire Rocky Mountain Range. The highest peaks, almost 100 of which are more than 10,000 feet high, are perpetually snowcapped, while the glacier-carved valleys and ravines glisten with rivers, lakes, and more than 750 varieties of plant life. Wildlife–including deer, elk, bighorn sheep, beaver, and golden eagles–is abundant. Presiding over the park is 14,256-foot Longs Peak.

The park is accessible from the east via Estes Park, a year-round resort town surrounded on three sides by Roosevelt National Forest. Trail Ridge Road winds through the park from Estes Park to the western entrance at Grand Lake, and rises over 12,000 feet to cross the continental divide, affording thrilling views. There are abundant opportunities for hiking, trail riding, birdwatching, and fishing within the park. The Hidden Valley Ski Area is within the park 10 miles west of Estes Park.

Anderson's Wonder View Motel and Cottages, *P.O. Box 427, Estes Park, CO 80517, (303) 586-4158* ☐ *Open all year* ☐ *3½ mi W of Estes Park via US 36, then right on High Drive Rd.* ☐ These cozy cottages, named for a wonderful view of Longs Peak, have contemporary, motel-style funishings, kitchenettes, fireplaces, baths, TV, large picture windows, front and rear doors, outdoor grill, and one to three bedrooms. Double-room motel units are also available. On the premises are a heated outdoor pool, lawn game areas, and a recreation room. ☐ **Inexp to Mod** ☐ **R, F, H, S**

Elkhorn Lodge, *Box 1560, Estes Park, CO 80517, (303) 586-4416* ☐ *Open May–Oct* ☐ *In Estes Park off of 34 W* ☐ Established as a homestead in 1874, Elkhorn Lodge is the oldest continuous business in Estes Park and was recently named a National Historic Landmark. The old lodge still houses guests, while the newer main building contains guest rooms, suites, a lounge, soda fountain, dining room, and candy kitchen. In addition, there are one- to five-bedroom rustic log cabins, closely grouped, without kitchen facilities. The Coachhouse Cocktail Lounge features steak dinners, dancing, and entertainment nightly. A heated outdoor swimming pool, shuffleboard courts, and game room are also on the 20-acre grounds. The lodge borders Fall River, with good fishing. Stables provide trail rides, steak rides, park trips, and hunting trips. □ **Inexp** □ **F, R, H, S**

Lane Guest Ranch, *P.O. Box 1766, Estes Park, CO 80517, (303) 747-2493* □ *Open June to Sept* □ *12 mi S of Estes Park via SR 7, watch for sign on left* □ Secluded in the woods among the peaks of Rocky Mountain National Park, Lane Guest Ranch offers daily trail rides into the mountains. Guests also explore the countryside on foot along the many hiking trails and on Jeep expeditions across the continental divide. The ranch has turned a variation on the standard breakfast and steak rides with an adult wine and cheese ride. Many activities for children, including hikes, cookouts, campouts and movies, are part of the daily schedule. Nighttime entertainment includes disco dances, bingo games, old movies, and fireworks displays. The MAP rates include breakfast in bed. Among the dinner specialties are the New York Steak buffet dinner and prime rib au jus with Hollandaise sauce. All dinners are served with homemade soups. Beside the heated outdoor swimming pool is a European-style cafe. For out-and-out relaxation there are a whirlpool bath and library with a fireplace.

Spread out around the property are the 23 guest cabins, each with shag carpet, controlled heat, private tub or shower, maid service, and picture windows taking in views of the mountain peaks. Numerous lakes and streams near the ranch provide a variety of trout fishing. **Mod to Lux** □ **R, F, H**

Longs Peak Inn & Guest Ranch, *Longs Peak Route, Estes Park, CO 80517, (303) 586-2110 (Summer) 586-3589 (winter)* □ *Open June to mid-Sept* □ *7 mi S of Estes Park on SR 7* □ This cluster of main inn and large guest lodges, set in a meadow at the foot of the mountain from which it takes its name, is essentially a family-oriented guest ranch. The modern lodges have double rooms and two-bedroom family units, many expandable with sofa beds and rollaways. All units are fully carpeted and have private bath. The AP or MAP rates include all ranch activities except horseback riding; weekly package rates include a horse.

Daily trail rides take you into Rocky Mountain National Park, and overnight pack trips can be arranged. Four stocked ponds are on the property, and many mountain lakes are nearby. Children 6 to 12 are supervised by counselors. A heated outdoor swimming pool and sun deck, a game room, family lounge, cocktail lounge, hiking trails, and outdoor game areas are among the facilities here. The glass-walled restaurant commands a stunning mountain view. □ **Inexp to Mod** □ **R, H, F**

Machins Cottages in the Pines, *P.O. Box 88, Estes Park, CO 80517, (303) 586-4276* □ *Open mid-May to late Sept* □ *3 mi SW of Estes Park via US 36, private road marked by sign* □ Secluded one- to three- bedroom cottages with picture-window views of the peaks of Rocky Mountain National Park, fireplaces, color TV, fully equipped kitchens, wood-paneled walls, and private baths. Each cottage has a private picnic area with charcoal grill, picnic table, and lawn furniture. **Inexp** □ **H, R, F**

McGregor Mountain Lodge, *P.O. Box 1969, Estes Park, CO 80517, (303) 586-3457* □ *Open all year* □ *3½ mi W of Estes Park on US 34* □ Overlooking Fall River Canyon, here are attractive one- and two bedroom frame cottages in a 10-acre wooded, natural setting, with lots of wood inside, as well as fireplaces, TV, private bath, patio or balcony, kitchenette, and room service. □ **Inexp to Mod** □ **H, R, S**

Rams Horn Cottages, *P.O. Box 1438, Estes Park, CO 80157, (303) 586-4338* □ *Open all year* □ *2 mi S of Estes Park on SR 36 & 66* □ On six wooded acres, these one- to four-room cottages have contemporary furnishings, kitchenette, and outdoor grill and picnic area; some have fireplaces. On the premises are a heated outdoor pool, game areas, and a playground. □ **Mod** □ **H, S, F, R**

GRANBY: Arapaho Valley Ranch, *P.O. Box 142, Granby, CO 80446, (303) 887-3495* □ *Open June 1 to early Sept* □ *16 mi NE of Granby off US 34* □ Nestled in magnificent country at the point where the South Fork of the Colorado River joins Lake Granby, this ranch is at the end of a secondary road just west of Monarch Lake and the continental divide. It generates its own electricity, maintains its own water system, and keeps a stock of staples, fishing tackle, and other necessities in its general store. Lodging is in 19 rustic cabins scattered among spruces and aspens. These have living room, equipped kitchen, bath, heat, fireplace, private porch, and one to four bedrooms. You can stay here on AP, MAP, or EP, but even if you do most of your own cooking, you can arrange to have occasional dining room meals or join the cookouts. Rates are modified depending on whether you want full-time use of a horse; if not, you can rent horses by the hour or day. Included in the rates are use of the heated pool, wading pool, outdoor game areas, a pretty pond with free rowboats stocked for children's fishing, and activities such as songfests, hayrides, and cookouts. Also included is all-day supervision of children over age four. Fishing (for trout, Kokanee, salmon, and mackinaw) is good in streams and in lakes Granby and Monarch. The ranch rents canoes, rowboats, and motorboats for the lakes, as well as fishing rods and backpacking equipment. Park trips can be arranged. □ **Mod** □ **W, F, R**

C Lazy U Ranch, *P.O. Box 378, Granby, CO 80446, (303) 887-3344* □ *Open June–Sept and mid-Dec–March* □ *From jct US 40, 3 mi W of Granby, take SR 125 3½ mi N to ranch* □ Successfully blended here are the hardy, outdoor lifestyle of a ranch with the luxury and facilities of a full-service resort. The 5000-acre spread is set in a valley near the headwaters of the Colorado River. Willow Creek flows through the ranch into adjacent Willow Creek Reservoir; both are good for fishing.

The two-story main lodge, a handsome lodge building, is set on a hillside and has porches on both stories, providing nice views of the ranch. Inside are the dining room and lounge; both have a large stone fireplace. The dining room fare tends toward steak and prime ribs, accompanied by fresh salads and vegetables and homemade breads and pastries. Also in the lodge are a bar and card room.

The lodge is flanked by two wings, each with three guest rooms. Larger units, including multi-bedroom family suites, are available in the sleeping cottages set around the ranch. Some units have fireplaces. All rooms are individually decorated, and the management pays careful attention to small details—for instance, a basket of fresh fruit is placed in every room each morning.

The principal summertime activity here is horseback riding, which is included in the AP rates. The ranch maintains a string of 130 horses, providing ample opportunity for you to find the appropriate mount. Two rides are scheduled every day but Sunday, and there's an all-day picnic ride on Thursday. The highlight of Saturday afternoon is the Shodeo—part rodeo and part horse show. Two Laykold tennis courts, skeet range, indoor racquetball court, a swimming pool with poolside service, whirlpool, sauna, and outdoor game areas are also provided. Next to the pool is Trail Peak Lodge, the center for evening entertainment, square dancing, Ping-Pong, and other indoor games. For children and teenagers, separate programs include riding, swimming, cookouts, and other activities, and babysitters are available.

In winter, cross-country skiing, with rentals and free instruction, is featured. Free bus transportation is offered to Winter Park, a full-service downhill ski area. □ **Lux** □ **R, F, S**

Drowsy Water Ranch, *Box 147, Granby, CO 80446, (303) 725-3456* □ *Open June 1 to Sept 14* □ 7 mi W of Granby on US 40 □ Named for the wide creek that winds through the property, Drowsy Water Ranch is set in high, semiarid country adjoining Arapaho National Forest. Log cabins for guests are set in a stand of aspen overlooking the creek and have spare but comfortable furnishings, pine paneling, and private baths. Single and double rooms in the main lodge are also available.

The weekly schedule here begins with riding instruction and shorter trail rides as well as Jeep expeditions into the backcountry. In the evening there's square dancing in the Teepee, the recreational building. Alternate activities include raft trips, sailing expeditions on nearby lakes, and golf at the 18-hole course on Grand Lake. Hayrides, campfires, and bingo are on the evening agenda. Children have their own program of activities. Saturday is rodeo day, with ranch guests participating in many of the events. Unscheduled activities include swimming in the outdoor heated pool or fishing in the creek and stocked ponds. In fall the ranch is a base for deer and elk hunting. All on-premises activities except overnight pack trips are included in the AP rates. **Mod to Lux** □ **R, F**

GRAND LAKE: Daven Haven Lodge and Cottages, *P.O. Box 528, Grand Lake, CO 80447, (303) 627-3381* ☐ *Open all year* ☐ *4 blocks S of Grand Lake Village on west shore of lake* ☐ Set on Grand Lake at the western entrance to Rocky Mountain National Park, this EP family resort has a year-round roster of activities. Fishermen and other waterbugs will take to Daven Haven. The private beach and the dock are just a few steps from the Lake Lodge, and sailboats and outboards are available for rent. Rocky Mountain National Park is a natural paradise for hiking, wildlife viewing, and climbing. Snowmobiling, ice skating, ice fishing, and cross-country skiing are the major winter attractions. Rental equipment is available.

There are two lodges, Lake Lodge and the Main Lodge, which is set back from the lake. Many rooms have mountain and lake views. All have private baths and TV. A dozen frame sleeping cottages among the pines surrounding the swimming pool are also available. The restaurant serves three meals daily and two cocktail lounges serve legal beverages. Daven Haven is only four blocks from the town of Grand Lake. **Inexp to Mod** ☐ **W, F, S**

GRANT: Tumbling River Ranch, *Grant, CO 80448, (303) 838-5981* ☐ *Open all year* ☐ *NW of Grant off of US 285* ☐ At an altitude of 9200 feet, surrounded by 12,000-foot peaks, the setting of Tumbling River Ranch is literally and figuratively breathtaking. The original homestead, built at the turn of the century, is now a guest house. The other ranch dwellings were built in the 1920s by a former mayor of Denver and by a daughter of Adolph Coors. Rooms in both the upper ranch house and the lower ranch pueblo, with its hand-carved beams and unusual Spanish-Indian decor, have fireplaces and private baths, as do the six cabins surrounding the upper ranch house. A sauna, lounge, and game room are in the upper ranch and the pool and corral are just outside. Family-style meals are served in the dining room in the lower ranch.

An AP week at Tumbling River begins with a riding orientation program in the corral. Guided trail rides of varying lengths leave the ranch every day except Sunday. Saturday is rodeo day, with competitions for adults and children. The children's program includes a trip to Silver Plume for an hour-long ride on a narrow-gauge train. There are two children's groups, ages three to six and seven to twelve, each with its own counselors, plus a teen program of hiking, riding, and entertainment. Raft trips on the Arkansas River and Jeep trips into the mountains in search of reminders of the mining era are part of the program. Fishermen have Geneva Creek, which runs through the ranch, as well as two stocked ponds on the property. Those looking for real native trout can pack into the mountains.

The winter program includes cross-country skiing on the many trails surrounding the ranch, and alpine skiing at Geneva Ski Basin, with two chairlifts and two Pomalifts, which is just a few minutes away from the ranch. ☐ **Mod to Lux** ☐ **R, F, S**

GUNNISON: Harmel Ranch Resort, *P.O. Box 944V, Gunnison, CO 81230 (303) 641-1740* ☐ *Open mid-May through Sept* ☐ *10 mi N on SR 135, then 6.5 mi E on Taylor River Rd.* ☐ On 300 acres on the Taylor River, this family ranch offers attractive accommodations in the main lodge and more than a dozen housekeeping cottages. Most people stay here on MAP, but family rates and EP are also available. In addition to restaurant meals, the riding program includes wranglers' breakfasts, chuckwagon dinners, and steak fries. Riding instruction, trail rides, moonlight rides, and hayrides are also offered. There is good fishing, and supplies, guides, and float trips are available. Also on the ranch are a swimming pool, playground, lawn game areas, and recreation room. An activities director arranges special events. ☐ **Mod to Lux** ☐ **R, F**

Waunita Hot Springs Ranch, *Gunnison, CO 81230 (303) 641-1266* ☐ *Open mid-May to Oct 1* ☐ *19 mi NE of Gunnison via 50, watch for sign* ☐ Besides the basic dude ranch facilities and activities, this ranch has the added attraction of a year-round natural hot springs pool. The pure water of the spring also provides all the hot water on the ranch. The hot pool may be just the thing after an all-day ride into Gunnison National Forest which surrounds the ranch. Less adventurous riders can stick to the trails that wind through the 480-acre ranch. Jeep trips to nearby mining towns, hayrides, canoe trips, and trail breakfasts are among the activities Waunita offers its guests. At night there are melodramas, singalongs, and cookouts. Fall visitors use Waunita as a base for deer and elk hunting and in the winter groups may reserve use of the ranch for snowmobiling and cross-country skiing.

Guests stay in the two-story white clapboard ranch house in rooms with private bath and either tub or shower. Family-style meals are served in the dining room. AP rates include all activities. ☐ **Inexp** ☐ **R**

KEYSTONE: Keystone Resort, *Box 38, Keystone, CO 80435, (303) 468-2361* ☐ *Open all year* ☐ *6 mi E of Dillon on US 6* ☐ Spreading out on either side of the beautiful Snake River, this is a well-designed, planned community with lodge, condominium units, and private homes all sharing extensive year-round resort facilities. The main center of activity is the village plaza on Keystone Lake, where Keystone Lodge offers 152 rooms and suites. Around the lake and along either side of the river are clusters of lodges with hundreds of condominium units sized from studios to four-bedroom apartments with kitchens. All units have huge windows, and many have extras such as fireplaces, balconies, or patios. For those who would prefer a more intimate, old-fashioned atmosphere, Ski Tip Ranch, less than a mile from the center, has some log buildings more than a century old; it houses up to 50 people and has its own dining and cocktail facilities. Units are rentable on EP and package plans, and a free shuttle bus connects all areas of this vast complex.

Keystone Lodge offers elegant dining in the lakeside Garden Room—varied continental cuisine, a good wine list, and Sunday brunch. Also here are a steak house, a lounge with entertainment, and a family restaurant on the lakeshore that offers buffet breakfasts and informal meals. The village also has a seafood restaurant, an English-style bar and restaurant with lake view, a delicatessen, an ice cream parlor, and a disco. Food and drink can also be had at the base of both ski areas, and Keystone Mountain has a summit restaurant. Less than a mile from the village plaza is the colorful and convivial Snake River Saloon, on the river, which also offers Sunday brunch.

The Sports Desk staff in the village plaza is the information center for Keystone action. They can tell you the location of the various swimming pools, saunas, and whirlpools. The Sports Desk arranges on-premises activities, such as guided nature walks in the village, as well as off-premises trips–Jeep tours of backcountry, guided mountain climbing, and bus tours to historic mining camps.

The Keystone Golf and Country Club centers on an 18-hole Robert Trent Jones course. The clubhouse is a beautiful Western log building, with huge stone fireplaces, that was originally a wealthy ranch house. This complex has tennis courts, swimming pool, pro shop, restaurant, bar, lounges, and many services. Behind Keystone Lodge is the John Gardiner tennis center, with 14 indoor and outdoor courts, pro shop, rentals, instruction, clinics, ball machines, and a full range of modern teaching aids.

Keystone Stables provide hourly rentals, lessons, stagecoach and hayride tours, trail rides, rides with cowboy breakfasts or Western barbecues, and rodeos.

Keystone Lake has rental canoes and paddleboats, and you can learn kayaking here. Sailboat rental and sailing cruises are available at Dillon Lake, five miles away. Whitewater rafting is offered on the Colorado River. For fishermen, Keystone Lake is stocked, guided tours are offered to backcountry lakes, and lessons in fly-casting and fly-tying are offered.

Also at Keystone are well-marked hiking trails (you can rent backpacks) and bicycle trails that include a lovely paved path along the Snake River (rental bikes are available). Chairlift rides up Keystone Mountain are run for summer sightseers. Mountain House (at the base of Keystone Mountain) runs a summer child-care program that includes day care for infants over six months old, evening babysitters, supervised activities for children (kayaking, cookouts, riding, etc.), and a teen program.

Winter sports focus on Keystone Mountain and Arapahoe Basin (five miles). Keystone has a vertical drop of 2340 feet, 27 trails, and 9 lifts. The longest run is three miles, and most of the trails are intermediary. Arapahoe has a 1640-foot vertical drop, 22 trails, and 5 lifts; the longest trail is 1½ miles; 40% of the trails here are for advanced skiers. Both areas have rentals, ski schools that offer everything from beginner's lessons to helicopter skiing, NASTAR, restaurants, and bars. Various types of lift tickets may be used at either area or at Copper Mountain and Breckenridge too. Mountain House winter child care includes activities for toddlers and ski school for children over age four.

The ice skating center is at the village plaza. At the opposite end of the lake is the cross-country center offering lessons, rentals, day and moonlight tours, a racing clinic, films, and demonstrations. Some 22 miles of trails are maintained in the village, and routes go out to scenic and historic areas. Snowcat tours and snowshoe rental are also available. ☐ **Mod to Lux** ☐ **W, R, F, S, H**

LOVELAND: Sylvan Dale Ranch, *2939 North Country Road 31 D, Loveland, CO 80537 (303) 667-3915* ☐ *Open June to early Sept* ☐ *9 mi W of Loveland on US 34* ☐ Sylvan Dale is not a dude ranch but an unpretentious 4000-acre working cattle ranch where the main lodge is set on the banks of the beautiful Big Thompson River. The 25 lodging units consist of double rooms and two- and three-bedroom units in the ranch house or in frame cabins scattered around the grassy grounds. All units are carpeted and have private bath. The AP rates include use of the heated outdoor pool, Laykold tennis courts, indoor and outdoor game areas, playground, supervised children's program that includes insights into a working ranch, and planned entertainment such as cookouts, square dances, and hayrides. Fishermen can wet their lines in three stocked lakes as well as in the Big Thompson. Horses here are extra, but breakfast rides, instruction, pack trips, and seasonal cattle drives are available. The Jessup family, who have run the ranch for over 30 years, are eager to impart a love of the landscape and an understanding of ranch life, and welcome your participation in ranch chores. ☐ **Inexp** ☐ **R, T, F**

LYONS: Peaceful Valley Lodge and Guest Ranch, *Star Route, Lyons, CO 80540, (303) 747-2582* ☐ *Open all year* ☐ *18 mi SW of Lyons. Take SR 7 to SR 72* ☐ Austrian-born Karl E. Boehm has given his guest ranch in the Rockies a distinctly alpine flavor. The lodges look Swiss, and little details like family crests add variety to the usual dude ranch appointments of pine paneling and elk antlers. Lodge rooms are basic motel-style with a choice of deluxe, featuring oversize beds, or smaller economy rooms. Private and shared bathrooms are available. You may also rent a rustic cabin in the woods. The two dining rooms enjoy views of the surrounding woods and mountains. Indoor family-style meals alternate with cookouts, picnics, and trail breakfasts.

The mountainous country around the ranch can be explored on foot with a hiking group or on horseback. Children and teenagers have their own programs of supervised riding and hiking. Native trout inhabit the pools of the St. Vrain River and the ranch's stocked pond. A special feature is the weekly Gymkhana–a cross between a rodeo and a horse show–at which guests can participate. AP rates include all activities.

The winter season is devoted mainly to cross-country skiing. Instruction in all levels of ski touring is offered, and six acres of snowmaking guarantee good conditions in the instruction area. Miles of trails above and below timberline are maintained, and the adventurous can blaze new trails after an initial ride from a four-wheel-drive vehicle. ☐ **Mod to Lux** ☐ **R, S, F**

MEREDITH: Horseshoe Bend Guest Ranch, *Meredith, CO 81642, (303) 927-3570* ☐ *Open all year* ☐ *31 mi E of Basalt on Frying Pan River Road* ☐ At 9000 feet above sea level, this 22-acre ranch sits in White River National Forest, with a small mountain lake behind the ranch house and the Frying Pan River out front. Guests stay in suites in the long log ranch house or in private log cabins with bare-log walls and heavy pine furniture. Family-style meals are served in the ranch house, and there are cookouts and fish frys. The AP rate includes breakfast rides, moonlight rides, picnic rides, free riding instruction, Jeep tours, nature hikes, and good trout fishing in the ranch's streams and lakes. The ranch rents rowboats and fishing gear. Available at extra cost are rafting on the Roaring Fork or Colorado Rivers and overnight pack trips. In the fall, the ranch is a base for deer and elk hunting, while the winter season is taken up with snowmobiling and cross-country skiing. ☐ **Mod to Lux** ☐ **R, F, S**

NATHROP: Deer Valley Ranch, *Nathrop, CO 81236, (303) 395-2353* ☐ *Open all year* ☐ *6 mi W of Nathrop* ☐ Deer Valley Ranch is set against the spectacular Chalk Cliff of 14,196-foot Mt. Princeton. Modest cottages with fireplace and kitchen and the larger year-round cottages are available on modified EP, while lodge rooms are rented on full AP, with family-style meals included. Clara and Parker Woolmington, who have owned and operated Deer Valley since 1955, emphasize that this is a Christian family guest ranch, with no alcohol on the premises. Guests explore the mountains on horseback or by Jeep (neither is included in the room rates). Expeditions to nearby ghost towns are arranged, and there are cookouts, wrangler breakfasts, and pack lunches. A tennis court, swimming pool, and game room are among the ranch facilities. Fishermen will find trout in the many high mountain lakes and streams nearby. In season, the ranch also arranges big-game hunts. ☐ **Inexp** ☐ **R, F**

SEDALIA: Lost Valley Ranch, *Rte. 2, Sedalia, CO 80135, (303) 647-2311* ☐ *Open all year* ☐ *15 mi SW of Decker* ☐ This friendly working horse and cattle ranch secluded in a valley of Pike National Forest is especially geared to families. Younger children enjoy a program of supervised riding, swimming, hiking, children's meals, and movies, while teenagers have their own programs of float trips, hay rides, and swim parties. Everyone can help out with the ranch chores, play tennis on one of the two courts, or take a trail ride. For fishermen there are miles of mountain trout streams. At night there are square dances, melodramas, cookouts, and other entertainment.

Lost Valley guests stay in rustic one- to three-bedroom cabins; the largest accommodates 10 people. Western meals are served in the dining room, at poolside, and in the meadows around the ranch. Western clothing and gifts are available in the ranch trading post. All activities except trapshooting and overnight pack trips are included in the basic rate. ☐ **Mod to Lux** ☐ **R, F**

VAIL: The name is practically synonymous with skiing. A case can be made for Vail as a year-round resort community with as much action in summer as in the ski season. On the other hand, most of the BMWs and four-wheel drive vehicles parked around the tennis courts in summer sport "Think Snow" bumper stickers. Vail is billed as the largest single-mountain ski resort in the country. The mountain has more than 10 square miles of skiing area serviced by 2 gondolas, 2 triple chairlifts, and 14 double-chairs. The miles of trails include terrain for all levels of ability, from beginners to experts. Powder buffs can take on three deep powder bowls.

Vail has a complete full-service village at the base of the slopes with motel, lodge, and condominium accommodations. Within the town are more than 70 restaurants and bars, more than 100 shops, and a movie theater. The Vail Institute presents cultural and musical events from early June to early September. The Vailfest, an annual carnival, is held at the end of September.

For a complete list of accommodations write to: Vail Resort Association, P.O. Box 1368, Vail, CO 82657, (303) 476-5677.

Black Mountain Ranch, *Crossroads Center E 209, 141 E Meadow Dr., Vail, CO 81657, (303) 476-1200* ☐ *Open all year* ☐ *40 mi from Vail via I 70, SR 131 N* ☐ This luxury AP guest ranch has a special focus on the winter season and a minimum of "Howdy, pardner" hokum. Accommodations range from contemporary private cabins with spacious living rooms, sun porches, kitchenettes, and bedrooms with private baths to lodge suites with lofts.

Winter activities covered in the regular rates include snowmobiling, cross-country skiing, ice skating, snowshoeing, and hay-sled rides. All equipment for these activities is provided, as is transportation to Vail and Steamboat for alpine skiing. At night live entertainment alternates with ski movies and old movies.

In summer Black Mountain takes on the aspect of a traditional guest ranch. Guests receive riding instruction if needed, and all the equipment they need for an overnight pack trip. Trout fishing in stocked ponds, Colorado River rafting, kayaking, canoeing and log rolling are among the other summer activities. Since this is a working ranch guests can help with the chores if they wish. Beer and wine are served with meals in the lodge. ☐ **Lux** ☐ **R, S, F**

WINTER PARK: In Arapahoe National Forest, 67 miles west of Denver, Winter Park is part of Denver's Mountain Park system. Open from mid-November through April, it offers snowmaking, patrol, ski school, rentals, and 13 double chairlifts serving 7 jumping hills and 51 trails for all levels of skiing ability. The base elevation is 9000 feet, the longest run two miles, and the vertical drop

2125 feet. Facilities include a rstaurant, bar, cafeteria, and nursery. In summer, a sightseeing chair operates, and guided hikes are offered.

About two miles north is Ski Idlewild, which operates from Thanksgiving through March and offers a double chairlift, Poma lift, school, restaurant, cafeteria, and nursery. Here there are over eight miles of cross-country trails. The ski shop has rentals for downhill or cross-country skiing.

Beaver Village, *P.O. Box 43, Winter Park, CO 80482, (303) 726-5741* ☐ *Ranch open about June to mid-Sept; ski facilities open about mid-Nov to late Apr* ☐ *1½ mi W of Winter Park on US 40* ☐ A guest ranch in summer and a ski resort in winter, Beaver Village nestles at the foot of Berthoud Pass. In the quaint, Swiss-style Main Lodge, paneled double rooms and suites come with double and twin beds, sleeping four people, while the roomier units in Lake Lodge have double and twin beds plus studio couches; rollaways are available for both lodges. The more expensive condominium apartments, with private balconies, fireplaces, and kitchens, have between one and four bedrooms. All units include maid service. Lodge units are rented on package plans that include meals in winter, and on packages that include meals, horses, and all ranch activities and amenities in summer. Condominiums are available in both seasons on either package plans or a lodging-only basis. Dining room meals include buffets and homemade breads, and the bar has entertainment.

In winter, free shuttle service takes you to the Winter Park ski area. Cross-country trails start outside the Main Lodge. Beaver Village has its own rental shop for both sports. Also offered are tubing, sleigh riding, whirlpools, saunas, a game room, and evening ski films and demonstrations.

In summer, the ranch has daily scheduled activities on horseback that include trail rides through magnificent country, cookouts, and options such as riding lessons or log rolling. On Saturday you may participate in the rodeo. Children aged four to twelve are split into two age groups with their own wranglers and counselors for trail rides, trail lunches, and other activities, and they have their own rodeo. For children under age four, babysitters are available at extra cost. If you don't like riding, you may choose the daily auto trips to scenic or historic areas instead. A weekly golf outing to Grand Lake Golf Course is also offered. Also on the ranch you'll find two tennis courts, an outdoor heated pool, whirlpools, sauna, hiking trails, and areas for outdoor games such as badminton and volleyball. Trout fishermen can fish in the private, stocked lake or in the mile of Fraser River that runs through the ranch. The ranch shop rents and sells fishing equipment, riding boots, and other gear. Evenings there are square dances, hayrides, films, and other activities. For extra charges, you may opt for whitewater rafting on the Colorado River, an overnight pack trip, or for children, an overnight camping trip. ☐ **Mod to Lux** ☐ **R, W, F, S, H**

Sitzmark Lodge/Sitzmark Guest Ranch, *Box 1239 W8, Hideaway Park (at Winter Park), CO 80450, (303) 571-5666* ☐ *Open all year* ☐ *Off US 40 just* *north of Hideaway Park* ☐ In architecture and furnishings, this resort, a ski lodge in the snow months, a dude ranch in summer, has a distinct Austrian accent. Double rooms in the main lodge sleep up to four people; cheaper cabin rooms sleep up to five people. Deluxe alpine chalets scattered among the evergreens have wood-paneled walls, kitchenettes, and free-standing fireplaces, and accommodate six or ten people in one- or two-bedroom units. During the ski season dorm rooms are available. The main lodge has a Ratskellar bar and an Austrian-motif dining room.

Summer AP rates include horseback riding and instruction, fishing in the ranch's private stream and lake, steak fries, cookouts, square dancing, a children's program (ages three to twelve), a heated pool, a sauna, and sightseeing trips. Float trips on the Colorado River, sailing, tennis, golf, overnight pack trips, and horse-drawn hayrides are available at an extra charge. In the fall and spring, when the ranch is geared to hunters, hikers, and fishermen, rates are EP and activities such as horseback riding are charged separately. Trap and skeet shooting and hunting guides are available. During winter, Sitzmark offers package plans that include breakfast and dinner, ski movies, use of the sauna, and free transportation to Winter Park. Cross-country skiing, tubing, sleigh riding, and snowmobiling are available. ☐ **Inexp to Mod** ☐ **R, S, F**

Connecticut

LAKEVILLE: Interlaken Inn, *Lakeville, CT 06039, (203) 435-9878* ☐ *Open all year* ☐ *On US 44 in Lakeville* ☐ This is a contemporary resort rather than an old New England inn. The modern buildings with wood exteriors are set on 20 acres alongside the unpronounceable Lake Wononskopomuc. The inn's 57 rooms and seven duplex town houses are appointed with bright, contemporary furnishings, color TVs, AM-FM radios, and private patios or balconies. Each town house also has a living room with fireplace, terrace, dining area, complete kitchen, large master bedroom, 1½ baths, a washing machine and drier.

The inn's dining room, Anthony's Vineyard, has colorful Danish decor, a view of the lake, and two fig trees growing beneath the skylights. The continental cuisine is complemented by an all-American wine list.

Interlaken has its own heated pool, sauna, tennis and paddle tennis courts, and game room. Docks and dining areas are on the lake, and adjacent to the inn is a tricky nine-hole golf course. ☐ **Mod** ☐ **W**

NEW PRESTON: Boulders Inn, *Rte. 45, New Preston, CT 06777, (203) 868-7918* ☐ *Open all year* ☐ *US 202 to New Preston and SR 45 to lake* ☐ Built as a private home in 1895, the Boulders has been an inn for some 35 years. The entire inn, including the six guest rooms, is furnished predominantly with antiques, including some standout pieces such as an 18th-century English sideboard. All rooms have private bath, and the rooms on the second floor overlook the lake and surrounding woods. The nine cottages, of contemporary design and furnishing, are set back into the hill immediately behind the inn, and each has a lake view from the window wall and private deck. All are winterized, have private baths, and some have fireplaces.

The restaurant, open for dinner, features dishes such as veal Marsala and four-herb chicken. In summer, dinner and cocktails are served on the terrace.

Guests may enjoy all the recreational activities of the lake–swimming, canoeing, sailing, and fishing–as well as the inn's own tennis court, game room, library, and lawn games. ☐ **Inexp** ☐ **W**

The Inn on Lake Waramaug, *New Preston, CT 06777, (203) 868-2168* ☐ *Open all year* ☐ *US 202 to New Preston and Sr 45 to lake* ☐ The main inn building was constructed as a home sometime between 1795 and 1815. Renovated and restored over the years, the building was moved to its present hilltop site by a team of oxen in 1860. The inn is now thoroughly colonial, with cherry and pine antiques and displays of silver, copper, and pewter—de rigeur for a New England inn. Resort amenities include a tennis court, a private beach on the freshwater lake, an indoor heated swimming pool, and a stern-wheeler, the "Showboat," a pontoon boat styled after a steamboat on which guests may cruise about the lake.

Overnight guests take their meals on the American Plan, and can select from a number of dishes on the buffet. The a la carte menu is more extensive.

The inn also has a variety of plans. For instance, if you are just passing through and stop here for dinner, you may use the pool, the sauna, and the showers for a small fee. ☐ **Mod to Lux**

SOUTHBURY: Harrison Inn, *Heritage Village, Southbury, CT 06488, (203) 264-8255 Open all year* ☐ *From I-84 in Southbury, take SR 67 to Heritage Rd.* ☐ Harrison Inn has the "modern-rustic" look—weathered cedar exteriors on the angular buildings, stone fireplaces, and high, beamed ceilings in the dining room. It is part of the Heritage Village complex, a tasteful, extensive mall of shops and restaurants done in the same style.

The Inn's rooms have cedar paneling, contemporary furniture, and classy features such as black marble bathrooms with brass fixtures.

The restaurant, the Timbers, is one of the best in the area. Its popular Sunday brunch features fresh carved corned beef, baked Virginia ham, blintzes, apple fritters, finnan haddie, chipped beef, sauteed chicken livers, fresh fruit, and more.

The inn has its own sauna, exercise room, and pool room. On the landscaped grounds, complete with stream and a bit of woods, are two lighted tennis courts and a heated pool. ☐ **Lux**

Delaware

FENWICK ISLAND: The Sands Beach Resort Motel, *Rte. 1, Ocean Hwy., Fenwick Island, DE 19944, (302) 539-7745* □ *Open Apr–Oct* □ *N of Ocean City, MD, on SR 1* □ Fenwick Island, with the Atlantic Ocean on one side and the bay on the other, is a recreational paradise for the whole family. It offers the unbeatable combination of sun, sand, and surf, unspoiled by excessive commercial development. The two-story Sands Beach Resort looks out across a wide, clean beach toward the Atlantic. The large, lifeguarded kidney-shaped pool is surrounded by a patio with chaise lounges for sunning. Guests also have free admission to the village's immaculate, uncrowded public beaches.

Keeping the needs of its family-oriented clientele in mind, the Sands offers various types of living quarters: fully equipped apartments and efficiencies and motel units with one or two double beds or two twin beds. Guests desiring extra space can arrange connecting rooms. All units have air conditioning, heat, maid service, a coffee warmer, and eight-channel cable color TV with in-room movie channel. □ **Inexp** □ **F, W**

Florida

AMELIA ISLAND: Amelia Island Plantation, *Amelia Island, FL 32034, (904) 261-6161* ☐ *Open all year* ☐ *On SR A1A, 20 mi E of I-95 at the Fernandina Beach-Yulee exit* ☐ Intelligently designed to preserve the natural beauty of the seashore, with impeccable furnishings throughout the inn and condominiums, and generously equipped with modern sports and spa facilities, Amelia Island outclasses most American resorts. Its four-mile beach is a smooth band of fine white sand, wide enough to get thoroughly isolated on. Behind the barrier of dunes is the beach club, whose focus is the large community pool, one of a dozen pools that dot the resort. Clustered together to avoid vegetation-encroaching sprawl, are the inn and condominiums, facing out to sea.

The 24 air-conditioned inn rooms have all the first-class amenities plus wetbar, refrigerator, and private terrace. The appointments of the one- to four-bedroom condominiums are more varied, but kitchens with dishwasher are the norm. Some suites even have their own secluded pool.

Other good reasons to spend time on this 900-acre haven include the 27 holes of seaside golf, designed by Pete Dye; the Racquet Park, with its 19 composition courts scattered among the trees, and the stable, with plenty of mounts ready to explore the island's trails. Not all of this resort is for sport, however. Another 900 acres have been given over to wildlife as a preserve. Boardwalks lead through swampy areas for observation.

Since all suites and inn rooms have at least a refrigerator, guests find little need to dine out. But when they do, they can choose between several fine dining and drinking areas in the beach club and clubhouse. Also here are shops, supervised children's programs, entertainment, and all the other facilities and services you'd expect of a full-service resort. ☐ **Lux** ☐ **T, G, R, F**

BOCA RATON: The Boca Raton Hotel and Club, *Camino Real, Boca Raton, FL 33432, (305) 395-3000* ☐ *Open all year* ☐ *½ mi E of jct Camino Real and US 1* ☐ Undisputably one of the top full-service resorts in the United States, Boca Raton is an exalted combination of Spanish mediterranean and space-age-modern architecture. The resort began in 1926 as a kind of millionaires club. It was called the Cloisters then, because, like a monastery, it has long porticoes surrounding a formal garden. The original structure, most of which remains, has a traditional Spanish exterior—pink stucco walls and a red-tile roof–and an exceptionally grand interior. For example, in the Cathedral dining room golden columns with carved stone capitals support an arched, coffered ceiling. The 26-story hotel tower, built in 1969, starkly contrasts with the original building. Of white concrete and smoke-colored glass, its futurist architecture makes it look a little like a home for extraterrestrials. Inside are bright, Spanish-inspired rooms and especially sleek bathrooms. The rooms in the original Cloisters are also contemporary but in a gracious style keeping with its architecture. All units at Boca Raton have the extra touches–feather pillows, a fresh copy of *The New Yorker* magazine, turndown service–that you thought were gone ages ago.

Sports facilities are first class, well maintained, and extensive. One golf course and 20 clay tennis courts are adjacent to the hotel. Three more golf courses and 20 tennis courts are a few miles away at Boca West, the resort community affiliated with Boca Raton. A bus shuttles guests back and forth between the hotel and community. A pro shop and teaching staff are on hand to meet the needs of both golf and tennis players. Two of the courses at Boca West were designed by Desmond Muirhead. The third, by Von Hagge and Devlin, is tight, tricky, and well trapped, as is the course at the hotel.

The hotel is perched at water's edge along the Intracoastal Waterway. Its marina, a calm, safe docking area with easy access to the Gulf Stream, has several deep-sea fishing boats and 21 slips with water, electricity, and telephone hook-up. Boca Raton has several pools, and guests who want to swim in the ocean can go to the hotel's Beach Cabana Club, a large semicircle of private cabanas facing the ocean. Each cabana has private dressing rooms, shower, and patio. The club has a dining area, pool, shops, and solarium.

The food and dining service is the pride of the resort. Both the Cathedral dining room and the Top of the Tower Restaurant, with its sweeping view of Florida's Gold Coast and the Atlantic Ocean, serve gourmet entrees and seafood dishes such as pompano papillote and prime rib of beef. ☐ **Lux** ☐ **G, T, F, W**

CAPE CORAL: Cape Coral Country Club Inn, *4003 Palm Tree Blvd., Cape Coral, FL 33904, (813) 542-3191* ☐ *Open all year* ☐ *From Cape Coral Blvd., take Palm Tree Blvd.* ☐ This inn is a low, modernistic building overlooking the golf course and pool. The 18-hole, 6827-yard golf course, designed by Dick Wilson, has many bunkered greens and lots of water. For warming up or casual play, there's also a par three executive course. The pro shop has cart and club rentals. Near the five Har-Tru tennis courts (two lighted) is the tennis pro shop. Lawn bowling and shuffleboard are also available.

The inn has several dining areas. The Terrace Dining Room specializes in seafood dishes such as filet of lemon sole, as well as traditional gourmet entrees such as coq au vin. The Sandtrap Coffee Shop serves less costly fare for breakfast, lunch, and dinner.

Guest rooms here come standard and deluxe. All are furnished in contemporary style and have two double beds, but the deluxe rooms are poolside. ☐ **Mod to Lux** ☐ **T, G**

CLEARWATER: Belleview Biltmore, *Belleair, Clearwater, FL 33517, (813) 442-6171* ☐ *Open Jan to mid-Apr* ☐ *From airport take SR 60 W to Alt US 19 S to Belleview Blvd. W* ☐ This traditional MAP resort hotel, built on the 19th-century European pattern, is often called "The White Queen of the Gulf," so enormous and grand is this welter of gables, chimneys, and dormer windows set on 625 acres on the Intracoastal Waterway. Built in 1896 by railroad baron Henry Plant, it was designed to lure the wealthy away from Florida's prospering east coast. The hotel remains noble and noted for gracious service. Perhaps the most impressive public room is the dining room. The high, vaulted ceiling, inset with stained glass skylights, is supported at each end with a long array of columns. Guest rooms are furnished with maple furniture–not grand, certainly, but very comfortable.

With three 18-hole golf courses, this *fin de siècle* resort has taken on a new identity as a golfer's haven. Two of the courses, designed by Donald Ross, are right at the doorstep. Laid out on the rolling terrain alongside Clearwater Bay, they are bordered by live oaks and date palms. The third course is a mile away. A pro shop and clubhouse are on premises.

On the south lawn are six Har-Tru tennis courts. The olympic-size pool, completely surrounded by a bath house with a balcony on the second story, is lined with mosaic tiles. Across the bay, a ten-minute complimentary ride on the hotel cruiser away, is the Cabana Club on Sand Key where guests can tan on the white-sand beach, swim in the Gulf, or enjoy a seaside meal or drink. Fishing boats, sailboats, and deep-sea charters are available. The many services include a children's hostess during spring vacation. ☐ **Lux** ☐ **G, T, W**

CRYSTAL RIVER: Plantation Inn, *P.O. Box 1116, Crystal River, FL 32629, (904) 795-4211* ☐ *Open all year* ☐ *On SR 44 SW of Crystal River* ☐ In accord with its name, this resort evokes the Southern manors of antebellum days: white columns support a Greek pediment, black shutters flank the windows. Public and guest rooms are contemporary, yet retain old-fashioned touches such as fireplaces in the dining room and a curving staircase in the lobby. The 100 guest rooms have separate dressing rooms and climate control. Seafood is the specialty in the dining room, and the bar has dancing and entertainment six nights a week.

The inn is on a virtual peninsula bounded by crystal-clear springs (great for scuba diving) that flow through lush vegetation into King's Bay. Crystal River is famous for its fishing, because here freshwater and saltwater species swim side by side. Manatee are frequently seen in winter. Guides, boats, takcle, scuba gear, waterskiing, and other services are available at the marina built into the inn's 1000-foot seawall.

The classy, tile-edged pool in front of the inn has a beautiful view of the winding waterway. Four all-weather tennis courts are at the water's edge, and ball machines and lessons are available. Behind the inn stretches the 18-hole golf course, designed by Mark Mahannah and dotted with lagoons and palms; a driving range and pro shop with cart rental are part of the facilities. AP, MAP, and packages are available. ☐ **Inexp to Mod** ☐ **W, F, G, T**

CYPRESS GARDENS: Grenelefe, *P.O. Box 143, Cypress Gardens, FL 33880, (813) 422-7511* ☐ *Open all year* ☐ *At jct SR 544 and SR 546, 5 mi E of Haines City* ☐ This beautifully landscaped 800-acre resort community is set around two fine 18-hole golf courses, one of which, the West Course, is considered to be among the finest in America. The tennis complex, with its 12 Laykold and Har-Tru courts and computerized ball-machine training system, is also recognized as one of the country's best. Both golf and tennis areas have their own pro shops and striking contemporary clubhouses. Besides all this, there are two large swimming pools, and a marina on 6500-acre Lake Marion.

The public buildings and condominiums here are nicely designed–modern lines created with stone, wood siding, and weathered shingles. Guest suites come in a variety of styles: kitchens, private patios, dressing rooms, and various combinations of bedrooms, living rooms, and dining rooms are available.

Incidentally, Grenelefe gets its name from the ballad of Robin Hood in which Little John disguises himself as "Reynald Grenelefe" to infiltrate the household of the Sheriff of Nottingham. ☐ **Lux** ☐ **T, G, F**

DAYTONA BEACH: The Inn at Indigo, *2620 Volusia Ave., Daytona Beach, FL 32014, (904) 258-6333* ☐ *Open all year* ☐ *US 92 at I-95* ☐ On the outskirts of Daytona Beach, this completely self-contained enclave on 600 acres escapes the racetrack noise and atmosphere of the famous beach. The modernistic, low buildings are nicely camouflaged behind date palms and palmettos. Each of the 150 guest units has a private porch or balcony overlooking the lawns and landscaped grounds. Family suites have kitchens. Buffets, Sunday brunch, gourmet as well as family dining, and a cocktail lounge are available.

The large L-shaped, Olympic-size pool has a three-meter diving board at the end of the long leg. For tennis players, there are eight lighted Laykold courts, a pro shop, and the services of an automatic ball machine and USPTA pro. The 6849-yard golf course, designed by Lloyd Clifton, has 87 sand traps (around landing areas as well as greens), 50 acres of lagoons, three miles of tropical rough, and not a single parallel fairway. From the white tees the course rates 70.20. A par-three course, driving range, pro shop, club, and PGA pro complete the golf facilities. ☐ **Mod** ☐ **G, T**

DESTIN: The Sheraton at Sandestin, *Rte. 98, Destin, FL 32541, (904) 837-2121* ☐ *Open all year* ☐ *On US 98, 8 mi E of Destin* ☐ Sandestin is spread over 440 acres of the Florida panhandle, on a narrow ribbon of land between Choctawhatchee Bay and the Gulf, with 1000 yards of that property directly on the Gulf and a bay shore too. Guests are housed in the comfortably furnished rooms in the six-story inn or in the 115 one- to four-bedroom contemporary villas. The latter are furnished in Mediterranean style and have washers, driers, equipped kitchens, and maid service. In addition to room service, an all-day grill, and a seafood restaurant, there is a bar with entertainment, and a social director is in charge of dancing and other activities.

Besides the eight lighted Rubico courts, Sandestin has grass courts. Teaching aides include ball machines and videotapes. Tom Jackson designed the par-72, 6700-yard golf course with lots of water and many traps, and on the grounds are driving range and putting green. Despite all this, water sports are the highlight–an outdoor pool with poolside service, a wading pool, a health club with sauna and steam bath, and a bay beach as well as the beautiful Gulf beach. Destin, once a commercial fishing area, has half-day, all-day, and over night deep-sea fishing for red snapper, grouper, and other fish. Trolling boats go after mackerel, dolphin, and marlin. All sorts of boat rentals are available, and you can surf fish off the beach, and waterski, sail, or learn scuba diving at the resort. The Sheraton also has a nature trail, bird watching, recreation room, lawn game areas, and all the usual services such as a children's program and airport transportation. ☐ **Mod to Lux** ☐ **W, F, G, T**

FORT LAUDERDALE: Built on and over several islets and peninsulas and criss-crossed by more than 200 miles of canals and waterways, Fort Lauderdale is sometimes called the "Venice of America." Since there is a canal in virtually every backyard, recreation naturally centers on boating and watersports. For the public, the state park beaches–Hugh Taylor Birch Park and John Lloyd Park–provide the best bargain. Be forewarned: these beaches are invaded by thousands of sun-and-fun seeking college students each spring. Holiday Park (on Sunrise Blvd.) has numerous tennis courts at inexpensive rates. These are the courts where Chris Evert learned to play.

Bahia Mar Hotel, *601 Seabreeze Ave., Fort Lauderdale, FL 33316, (305) 764-2233* ☐ *Open all year* ☐ *On SR A1A* ☐ This resort is surrounded on three sides by its huge complex of docks and the Intracoastal Waterway. On the fourth side, across A1A, is a seven-mile strip of public beach on the Atlantic Ocean with picnic tables under palm trees. The focus is the marina; with slips for 350 yachts, sales and rentals of all sorts of marine equipment, it is one of the largest and best-equipped marinas in the country. The charter fleet alone numbers 24 boats; drift fishing, yacht charter, and waterskiing equipment and drivers are all available. The Pro Dive Shop gives snorkeling and scuba lessons as well as courses in underwater photography and tropical fish collecting. *The Jungle Queen,* an excursion boat resembling a Mississippi paddle-wheeler, sails on sightseeing and dinner cruises.

The Bahia Mar buildings are connected to each other and to the beach with pedestrian walkways elevated above the parking areas and highways. The walkways also connect to the outdoor pool set on the roof of the 20-store shopping mall. The 300 EP rooms and suites are in a low building overlooking the marina and in a tower rising 14 stories above the pool deck. More than 200 rooms have a terrace or balcony. Formal and family-style dining with garden or water views, lounges with and without entertainment, and services such as babysitting complete the complex. Golf and tennis can be arranged nearby. □ **Lux** □ **F, W**

Marina Bay, *2175 State Road 84, Fort Lauderdale, FL 33312, (305) 791-7600* □ *Open all year* □ *From SR 84 at I-95, take Marina Bay Dr.* □ In a water-obsessed community such as Fort Lauderdale, where the value of waterfront property is skyrocketing and dock space is always in demand, somebody would have had to build a place like Marina Bay sooner or later. Out near the airport, these lodgings are in rows of rectangular "houseboats" or, rather, in floating lodges, since they aren't made to be navigated away from the docks. The two-story units contain 60 double rooms and 75 suites that are much like luxury hotel accommodations except that they rock gently. Some suites have kitchens, whirlpools or steambaths. Beside the docks is the Storehouse, a clubhouse and restaurant. It was constructed of old wood to resemble a ramshackle warehouse, complete with tin roof, lobster pots, and other nautical paraphernalia. The stylish two-tier restaurant serves local specialties such as conch and alligator steak, and has entertainment. Outside is a heated pool with saunas, whirlpools, and sundeck, and near that a pleasant little landscaped lagoon with tropical fish. Adjacent is the tennis complex with pro service–nine Laykold courts (three lighted), four clay courts, practice alley, and ball machines. □ **Lux** □ **T**

Pier 66 Hotel and Marina, *17th Street Causeway, Fort Lauderdale, FL 33316, (305) 524-0566* □ *Open all year* □ *On SR A1A* □ The phrase "hotel and marina" describes this place, but doesn't do it justice. The hotel is a lone 17-story tower in this land of low rise. On top of the tower, like a panama hat, sits the slowly revolving, circular, Pier Top Lounge. Spread beneath the tower are 22 acres of palm-studded grounds, the Intracoastal Waterway, and the docks of the marina–its 134 boat slips providing dockage for boats as large as 150-foot yachts. Visiting sailors enjoy all the recreational facilities of the hotel as well as a marine store, shower and laundry facilities, customs clearing, and an official weather display station. Deep-sea fishing charters are available at the dock, as well as rental boats, scuba, waterskiing, and cruise services.

Rooms in the tower, from studios to executive suites, all have balconies overlooking the marina or the courtyard containing the swimming pools, two lighted tennis courts, and the nine-hole pitch-and-putt golf course. In addition to the 156 units in the tower, 102 rooms and suites are in a two-story motel section which borders the courtyard. Ground-floor units have enclosed lanais, and many second-story units have balconies.

Besides the rooftop lounge, Pier 66 offers several dining and drinking areas. At the conference center, near the marina, the Panorama dining room serves French and American cuisine in an elegant setting. ☐ **Lux** ☐ **F**

FLORIDA KEYS: This thread-thin archipelago stretches from the tip of Florida over 100 miles southwest through the Gulf. South of Key Largo there is only one road down and back, US 1. The two-lane highway is built on the roadbed of Henry M. Flagler's ambitious East Coast Railway, which connected the Keys to the mainland from 1912 until 1935 when the railway was washed out by a hurricane. The highway joins island to island with 42 bridges, and a pipeline runs alongside to supply the Keys with fresh water.

Anyone interested in marine life should make John Pennekamp State Park in Key Largo the first stop on the highway. The "land" the park protects is underwater–a spectacularly colorful 75-acre live coral reef several miles off shore. The fish, the shipwrecks, and the 40 species of coral can be seen from glass-bottom boats or with snorkeling or scuba equipment.

Tavernier, the next town, is good for birdwatching and has Harry Harris County Park for picnicking. A little farther on, at Plantation Key, is a museum of treasure salvaged from Spanish galleons and other wrecks. Two more coral gardens, both smaller than Pennekamp, are south of Islamorada, one off Upper Matecumbe, one off Lower Matecumbe. Long Key State Park at Layton has swimming, scuba diving, fishing, camping, and picnicking. On Grassy Key is Flipper's Sea School, which presents daily porpoise shows.

The keys south of Grassy Key are generally known as the Lower Keys. Marathon, the biggest city after Key West, has numerous marinas with guides available for light tackle fishing as well as deep-sea boats for charter. This bustling, extensively developed city also offers an 18-hole golf course, an airport, and a complete array of accommodations and restaurants. From Marathon to Key West, spear fishing is allowed one mile offshore.

The next landmark is Seven Mile Bridge at Bahia Honda Key, which has a state recreation area with ocean and Gulf swimming, fishing, boating, and camping. A few miles along is Big Pine, the largest of the islands and home of the Key deer, a miniature species that stands only about 2½ feet high at maturity. Much of the key is a preserve for the deer. Six miles from the highway at marker 25 is Summerland Orchid Gardens with over 65,000 exotic plants. South of Sugarloaf some of the keys have become suburbs of Key West, but many of the small islands remain undeveloped.

Long before the Cuban missile crisis, Key West was an important military outpost, first against pirates and then, surprisingly, against the confederacy. Fort Zachary Taylor and the Martello Towers, which still stand on the south shore, were manned throughout the Civil War by Union troops. Fishermen and marine salvagers from the Bahamas and New England settled the island, and the fishing and shrimping industry is still important, despite the influx of tourists. Besides its quaint fishing village atmosphere, Key West is famous for the writers and artists who have lived there. The Audubon House (Green and Whitehead Street), is open for tours and has some of his work on exhibit. Hemingway's home (Olivia and Whitehead Streets) is also open. Many of the city's sights can be seen from the Conch Tour Train that leaves daily from two stops, 303 Front St. and 3850 N. Roosevelt Blvd. Public beaches, fishing piers, docks, charter boats, boat rental, reef spear-fishing, and glass-bottom cruises over the reefs are available.

For more information on the Keys contact the Florida Upper Keys Chamber of Commerce, Key Largo, FL 33037 and the Lower Keys Chamber of Commerce, Big Pine Key, FL 33043.

Cheeca Lodge, *Islamorada, FL 33036, (305) 664-4651* □ *Open all year* □ *Off US 1 at marker 82* □ The white, four-story, EP lodge, flanked by villas to the north and south, is just steps away from the 1100-foot-long, palm-shaded beach and the 525-foot-long fishing pier. Between the lodge and US 1 is the beautifully landscaped sports complex. The nine-hole executive golf course was designed by the company that builds all of Jack Nicklaus's championship courses. It shares a pro shop with the four all-weather tennis courts. Also in this area is the game room, playground, snack bar, and large family pool; a smaller pool, for adults only, adjoins the lodge.

Some of the light and airy lodge rooms have balconies facing the ocean or golf course. Each of the villas has a screened porch and three rooms that can be rented individually, with or without kitchenette.

The dining room, featuring fresh seafood, has large picture windows overlooking the ocean. Drinks are available at the pool bar or the intimate, trophy-hung and wood-paneled Light Tackle lounge, which also has regularly scheduled live entertainment. □ **Lux** □ **T, W, F, G**

Marriott's Casa Marina Inn, *1500 Reynolds St., Key West, FL 33040, (305) 296-3535* □ *Open all year* □ *Off Atlantic Blvd, near the West Martello Tower* □ One of the great Flagler resorts, the seaside Casa Marina was the last stop on his East Coast Railway and the last hotel in his chain, completed after his death. As with most of Flagler's hotels, the style is Spanish Mediterranean. Over the years this distinguished hacienda, with its arches and porticos, fell from its stature as a watering hole for the rich and was used by the Army and Navy as a barracks and as the site of a Peace Corps conference. Now, once again a full-service resort, all its 250 rooms and suites have modern hotel amenities: private bath, color TV, telephone, and message lights; most have balconies.

The inn is set on six and one-half acres of tropical gardens and lawns along the water. In the courtyard facing the 1100-foot beach are the large pool and the Sun-Sun Hut, a snack pavilion serving hot and cold sandwiches, soft drinks, beer, and rum punch. The resort also has lighted tennis courts, game room, whirlpool, and sauna.

Henry's restaurant features continental cuisine and Key West specialties, and the Calabash Lounge has nightly dancing and entertainment. In addition, the resort offers theme parties such as a "shipwreck barbecue" or a "conch fiesta" on the beach. ☐ **Lux**

Sugar Loaf Lodge, *Box 148, Sugar Loaf Key, FL 33044, (305) 745-3211* ☐ *Open all year* ☐ *At the 17 mi marker on US 1* ☐ This resort's most unusual feature must be its pet porpoise, which puts on daily shows in its fenced-in cove in front of the restaurant. In other ways Sugar Loaf is more typical of a Florida resort–a marina next door provides boat rentals and fishing charters, the restaurant serves such dishes as fried shrimp and prime rib, and on the grounds are two tennis courts, pool, shuffleboard, and miniature golf.

The one- and two-story lodge contains 55 air-conditioned rooms all with color TV, 10 with kitchenettes. One special two-room suite with four double beds is especially good for families. Nearby is a 3000-foot airstrip. ☐ **Mod to Lux** ☐ **F**

HOMOSASSA SPRINGS: Six million gallons of water bubble up every hour from this 55-foot deep spring. This water, which flows into the Homosassa River and on to the Gulf of Mexico, attracts more than 34 species of saltwater and freshwater fish. To reach the spring, you board a boat at the main entrance (on US 19/98) for a 17-minute scenic ride. From the boat dock, trails wander through tropical woods of moss-draped like oak, sabal palms, and wild orchids. Peacocks and flamingoes often cross the trails, and imported and domestic waterfowl wander about. Tame deer, sea lions, otter, and alligator feeding are other attractions. At the spring is an observatory where thousands of fish can be seen through the underwater windows.

Sheraton-Homosassa Springs Inn, *P.O. Box 8, Homosassa Springs, FL 32647, (904) 628-4311* ☐ *Open all year* ☐ *On US 19/98, S of jct SR 490* ☐ A modern blend of wood and stone, this 104-room Sheraton offers easy access to the main entrance of Homosassa Springs. The two wings of the hotel are set in a landscaped area alongside a scenic canal. A covered walkway and a foot bridge across the canal lead from the hotel to the entrance. The hotel's pool, sun terrace, and children's playground overlook the canal, but the tennis courts are on the other side of the buidlings, by the parking lot. Guests take the short walk across the footbridge to eat at the Springs restaurant at the entrance to the springs. ☐ **Inexp**

Riverside Villas Resort Motel, *P.O. Box 258, Homosassa, FL 32646, (904) 628-2474* ☐ *Open all year* ☐ *On SR 490, 3 mi W of jct US 19* ☐ This resort is sent on a tree-shaded promontory on the beautiful Homosassa River. Strictly speaking, the "villas" are contemporary motel units with balconies overlooking the river. All units are air-conditioned and equipped with color TV and wall to wall carpeting, and half of the units are efficiencies with kitchenettes.

Sidewalks border the water's edge, and floating docks extend out into the water. The marina can handle 72 boats and has launching ramps, enclosed cleaning area, fishing supplies, boat rental (including paddle boats and canoes), and guide services. The catches here are bass, redfish, sea trout, and tarpon. Also on the grounds are all-weather tennis courts, a swimming pool, and bicycle rental.

The Yardarm restaurant, in nautical decor, has picture windows all around, giving diners superb views of river activity. The menu features seafood dishes, and they will cook your catch. The Ship's Lounge, also overlooking the water, has live entertainment many nights.

The Riverside runs another restaurant, the Crow's Nest, on an island six miles away, where the river flows into the Gulf of Mexico. You are ferried by boat, and the scenic ride takes 45 minutes each way. The menu includes fried local fish or barbecued chicken or ribs and Key lime tart, and is very reasonably priced. ☐ **Inexp** ☐ **F**

HOWEY-IN-THE-HILLS: Mission Inn and Country Club, *Box 441, Howey-in-the-Hills, FL 32737, (904) 324-3101* ☐ *Open all year* ☐ *From Leesburg, take US 27 4 mi S to SR 48, then 7 mi E to club* ☐ True to its name, this family owned and operated resort is done in a style that would make Spanish conquistadors and missionaries feel at home. Red-tile roofs, slender Moorish archways, fountains, wrought iron chandeliers, beamed ceilings, and other architectural details contribute to the Old World atmosphere. The 100 double rooms, in low lodges scattered on the landscaped grounds, are spacious. Four large suites with bedroom, living room, and kitchenette are available. All units have balconies. The El Conquistador dining room specializes in beef and seafood, and the cocktail lounge has dancing and entertainment. EP and package plans are available.

The grounds spread over 300 nicely landscaped acres containing five lakes. The 18-hole golf course is a par-72, measuring 6709 yards. Designed by 1926 by Capt. C. E. Clark of Scotland, it is considered one of Florida's classic courses. The course here has a pro shop with rentals. Instructional aides from ball machine to videotape analysis are offered to adult tennis players as well as children. Each of the four all-weather courts is lighted. On Harris Lake is a marina with boat rental, waterskiing, and fishing supplies. Bass and speckle perch is the catch, and the lake was the site of the U.S. Bass Open. Behind the main building is a heated swimming pool surrounded by a patio. The inn can arrange horseback riding at nearby stables, and some guests like to canoe on the extraordinarily beautiful Juniper Run in nearby Ocala National Forest. ☐ **Lux** ☐ **G, F, W**

JENSEN BEACH: Frances Langford's Outrigger Resort, *Sewall's Point Rd., Jensen Beach, FL 33457, (305) 287-2411* ☐ *Open all year* ☐ *Off SR 707, 15 mi S of Ft. Pierce* ☐ This motel-marina is a perfect spot for those on the Intracoastal Waterway to drop anchor. The well-protected boat basin, minutes from the St. Lucie inlet, has 17 slips (for boats up to 120 feet) with telephone, water, electricity, cable TV connections, showers, and fuel. At the marina guests may charter deep-sea fishing boats to try the Gulf Stream waters off the coast—prime sailfish and dolphin territory. Those after bluefish, pompano, and snook can find them nearer the shore.

Each of the villas here has its own carport, sunbathing patio, screened-in porch, and full kitchen. All of the units have cable TV and air conditioning. The villas surround a grassy central area containing the heated kidney-shaped pool and the 18-hole putting green. Right at the water's edge, overlooking the Indian River, is the Outrigger restaurant, straight out of "South Pacific" with a thatched roof and Polynesian food. ☐ **Mod** ☐ **F**

KEY BISCAYNE: Key Biscayne Hotel and Villas, *701 Ocean Dr., Key Biscayne, FL 33149, (305) 361-5431* ☐ *Open all year* ☐ *From US 1, take Rickenbacker Causeway 7½ mi SE* ☐ Between the tough Key Biscayne course down the road and the hotel's own pitch-and-putt course and putting green, golfers can while away quite a few hours sharpening their game. And those whose idea of a vacation involves water will appreciate the large pool and palm-shaded beach where they can swim, bask in the sun, or rent sailboats. Tennis players also have a fair shake: a tennis center with 10 courts.

Unlike recreation, the choice of oceanside accommodations is simple—one of the 100 hotel rooms or 75 villa suite. All have air conditioning and color TV, but the one- to three-bedroom villas have kitchenette.

In summer, both a children's program and MAP are available. In winter, dancing and other entertainments are regularly offered. ☐ **Lux** ☐ **T, W, G**

Royal Biscayne Beach Hotel and Racquet Club, *555 Ocean Dr., Key Biscayne, FL 33149, (305) 361-5775* ☐ *Open all year* ☐ *From US 1 or I-95, take Rickenbacker Causeway to Ocean Dr.* ☐ Right on the ocean, the long three- and four-story lodging wings with 200 rooms and suites surround palm-shaded courts with a large freshwater pool and broad sun terraces. All of the rooms are airy and cheerful; most have screened porches, and some have kitchenettes. The posh Caribbean dining room has dancing, and there is also terrace dining. Informal lunches and cocktails are at the oceanside Racquet Club. The bar has entertainment. Set back from the water are 10 Acryflex tennis courts (four lighted) supervised by a pro. Nearby is the Key Biscayne Golf Course, a beautiful and difficult 18-hole course. At the beach is another pool, plus cabanas and sailboats. Deep-sea fishing can be arranged. ☐ **Lux** ☐ **T, G**

Sonesta Beach Hotel and Tennis Club, *350 Ocean Dr., Key Biscayne, FL 33149, (305) 361-2021* ☐ *Open all year* ☐ *From US 1 take Rickenbacker*

Causeway 7 mi SE ☐ This hotel is easy to recognize: it looks like a pyramid with the top cut off. On the ocean side are the pool, the sun terrace, and of course the beach, dotted with the colorful sails of the catamarans and with the thatched beach umbrellas, looking like so many oversize Chinese hats. Behind the hotel are the 12 Laykold tennis courts. The teaching staff has videotape and ball machines and holds weekly clinics.

All of the rooms have private balconies and color TV. A few villa suites with kitchen are available. Among the hotel's package programs is one for families, featuring the "Just Kids" program that has supervised activities for children 5 to 13. It includes trips to Crandon Park Zoo, the Miami Seaquarium, and other local sights, as well as organized sports and evening movies at the hotel.

The two main dining areas are the Rib Room, with steaks, seafood, and live music, and the Two Dragons, with Chinese and Japanese cuisine. ☐ **Lux** ☐ **T, W**

LEHIGH: Lehigh Resort, *225 E. Joel Blvd., Lehigh, FL 33936, (813) 369-2131* ☐ *Open all year* ☐ *From Ft. Myers, take SR 80, SR 82, or Colonial Blvd. E 12 mi* ☐ This resort features two 18-hole golf courses, both designed by Mark Mahannah. The par-71 course is rated at 69 and has quite a few water hazards. The Mirror Lakes Course, about three miles south, is slightly longer (6722 yards) and tougher (par 72) The racquet club has four lighted Plexipave courts. Another lighted court is near the country club building, where the golf and tennis teaching pros and pro shops are found.

Accommodation is in one- and two-story motel units, most of which face the 20-meter swimming pool, and most of which have private balconies or patios. Four dining rooms include poolside grill and clubhouse restaurants. Of the three cocktail lounges, two have entertainment, and there is a Spanish-style supper club. On the grounds are driving ranges, putting greens, practice traps, mini-golf, outdoor game areas, a nature trail, and bicycle rental. A social director organizes activities for adults and children, and babysitters are available. One unusual feature is the folk and square dancing included in the vacation packages, where nationally known callers and instructors hold classes and dances. ☐ **Mod** ☐ **G, T**

LONGBOAT KEY: The Colony Beach and Tennis Resort, *1620 Gulf of Mexico Dr., Longboat Key, FL 33548* ☐ *Open all year* ☐ *On SR 789 5 mi S of Longboat Key* ☐ With a fine oceanfront setting, classy villas, and one of the best ratios of tennis courts to suites–one court for every 10 suites–the Colony qualifies as one of the top tennis resorts in the country. Yet the atmosphere here suggests a relaxed vacation resort, not a tennis camp. You are encouraged to take the tennis clinics and private lessons but are free to idle on the beach or by the pool. Besides the 21 all-weather tennis courts (some lit), the Colony boasts a health club with sauna, steam room, and whirlpool. An all-day program of children's activities offers crafts workshops, supervised games, and cookouts.

All accommodations here are furnished in bright contemporary style, The two-bedroom duplex apartments have two baths, sun deck, a dining and living area, and an equipped kitchenette. Guests can order drinks, salads, and sandwiches at the Seagrape Patio; the Beach Bistro serves more gourmet, and more pricey, dishes.

Across Sarasota Bay are the Ringling Museums, which include Ringling's palatial residence, an 18th-century Italian theater, a museum of art, and a museum of the circus. ☐ **Lux** ☐ **T**

The Field's Buccaneer Inn of Sarasota, *595 Dream Island Rd., Longboat Key, FL 33548, (803) 383-1201* ☐ *Open all year* ☐ *From Sarasota take SR 780 (Ringling Causeway) to Lido Key, then 10 mi N on SR 789* ☐ A high point of this waterfront motel is the restaurant which specializes in ribs and steaks roasted or barbecued "pirate style," which means up to eight hours over hickory coals. (The French word *boucaner,* from which "buccaneer" came, means to cook slowly over the coals. Pirates were called buccaneers because they would steal cattle and smoke the meat over the coals to preserve it.)

The pool terrace is nicely bordered by tropical flowers and has a good view of Sarasota Bay. Two of the four Laykold tennis courts are lighted, and a pro is on hand to teach guests. The Buccaneer has its own marina with electrical services at each slip. All of the 25 guest rooms and suites have either refrigerator or kitchen. ☐ **Mod to Lux** ☐ **T**

MARCO ISLAND: Marco Beach Hotel & Villas, *Marco Island, FL 33937, (813) 394-2511* ☐ *Open all year* ☐ *Off SR 92 on Collier Blvd.* ☐ This unusually attractive beachside resort offers three types of accommodations—rooms in the hotel tower, in the two-story wing housing lanai suites, or in the clustered villas. Each of the one- and two-bedroom lanai suites has a kitchenette, living room, *two* color TVs, and a private balcony or patio overlooking the pool or beach. Each villa has two bedrooms with twin beds, two baths, a kitchen, and a living room with dining area. The villas are modern single-story structures in a garden-like area around the nine-hole pitch-and-putt course and the second pool.

The three and one-half miles of beaches on Marco Island are known as a paradise for shell collectors and birdwatchers. At the water's edge, by the main hotel, Sunfish and Hobie 14s are available for hourly and daily rental. Across South Collier Boulevard are the 15 Har-Tru courts (hourly fee) and the pro shop. Near the villas is a Laykold court, and two courts at the Marco Island Country Club may be used by guests. Guests may use the 6800-yard golf course at this country club or the 7000-yard course at Marco Shores Country Club.

Six restaurants are on the premises, from the elegant Marco Dining Room, which features dinner dancing, to the informal Quinn's Patio, a beachfront bar and restaurant offering fresh shucked clams and oysters.

At this resort there's always some kind of program or activity going. Guests can polish up their tennis game with pro instruction and ball machines, take a seminar in sailing, backgammon, or shell collecting, or join the inspirational meditation class. The supervised children's program for ages 4 to 10 runs daily from 1 to 5 and 6 to 9. ☐ **Lux** ☐ **T, W**

Marriott's Marco Beach Hotel & Villas, *Marco Island, FL 33937, (800) 228-9290* ☐ The pride of this complex is the wide, dazzling-white, 3½-mile beach with no other buildings around and an unobstructed view of the Gulf. This hotel is a 528-room high rise that gives every room a private balcony. Eleven stories up are one- and two-bedroom penthouse suites. Flanking the main building and clustered around twin outdoor pools are 75 villa units with one or two bedrooms. Some units have view of the Gulf, kitchenettes, or oversize beds. The main dining room, with continental cuisine and flaming dishes, is rather grand. The Voyageur dining room overlooks one of the pools and the golf course and specializes in Sunday buffets with international menus. An outdoor cafe is set under palms. A couple of the lounges have dancing or entertainment.

Near one of the pools is a nine-hole pitch-and-putt course, and the hotel can arrange for you to use the 18-hole championship course at Marco Shores Country Club. Sunfish and Hobie Cats and a fishing dock are available, and the hotel will arrange boat tours, charter fishing, and fishing guides. On the grounds are 15 Har-Tru courts (lighted), and tennis instruction is offered. For exploring you can rent bicycles or motorbikes. Also offered are supervised play programs for children aged four to ten, service desks, airport transportation, and shops. ☐ **Lux** ☐ **W, F, G, T**

MIAMI: Coconut Grove, *2649 S. Bay Shore Dr., Miami, FL 33133, (305) 858-2500* ☐ *Open all year* ☐ *Bay Shore Dr., S of jct US 1 and I-95* ☐ This 20-story luxury hotel boasts 200 modern, individually decorated rooms, all with air conditioning, color TV, and telephone (some even in the bathrooms). The suites have kitchen or refrigerator, and a number of deluxe suites feature such extras as a whirlpool, a wetbar, or a mirrored ceiling.

Although not a sports resort, Coconut Grove does have two tennis courts (lighted) and a pool. Also on premises are a men's and women's sauna, a boutique, and a gift shop. Across the street is the Dinner Key marina and the hotel's 53-foot yacht, *Horatio,* available for charter to guests.

Similarly named is the 20th floor dining room, Horatio's, with its long, arched ceiling of leaded glass and its wood-paneled walls. It has a splendid view of the bay and fine continental food. Cafe Brasserie, done in a fin de siecle opulence and hung with a profusion of plants, is for a cup of coffee or a quick lunch. ☐ **Lux**

Doral Country Club, *4400 NW 87 Ave., Miami, FL 33166, (305)532-3600* ☐ *Open all year* ☐ *Off SR 826, 5 mi S of jct US 27* ☐ This exclusive enclave, set on 2400 acres eight miles from central Miami, is for pampered vacations. You enter the gates into the beautifully landscaped heart of the complex–a pretty lake with a waterfall and the huge outdoor swimming pool set in gardens and shaded by date palms. Ringing this oasis are ten sparkling-white, two-story guest lodges and the circular, glass-fronted clubhouse that contains dining and drinking areas, lounges, and the superbly equipped pro shop. Surrounding all this are the recreational facilities, an enormous buffer zone against the outside world. The 650 accommodations here are all first-rate. Most units have a private terrace, all have amenities such as refrigerators, and many have dressing rooms, sitting rooms, walk-in closets, or other extras.

Avid sports people should note that the rates here, which include EP, MAP, and package plans, include free use of facilities–not even the usual greens fees are levied. You have unlimited golf on the *four* 18-hole courses, dotted with lakes and palms and including the famed Blue Monster. In addition there are a nine-hole, par-three course, three putting greens, and a driving range. For tennis there are 19 courts, some lighted, and a pro. The health club offers saunas and a solarium. On the grounds are 18 lakes, one with a beach and some stocked with bass for fishing, as well as courts for basketball and other games, areas for cards and other indoor games, beauty and barber shops, and shops for sundries. The range of services included supervised play for children over holiday periods.

In addition, a free bus shuttles you to the sister resort, Doral-on-the-Ocean, 4833 Collins Ave., Miami Beach, FL 33140, (305) 532-3600. This 420-room tower is on the ocean and offers a beach with thatched umbrellas and cabanas. Guests at either Doral may use facilities at the other resort. ☐ **Lux** ☐ **G, T, F**

MIRAMAR: Seascape Resort and Conference Center, *Miramar Beach, Destin, FL 32541* ☐ *Open all year* ☐ *On US 98, 5½ mi E of Destin* ☐ Besides a Joe Lee 18-hole golf course and 8 tennis courts (six Rubico and two all-weather), Seascape's 220 acres embrace 2000 feet of Gulf shoreline and three freshwater lakes stocked with bass. Set around the course and courts are the one-, two-, and three-bedroom condominiums with wood exteriors and shake roofs. Privately owned, most have been professionally decorated. All have color TV, telephone, air conditioning, kitchen, and private patio or balcony.

Of the resort's two restaurants, the State Room is the more formal, serving continental food, including a special menu of Basque dishes. The Galley grill, open for lunch, serves sandwiches and quick meals. In addition to the dining areas, the clubhouse contains saunas, whirlpools, and exercise room. Guests can take a pedal boat out on a lake or swim or surfcast in the Gulf. ☐ **Mod to Lux** ☐ **T, G, F**

MOUNT DORA: Lakeside Inn, *Box 175, Alexander St., Mt. Dora, FL 32757* ☐ *Open mid-Dec to mid-Apr* ☐ *Off US 441* ☐ The oldest part of this restful, rambling lakeside hotel dates from 1882. Conceived in the Grand Hotel tradition, it offers 111 AP rooms and suites (22 units share baths) in three-story pink buildings with green peaked roofs. The buildings are set endwise on the six-acre shore so that most rooms overlook the lawns and swimming pool rather than the lake. Public and guest rooms have an old-fashioned coziness and some antiques. The restaurant serves local specialties such as peanut soup and orange chiffon pie, and there is a bar. On the lake is a sand beach and dock with boat rental. Fishing is good, especially for largemouth black bass. Mount Dora has the third largest lawn bowling club in Florida. Golf and tennis are available nearby. ☐ **Inexp to Mod** ☐ **W, F**

NAPLES: La Playa Beach & Racquet Inn, *9891 Gulf Shore Blvd., Vanderbilt Beach, Naples, FL 33940, (813) 597-3123* ☐ *Open all year* ☐ *Take US 41 5 mi N of Naples to Vanderbilt Beach Blvd.* ☐ This modern, low-rise hotel is right on the water–only a few feet of lawn separates it from the sandy beach on the Gulf–and to rear of the hotel, just beyond the tennis courts and pool, are the attractive waterways of the Vanderbilt lagoon. Every room has a private balcony or terrace overlooking either the Gulf or the lagoon. Rooms are decorated in a South Seas theme, with some rattan furniture, Gaugin prints, and so on. The hotel has three one-bedroom suites, two on the Gulf side and one on the lagoon side. Although the inn has a continental restaurant, rooms with kitchens are available. The heated pool is set in a sun terrace bordered by palm trees and other tropical plants. A high stone wall encloses it for privacy. The four tennis courts are just beyond the pool. ☐ **Mod** ☐ **T**

ORLANDO: John Newcombe's Vacation Resort, *P.O. Box 2527, Orlando, FL 32802, (904) 394-6171* ☐ *Open all year* ☐ *On US 27, 1 mi N of jct. US 192* ☐ Here the idea is tennis. Seven of the 11 Laykold and six Har-Tru courts are lighted, and the tennis program offers pro instruction for all levels with video replay and other modern aids. There is a well-stocked pro shop. Most of the attention has been lavished on the courts. Two-story rectangular housing units, surrounded by parking space, offer 200 motel-style rooms and suites. Also on the grounds are a swimming pool, restaurant, bar, game room, shuffleboard courts, playground and laundromat. ☐ **Mod** ☐ **T**

PALM BEACH: The Breakers, *Palm Beach, FL 33480, (305) 655-6611* ☐ *Open all year* ☐ *On US A1A, in town* ☐ As long as this truly palatial hotel stands guard at the end of its tree-lined private drive, Palm Beach cannot be said to have lost its class. Though condominium towers are beginning to crowd the town from the south, Worth Avenue is still the most opulent of Main Streets, U.S.A. Cool white Bentleys and Rolls-Royces park by the side of the palm-shaded street lined with boutiques and galleries. Presiding over the scene, a lawn-protected walk away, is The Breakers, America's best try at a perfect match of renaissance magnificence and dignity.

Unbelievably, The Breakers was originally built as Flagler's *second*-class resort in Palm Beach, to catch the overflow from the Royal Poinciana, which was, until the twenties THE resort hotel on the East Coast. But changing tastes and an Act of God (a hurricane) led to that mammoth hotel's demolition in 1934. This left The Breakers, America's richest orphan, to fill its place. The stucco and concrete building is all imposing symmetry; two flag-topped bell towers flank the arches and Ionic columns of the handsome portico, and two C-shaped wings branch off the facade. Step inside and you find yourself in an enormous lobby, a grand hall with vaulted arches, hand-painted frescoes, and floor to ceiling windows, furnished in a style suggestive of the Italian villas after which the building was modeled. The 526 double rooms and 41 parlor suites all have views of either the ocean or the gardens; they are equipped with color TV, air conditioning, and direct-dial telephones. Included in the rates are breakfast and dinner in the hotel's Florentine dining room. There are also restaurants and lounges at the beach and poolside, and at the golf courses.

The 18-hole Ocean Course adjoins the hotel, and the newer, longer West Course, designed by Willard Byrd, is in West Palm Beach. There are two swimming pools as well—one outdoor saltwater and one indoor freshwater—and a private beach where you can rent a cabana. The needs of tennis buffs are supplied by 12 courts. Other activities include shuffleboard, putting green, lawn bowling, and bicycling around the grounds and town. The hotel can make arrangements for deep-sea fishing expeditions, or will help you rent a boat for waterskiing, snorkeling, or sailing. At night there's entertainment, dancing, and cocktails in the ocean-view Alcazar Room. And of course, the Breakers offers everything from massage to supervised children's activities to an airport bus, in addition to topnotch, unobtrusive service. ☐ **Lux** ☐ **G, T**

PENSACOLA: Perdido Bay Inn, *1 Doug Ford Dr., Pensacola, FL 32507, (904) 492-1212* ☐ *Open all year* ☐ *16 mi W of Pensacola on SR 292* ☐ The only thing between this complex of 15 one-story villas and the Gulf of Mexico is the driving range and golf course—6871 yards of Bermuda and Tifton grasses—designed by William Amick. Built of brick and cedar with shingle roofs, each modern villa is laid out like a capital "H"—two bedrooms form each leg, and a studio with fireplace (in seven units only), kitchenette, and sofa beds is the cross-tie. Every room has a private entrance, color TV, air conditioning, and private bath. Rooms can be rented in any combination. A golf package (three night minimum) including breakfast, dinner, greens fees, and gratuities is also available. This 1000-acre resort also offers two all-weather tennis courts, a 1600-foot sand beach and a cabana, a restaurant and lounge with regularly scheduled dancing and entertainment, a social director, movies, barbecues and box lunches, and swimming and wading pools. ☐ **Mod** ☐ **G, W**

PONTE VEDRE: Sawgrass, *P.O. Box 600, Ponte Vedre Beach, FL 32082, (904) 285-2261* ☐ *Open all year* ☐ *From Jacksonville take I-95 to J. Turner Butler Blvd. E, then SR A1A S* ☐ Sawgrass is a private resort community with 1100 acres of trees, waterways, and beachfront. The community is carefully arranged around the excellent 18-hole golf course and several freshwater lakes. About 300 acres surrounding the development is a wildlife preserve, harboring shore and marsh birds, deer, and alligator. You may rent accommodations ranging from beachfront studios to four-bedroom villas. All come with fully equipped kitchens, washer/drier, and color TV; since each villa is privately owned, each is individually decorated to the owners' taste.

Recreational facilities are superb. The 18-hole golf course, designed by Ed Seay, is among the top courses in the country. From the championship tees it measures an awesome 7111 yards, but it can be played as short as 5090 yards. The most spectacular holes are the last four, which play along and across a series of bass-filled lagoons. The Tournament Players Championship is held here during March. The clubhouse has pro shop and an elegant restaurant and lounge. The practice area has more than 20 stations, a bunker, and putting and chipping greens. The 12 Har-Tru clay courts (two lit for night play) have been the site of the Sawgrass Professional Tennis Match. At the Racquet Club there's a large pavilion for viewing matches and a pro shop. Adjoining is a large freshwater pool with children's pool, and a children's playground. Supervised children's programs are available. At the Beach Club, you may use the olympic-sized pool or the "adults only" pool, enjoy a cocktail in the Oasis Bar, or have a cookout on the beach. ☐ **Lux** ☐ **T, G**

ST. PETERSBURG: Don Cesar Beach Resort Hotel, *3400 Gulf Beach Blvd., St. Petersburg Beach, FL 33706* ☐ *Open all year* ☐ *From St. Petersburg, take SR A19A S to Gulf Blvd.* ☐ Looking like a castle from Moorish Spain, the pink Don Cesar is right on the Gulf of Mexico; only the large sun terrace surrounding the heated pool lies between the hotel and the white sands of the beach. Costing $1,150,000, the Don was finished in 1928. It was built to last: the walls and floors are of Belgian concrete 10 inches thick. Guest room furniture was custom made, and the wooden ceiling beams for each room were individually cut because no two rooms have the same dimensions. In the public rooms the chandeliers are Austrian crystal and many of the antique furnishings and fixtures are from Spain and Mexico. The hotel is now on the National Register of Historic Places. Restaurants include a bistro in a greenhouse with stained-glass skylights, a charming cafe, the Beachcomber grill, and the opulent King Charles with notable cuisine. Entertainment is available in several of the lounges.

The Beach and Tennis Center, with cheerful, airy lounges, is headquarters for most activities. Here is a health club for whirlpool, sauna, or massage. You can rent bicycles, beach cabanas, sailboats, jet skis, take scuba lessons, or take two-hour or all-day cruises. Two lighted tennis courts here are supervised by a pro. Pier and drift fishing or rental of party or charter boats can be arranged. Just Kids Limited is an all-day program of supervised activity for ages 5 through 12. You also have complimentary transportation and privileges at the 18-hole Mark Mahannah golf course, less than a mile away at Isla del Sol country club. A tour and transportation desk arranges off-premises sightseeing. □ **Lux** □ **T, G, F, W**

SANIBEL AND CAPTIVA: These two islands are connected to the mainland by a toll causeway and to each other by a bridge. Because the islands are small, everything is close by. There are marinas, a fishing pier, and boat tours. Some people consider the Gulf-side beaches prettier, and others prefer those on the bay side. Sanibel is the bigger island and has a wider range of accommodations. Captiva is so narrow a duffer could easily drive a golf ball across it at some points, and it has only one place to stay, the South Seas Plantation resort. The islands are probably most famous for the profusion of rare shells that wash up on the miles of Gulf beaches. Shelling guides are available to help you sort out the cockles from the welks. Another attraction is the J. N. "Ding" Darling National Wildlife Refuge on Sanibel. It is so beautiful that lots of people uninterested in alligators or roseate spoonbills go there just for the tropical scenery.

Incidentally, Sanibel got its name from the English pronunciation of the Spanish original, "Santa Isabella," after the queen of Spain. Its sister islet is called Captiva because it was believed that pirates held their female captives there, under the guard of squaws and eunuchs, so their crews would not molest them and thus decrease their ransom price.

Island Inn, *Sanibel, FL 33957, (813) 472-1561* □ *Open all year* □ *From causeway, take Periwinkle Way (SR 867) to the corner of Island Inn Rd. and W. Gulf Dr.* □ This seaside resort consists of 3 lodges and 6 cottages. Accommodations run from single rooms with refrigerator up to the three-bedroom Zeeryp Cottage with living room, dining room, enclosed porch, kitchen, two bathrooms, and car port. The largest of the lodges, Kimball, has a fireplace and library in the lounge and a sundeck on the second floor. All units, except those in Matthews Lodge and Matthews Cottage, have color TV; most are air-conditioned. In winter the inn runs on MAP. In summer the dining room is closed, and maid service is extra in the cottages. The inn has tennis courts, shuffleboard, and a small circular pool set in a sandy, palm-studded courtyard on the beach. □ **Mod** □ **W**

South Seas Plantation, *P.O. Drawer 646, Captiva Island, FL 33924, (813) 472-5111* □ *Open all year* □ *17 mi NW of causeway* □ This resort is called a plantation because at one time Captiva was producing 90 per cent of the Key limes sold to northern markets. Agriculture had begun to falter on the island by the 1930s when the newly introduced copra crop failed to materialize as quickly as the owners planned. The workers' cottages were rented to sportsmen and from then on Captiva flourished as a resort.

At the very tip of the island is the resort's attractive nine-hole golf course overlooking the Gulf and bay. Alongside the course, near the bay shore, are the swimming pool and the Plantation House. Designed to resemble a Southern manor, this elegant building houses the most reasonably priced guest rooms. Each has a living room/bedroom, balcony, private bath, and dressing room.

The deep-sea yacht basin is nearby, also on the bay side. Tucked inside this little harbor is 3400 feet of dockspace for boats up to 100 feet in length. Near the docks are the King's Crown, which serves fine food, and Captain Al's, a pub-style eating and drinking establishment. The resort villas are also in this area. These are efficiency apartments with one or two bedrooms. Across the yacht basin are the marina villas, slightly larger lodgings, each with two bedrooms and two baths. These villas share a private pool. Stretching south along the Gulf shore are the bigger homes: the Gulf cottages, beach homes, and villas. These have from one to four bedrooms and full kitchens, and they share private pools and tennis courts. The homes and cottages also have washers and dryers.

At the southern end of the island are Chadwick's Restaurant, featuring continental and seafood dishes, and the one- and two-bedroom tennis villas overlooking the main tennis complex. All told there are 22 Laykold courts at the resort, with one court lit for night play, and a clubhouse and pro shop.

Charter boats and guides are available at the marina for deep-sea fishing for tarpon, snook, grouper, and redfish. The Steve Colgate offshore sailing school offers week-long introductory sailing, cruising, and racing instruction aboard 27-foot Olympic Solings and Pearson 33s.

All the development and construction on this little island has unfortunately made the housing seem a bit urban. Nevertheless the glorious beaches and fine facilities make this an exceptional retreat. ☐ **Lux** ☐ **T, G, W, F**

The West Wind Inn, *P.O. Box F, Sanibel, FL 33957, (813) 472-1541* ☐ *Open all year* ☐ *On W. Gulf Dr. NW of Island Inn Rd.* ☐ Like a huge "U" with the open end facing the Gulf, these two-story lodges embrace a nicely landscaped grass courtyard with a palm-shaded pool at its center. All rooms have recessed private balcony, telephone, color TV, and daily maid service. Efficiencies are available. On the grounds are two tennis courts, putting green, lawn bowling, and shuffleboard. The lodge does not have a restaurant, but one is a few blocks away. ☐ **Mod to Lux** ☐ **W**

SEBRING: Harder Hall Golf and Tennis Resort, *Golfview Dr., Sebring, FL 33870* ☐ *Open mid-Nov to early May* ☐ *On SR 17, 2 mi N of jct US 27* ☐ Harder Hall is a grand old resort done in classic mission-style with red-tile roof, enclosed courtyards, high archways, and pink stucco exterior. The resort's 300 acres lie alongside Lake Jackson, giving guests ample opportunity to swim, fish, waterski, or sail. Catamarans, as well as fishing and waterskiing boats, are available for rental. The resort is well known for its golf. Its private course, right outside the door, is a 6800-yard, par-72 Dick Wilson course, which passes a number of small lakes. The golf program utilizes instant video-replay, movies, demonstrations, and stop-action photography, as well as supervised practice and play. The ratio of students to teachers is excellent: four golfers to one pro. An equally intensive tennis program, on 12 acrylic courts (5 lit), is supervised by a team of pros. Ball machines and replay video are regularly used. Facilities include a pool and a health club with sauna, exercise room, steam room, and solarium. Dining is at the Royal Palm room or the golf clubhouse. A live orchestra provides cocktail and dinner dancing six nights a week. Rooms have been done in bright contemporary colors and a few of the rooms have balconies. ☐ **Mod to Lux** ☐ **T, G**

TALLAHASSEE: Killearn Golf and Country Club, *100 Tyron Ci., Tallahassee, FL 32308, (904) 893-2186* ☐ *Open all year* ☐ *Take Thomasville Rd. exit off I-10, then SR 61 N to signs for club* ☐ This 300-acre country club has a full range of recreational facilities. The original country club, a contemporary-style building, houses 15 doubles and four suites, and the newer inn has 20 more doubles. Some rooms have kitchenettes or balconies. All have color TV and climate control.

The golf course, designed by William Amick on rolling terrain, is site of the Tallahassee Open each spring. The course totals 27 holes; each 9 holes are a par 36. The country club has eight courts: four Har-Tru and four lighted Laykold. Handball and racquetball courts and a practice wall are adjacent to the tennis courts. Both a tennis and golf pro shop are on premises. The "L" shaped, olympic-size pool has championship diving boards, lounging area, and a rustic wooden pavilion.

The main dining room has a gourmet dinner menu with specialties such as New Zealand lobster tail. Breakfast and lunch are served in the country club's restaurant and lounge. ☐ **Lux** ☐ **G, T**

TARPON SPRINGS: Innisbrook, *P.O. Drawer 1088, Tarpon Springs, FL 33589, (800) 282-9813* ☐ *Open all year* ☐ *3 mi S of Tarpon Springs on US 19* ☐ This beautifully landscaped resort has some of the finest and most extensive golf and tennis facilities in the country. A guest can play 18 holes of golf every day for a week and play each hole just twice. Of its three golf courses, the Island Course and the 27-hole Copperhead Course are the most challenging. The par-70 Sandpiper Course is shorter and puts a premium on accuracy. Each of the three courses has a clubhouse with dining room and lounges. Tennis fanatics will find 12 Har-Tru and five Laykold courts and a professional teaching staff aided by the most modern video taping equipment.

The 25 attractive lodges, secluded under trees bordering the fairways, contain one- and two-bedroom suites. Although these are privately owned and therefore furnished to the owners' tastes, all have color TV and private patio or balcony. Most have kitchenettes.

Also within Innisbrook's 1000 acres, wooded with pine and citrus trees, are five swimming pools, sauna, fishing lakes, bicycling, and a wildlife preserve. Children's counselors supervise a complete activities program that includes tennis, swimming, sailing, and fishing. There is shuttle service to the nearby Gulf beaches. ☐ **Mod to Lux** ☐ **G, T, F**

WALT DISNEY WORLD: Walt Disney World, *P.O. Box 40, Lake Buena Vista, FL 32830* ☐ *Open all year* ☐ *On US 192, near jct I-4* ☐ This *is* a world unto itself, or rather several worlds. A well-organized army of idealistic real estate planners has transformed 43 square miles of central Florida into a utopia for grown-ups as well as children. Active adult vacationers who couldn't care less about Mickey Mouse will find luxury accommodations, 54 holes of championship golf, and three groups of tennis courts. Of course, the main attraction is the Magic Kingdom, as the amusement park is called, with its fairy-tale castle, Frontierland, Fantasyland, Tomorrowland, and other theme lands. The Disney people are unexcelled in making mock-up versions of steamboats, hippos, spaceships, haunted houses, submarines, and other exciting attractions. Altogether, Disney World has more than 45 attractions and 70 restaurants and shops. All of it is good fun and cleverly done.

A fine time can be had at Disney-run resorts just outside the Magic Kingdom. These resorts center on three lakes: closest to the Magic Kingdom is Seven Seas Lagoon, circled by the sleek monorail. Starting at the entrance to the Magic Kingdom, the monorail runs by Golf Resort, stops at Polynesian Village Resort, and then stops *inside* the giant A-frame of Contemporary Resort before coming back to the Magic Kingdom. East of the lagoon is Bay Lake, and south of that is Lake Buena Vista. No matter which resort area you choose, you may take advantage of the golf, tennis, boat rental, horseback riding, or dining and drinking facilities of the other resorts.

Golf Resort is composed of low, shingled lodges of wood and volcanic stone set beside the Magnolia and Palm golf courses—two par-72, 18-hole courses designed by Joe Lee. The Palm is among the top courses in the country. Shorter and tighter than the Magnolia course, it has many hazards and traps. Lined by more than 1000 magnolia trees, the Magnolia course has elevated tees giving a commanding view of each hole. Most of the 151 rooms here have balconies overlooking the fairways or the swimming pool. Each room is decorated in autumn hues of burnt orange, gold, and brown, and has air conditioning, two queen-sized beds, and a fold-out bed. The Trophy Room restaurant has buffet breakfasts, sandwiches, late-night fondues, and such dinner entrees as veal scallopine and red snapper. Also at Golf Resort are putting greens, driving ranges, heated swimming pool, two lighted tennis courts, and a golf studio offering instruction, swing analysis, videotaping, and equipment check.

Nine "longhouses" decorated in a South Sea Islands style contain the 636 guest rooms of *Polynesian Village.* In the center of the complex is the Great Ceremonial House, an expansive building which resembles a royal Tahitian assembly lodge. The resort has a lush setting amid tropical gardens at the palm-lined shore of Seven Seas Lagoon, which has a marina, beaches, a large, boat-shaped swimming pool, and a smaller swimming pool with a waterfall and curving slide. Five lounge and dining areas plus several small bars provide a full range of food and drinks and entertainment. Chicken Pago Pago and Toa Nui are just two of the exotic dishes regularly featured, and luaus are held nightly.

With 1046 guest rooms in three buildings, the *Contemporary Resort* is the flagship. Flanked by two widespread, three-story lodging wings is the most impressive building–the A-frame tower through which the monorails silently glide. Over 100 feet high and 450 feet long, the lobby is startlingly spacious. On top of the tower, set on the plateau made by its truncated peak, is the glass-enclosed Top of the World Restaurant, the entertainment center featuring an orchestra and nightly floor shows. Contemporary Resort, set between Seven Seas Lagoon and 450-acre Bay Lake, is closer to the shore of Bay Lake. In fact, the marina and one of the two swimming pools extend out onto this lake. For tennis buffs, the resort's racquet club has six courts for day and night use.

The only other place to stay on Bay Lake is a camping and trailing area called Fort Wilderness where, if you don't bring your own shelter, you may rent Terry trailers that sleep up to six people and are fully furnished right down to the housewares and linens. A number of creeks and streams crisscross the large, wooded site, providing secluded spots for fishing and canoeing. Horseback riding and bike rental is also available here. Directly on Bay Lake in the Fort Wilderness section is River Country, a novel playground area with rope swings and a 260-foot water slide that glides swimmers over fiberglass rocks and drops them in an old swimming hole.

Lake Buena Vista is most distant from the Magic Kingdom, but like all the resorts, it is easily accessible by the complimentary motor coaches that regularly run between all major areas of Disney World. Here is the *Lake Buena Vista Club,* a chalet-style country club overlooking an 18-hole championship golf course, a swimming pool, tennis courts, and a marina. Guest accommodations are in three types of villas: Treehouse, Vacation, or Fairway villas. The treehouse villas are two-bedroom, two-story units giving you a treetop view of the golf course from the second story. Vacation villas, on the shores of the lake, are plush one- or two-bedroom units that are good for families. The Fairway villas have up to three bedrooms and a living, dining, and kitchen area. All villas come complete with kitchen appliances, dishes, flatware, utensils, color TV, and central air conditioning. The two main attractions on Lake Buena Vista are the *Empress Lillie,* an enormous stern-wheel steamboat that serves as a restaurant, and Disney World Village, a shopping mall with about 30 shops and restaurants.

Disney World resorts offer package plans and many special events and services, but some facilities can become overcrowded. Reservations for some of the posher restaurants should be made up to 30 days in advance. For babysitters, it is suggested you reserve up to 60 days prior to arrival. Reservations are also required for supervised play programs for children aged three through nine at Mouseketeer Clubhouses at Contemporary Resort and Polynesian Village.

Also near Lake Buena Vista, but not managed by Disney World, are Dutch Inn Resort, Howard Johnson's, TraveLodge Tower, and Hotel Royal Plaza. If you book with any of the Disney resort accommodations and they are all full, you will automatically be booked into one of the above hotels unless you indicate otherwise on your reservation form. Forms and other information on accommodations can be obtained by contacting Walt Disney World Central Reservations, P.O. Box 78, Lake Buena Vista, FL 32830, (305) 824-8000. ☐ **Lux** ☐ **G, T, W, F, R**

Georgia

BIG CANOE: Big Canoe Resort, *Big Canoe, GA 30143, (404) 579-3333* ☐ *Open all year* ☐ *From US 19 in Dawsonville, take SR 53 W to Yellow Creek Rd. N to resort* ☐ A resort community in the Appalachian foothills, Big Canoe operates on the European Plan and offers modern one-, two-, and three-bedroom homes and condominiums, varying from modern versions of the log cabin to round houses to unusual "Treetop" villas–elevated octagonal, glass-walled two-bedroom units. The homes, handsomely designed of wood and stone, are laid out with attention to details–such as underground power and telephone lines. Features include attractive homelike furnishings, fireplaces, daily maid service (no room service), fully equipped kitchens, color TV, sun decks, and beautiful views.

Within the resort's 5460 acres are three lakes and a complete array of recreational facilities. The 18-hole golf course, curving around Lake Sconti, was designed by Joe Lee and includes pro shop, putting green, and driving range. The racquet club (extra fee) has 10 tennis courts (including two indoors), as well as squash and platform tennis courts. On Disharoon Lake is the swim club, complete with beach facilities, two pools, and a 200-foot-long, slippery-rock water slide. Other facilities include stocked ponds and streams for fishing, hiking and jogging trails, a nature preserve and birdwatching, Jeep trails, paddleboats and canoes. Parents will find a babysitter list, a playground, and summer children's day camp and other programs.

The Chimneys restaurant has salads and hearty country dishes such as barbecued pork and fried chicken. No alcoholic beverages are sold, so guests brown-bag liquor into the restaurant and clubhouse, where setups are provided. An authentic country store, log cabin, and history museum are also part of Big Canoe. ☐ **Lux** ☐ **T, G, W**

CHATSWORTH: Cohutta Lodge, *5000 Cochise Trail, Chatsworth, GA 30705, (404) 695-9601* ☐ *Open all year* ☐ *9 mi E of Chatsworth on US 76* ☐ This attractive European-plan lodge, two rustic but contemporary buildings, is set on some 300 acres atop Fort Mountain overlooking Chattahoochee National Forest. From the balconies of this mountaintop perch, guests can scan Bear Pen Cove valley and the range of heavily forested mountains surrounding it. Inside, the public areas and guest rooms have modern appointments with woodsy accents such as the massive timber columns and ceiling beams in the dining room and the stone fireplace in the lounge. All guest rooms have color TV, telephone, and air conditioning. The restaurant specialties include prime rib, trout, barbecued flank steak, and peanut pie. As this is a dry county, guests brown-bag their drinks and the restaurant provides setups.

The lodge offers a heated pool, providing winter and summer swimming, two all-weather tennis courts, archery, horseshoes, and hiking trails. In winter there is tobogganing, ice skating, and some cross-country skiing. In the craft village one-half mile away, you may rent a horse for trail rides, and several good trout streams are nearby. Fort Mountain State Park and Carters Lake have swimming and boating. □ **Mod**

CLAYTON: Kingwood Inn and Country Club, *P.O. Box 725, Clayton, GA 30525, (800) 841-6266* □ *Open all year* □ *Take US 441/23 to Clayton, then US 76 2 mi E* □ This attractive complex in the foothills of the Blue Ridge Mountains offers a lodge with 44 double rooms and suites with the usual amenities (color TV, telephone, balcony), as well as two- to four-bedroom condominiums with balconies, stone fireplaces, and fully equipped kitchens. Activities center on the Clubhouse, which contains the registration desk, lounges, and dining room. Constructed of natural stone and hand-hewn beams, the dining room is aristocratic, with high-backed chairs and warm colors. In daylight the floor-to-ceiling windows look out on the pond and golf course; at night the chandelier lighting is subdued. Just outside is the Robert Trent Jones championship golf course and the pro shop; farther down the sloping lawns is a cloverleaf-shape swimming pool; and nearby are two tennis courts.

A special feature here is the Chrysalis, a low-slung, wood-and-glass building housing an indoor pool, saunas, whirlpools, exercise equipment, beauty salon, lounge, and a dining room with low-calorie menu. You can enroll here for four- and seven-day programs aimed at reducing stress, losing weight, or just for general conditioning; these package plans include room and board, a physical examination, stress test, biofeedback therapy, and other activities and facilities.

Kingwood can arrange rafting on the Chattooga River, tell you about fishing, riding, and hiking in the area, or arrange for day care for children. □ **Lux** □ **G**

JEKYLL ISLAND: Guale-Creek Indians used this small, windswept barrier island as site of their New Year's bonfire, into which they pitched every old possession they could bear to part with; the more an Indian threw in, the greater his prestige. French explorers came in the 16th century, Spanish missionaries in the 17th, English pirates in the 18th, and then cotton planters ruled it until Union troops burned the plantations in 1862. In 1886 it was bought by New York financiers and became an exclusive vacation retreat for Rockefellers, Goulds, Morgans, Vanderbilts, and Pulitzers. They built extravagant "cottages"—Rockefeller's had 25 rooms and 11 baths, Edwin Gould's had a swimming pool in the foyer—but none of the houses had a kitchen because the families brought along the Delmonico chefs to prepare meals at the clubhouse. It is said that in 1900 the inhabitants of the island were worth a sixth of the world's wealth.

World War II disrupted this idyll, and in 1947 Jekyll Island was bought by the state of Georgia as a historical resort park. All nine miles of snowy-white, palm-fringed beaches are open to the public. There's a public fishing pier for saltwater casting, and charter boats are available at the marina. Other facilities include tennis courts, three 18-hole and one 9-hole golf courses, an indoor olympic-size pool called Aquarama (which is air-conditioned in summer and heated in winter), bicycle rental, and shops. The millionaires' vacation complex is being preserved, and many of the homes may be toured. With moderate rates and so many activities, Jekyll Island is a good value for your vacation dollar. The island is reached by causeway by taking US 17 south from Brunswick and SR 50 east.

The Buccaneer, *85 Beachview Dr., Jekyll Island, GA 31520, (912) 635-2261* ☐ *Open all year* ☐ *S on Beachview Dr.* ☐ Although its architecture is undistinguished—low-rise, modern-motel style—the Buccaneer is nicely appointed in contemporary style. Besides the regular bedrooms, rooms with kitchenettes, two- or three-bedroom penthouses, and apartments are available. All rooms have two double beds, color TV, direct-dial telephones, air conditioning, and private balcony or terrace. Both the lounge and the restaurant are done in a bright Caribbean decor with bamboo furniture and ceiling fans. The motel, right on the beach, has its own pool and playground. Cocktail parties, buffets, game nights, oyster roasts, dances, and other guest entertainment are scheduled during the fall and winter. ☐ **Mod**

Jekyll Island Hilton Inn, *Jekyll Island, GA 31520, (912) 635-2531* ☐ *Open all year* ☐ *From SR 50, take Beachview Dr. N* ☐ This beachside Hilton consists of an inn and swimming pool flanked by a platoon of two-story accommodations buildings totaling 192 rooms and 84 one- and two-bedroom units with kitchen facilities. Whitewashed walls and red roofs give the whole an appearance of Spanish-influenced tract housing, and buildings are placed so that only a third of the units face the ocean; some units have balconies or patios, however.

In the inn is the restaurant, St. Andrews Landing, with glass-walled views of the pool and the ocean beyond. The menu features Georgia seafare. The lounge area, Misty's, has ocean views as well as full-service bar and nightly entertainment. Also available are poolside service and a group outdoor cooking patio.

The recreation center has game areas, playground, and bicycle rental, and seasonal supervision of children. ☐ **Mod to Lux**

Sheraton-By-The-Sea, *1175 N Beachview Dr., Jekyll Island, GA 31520, (912) 635-2521* ☐ *Open all year* ☐ *N on Beachview Dr.* ☐ In a successful attempt to avoid the motel look, this Sheraton offers clusters of one-, two-, and three-bedroom apartments in low, gray buildings that blend into the beautifully landscaped grounds. All units offer climate control, living/dining areas, fully equipped kitchen, and private balcony or patio (or both) that may face the pool, a flowered courtyard, or the ocean. Some units are duplexes, with bedrooms upstairs. Interiors, too, avoid the motel look, with tasteful contemporary furnishings.

The restaurant decor is restrained and attractive, and the menu features fresh salads and seafood. The cocktail lounge has entertainment. On the grounds are a swimming pool, wading pool, boutique, and laundromat. A boardwalk leads to the beach. Babysitters are available, and in summer there are supervised programs for children. Several package plans (such as for golfers, tennis players, or honeymooners) are offered. □ **Lux**

The Wanderer, *711 Beachview Dr., Jekyll Island, GA 31520, (912) 635-2211* □ *Open all year* □ *N on Beachview Dr.* □ Stretched out along the shore to give half the rooms an ocean view, this 178-unit, two-story motel offers little razzle-dazzle but is a good value for budget-conscious families. All double rooms have TV, telephone, air conditioning, and either balcony or patio, and are priced lower if you will do without an ocean view. Kitchenette rooms, all of which are oceanside, have two double beds and a studio couch and will sleep six.

The Wanderer has a swimming pool, wading pool, restaurant, cocktail lounge, seasonal entertainment, and a babysitter list. □ **Inexp**

PINE MOUNTAIN: Callaway Gardens, *Pine Mountain, GA 31822, (404) 663-2281* □ *Open all year* □ *On US 27, 20 mi N of Columbus* □ The late Cason Callaway was a golfer as well as an industrialist. So, when he decided to develop a vast tract of over-cultivated, hilly farmland into a model resort and recreation area he made sure that it had a round of golf that takes more than a few hours. In fact, it now takes more than a day. Of the three 18-hole golf courses, Mountain View is probably the toughest, followed by Lake View, and then Garden View, the newest course. The nine-hole Sky View is ideal for a brisk warm-up or family golf. The courses are served by two clubhouses complete with equipment sales, rentals, and food service.

There's much more here than just golf, though. For one thing, the 2500-acre resort contains some of the finest botanical gardens and greenhouses in the country. Thirteen miles of roads wind through formal gardens and forests, and special outdoor plots have seasonal displays of vegetables and flowers.

The resort and gardens are set around 13 man-made lakes, including the 175-acre Mountain Creek Lake, stocked with bream and largemouth bass. Wide, white-sand beaches and boat rentals of all types make these lakes a watersports enthusiast's dream. In addition, 18 tennis courts, bridle trails, a 1000-acre quail hunting preserve with rental dogs and guides (October through March), and a summer program featuring daily circus performances

(June only) and waterskiing exhibitions make this a truly outstanding and unusual resort.

Guests stay in the 365-room inn or in 175 one- or two-bedroom air-conditioned cottages with kitchenettes. EP and MAP are available. The Plantation Room at the inn and the Gardens Restaurant feature home-style Southern cooking, and the Gardens Country Store sells food specialties prepared from produce raised at the gardens.

The inn complex also provides a bar, two swimming pools, a wading pool, lawn games and indoor recreation room, a kennel, beauty and barber shops, and stores. A babysitter list and supervised play are also offered. ☐ **Mod to Lux** ☐ **T, G, W, F**

ST. SIMONS ISLAND: King and Prince Beach Hotel, *P.O. Box 798, St. Simons Island, GA 31522, (912) 638-3631* ☐ *Open all year* ☐ *On the SE corner of the island, 201 Arnold Rd.* ☐ Built in 1939, this quietly elegant hotel has a slightly old-fashioned air and is noted for old-fashioned Southern hospitality. Its whitewashed, interconnected one- to three-story buildings in Spanish colonial style spread along the shore and are surrounded by palm-dotted lawns. Public rooms have fine paneling and are carefully furnished with antiques and reproductions.

The hotel operates on European Plan. The 99 guest rooms are individually decorated, and many have matching wallpaper and curtains. All have color TV, most have private patio or balcony, and some have kitchenettes. There is a romantic honeymoon suite in a round tower with ocean views.

The Delegal dining room, overlooking the ocean (flood-lit at night) has arched windows with stained-glass scenes at the top, illustrating episodes from local history. Decorations here include brass chandeliers imported from England, period furniture, and oil paintings of King Edward and Prince Albert, for whom the hotel was named. Specialties include deviled crab, native shrimp creole, and country ham with red-eye gravy. The Frederica Tavern, also facing the water, has hatch-top tables, flagstone floors, and 18th century memorabilia.

Guests here play at the St. Simons Island Club, which features an excellent 18-hole course designed by Joe Lee. The hotel has three tennis courts, sailboat rentals, swimming pool, and horseback riding, and will arrange charter fishing and historic tours. ☐ **Lux** ☐ **T, G, R, W**

SAVANNAH: Sheraton Savannah Inn & Country Club, *612 Wilmington Island Rd., Savannah, GA 31410, (912) 897-1612* ☐ *Open all year* ☐ *US 80, 10 mi E of Savannah* ☐ One of the South's grand old hotels, renovated by the Sheraton chain at a cost of $2 million, sits among palm trees and live oaks and overlooks the Wilmington River. Public rooms evoke turn-of-the-century nostalgia with decorated columns, crystal chandeliers, and oil portraits. The guest rooms–210 doubles and 17 suites in the 8-story hotel plus kitchenette cottages–were decorator-designed with matching wallpaper and bedspreads. Some rooms have four-poster beds or other touches in addition to modern amenities such as wide-screen color TV.

Local seafood and fine wines are available in the elegant Emerald dining room, while steaks, entertainment, and dancing are featured at the livelier Oliver's, down by the river. You can also have drinks in the inn's bar or a lunch at the Clubhouse on the golf course.

The olympic-size swimming pool has a sun terrace. At the Spanish-style cabana you can fish from the dock or launch a sailboat ride. The golf course–tropical, imaginative, and testing–was designed by Donald Ross and hosts the Michelob Open and Georgia State Amateur tournament. Tennis instruction is available at the five all-weather tennis courts (three lit). The inn can also arrange horseback riding, charter fishing, or supervision of children. Here is a relaxing base from which to tour the wonderful National Historic District of Savannah, with landscaped squares, waterfront promenades, and more than a thousand restored buildings and museums. ☐ **Mod** ☐ **G, T**

SEA ISLAND: The Cloister, *Sea Island, GA 31561, (912) 638-3611* ☐ *Open all year* ☐ *From US 17 in Brunswick take causeway 10 mi to island* ☐ One of the country's foremost resorts and certainly one of the most beautiful, The Cloister is a study in good taste and extra-cordial service. As might be expected from the name, the oldest buildings here have the look of a Spanish monastery, complete with red-tile roofs and arched porticos. The resort extends back from the sea across 12,000 landscaped acres with formal gardens, fountains, palms and moss-draped live oaks, marshes, and meadows.

The variety of accommodations is too complex to describe here (more than 500 units available), but The Cloister operates on American Plan and offers everything from simple twin-bedded rooms to luxury suites (one to four bedrooms) in the hotel and in clusters of guest houses. All units have private bath and air conditioning, and many have balconies, patios, ocean views, or other extras. Rates vary not only by season but by size and location of unit and length of stay. Special rates apply to children and for golf, tennis, honeymoon, and other packages. In lieu of tipping, a service charge is imposed. The resort also rents by the month many of the privately owned two- to five-bedroom houses in the complex.

Dining varies from candlelit indoor dinners to outdoor terrace lunches, with meal service at the clubrooms and beside the pool. Outdoor buffets, cookouts, and dinner dances are also scheduled. Cocktail lounges are in the hotel and at the beach and golf clubs.

A major attraction is the five-mile, snow-white beach, with good swimming. But there is also a beach club with freshwater pools for swimming, diving, and wading.

Most other sports here involve extra fees (unless they are included in a package plan). Golf–some of the best 36 holes in America, with beautifully kept greens and imaginative water hazards–is nearby on St. Simons Island. Station-wagon shuttles leave hourly for the clubhouse, which offers a pro, putting green, caddies or carts, rental and repair, and other services. Tennis buffs will find 18 clay courts plus instruction and services. The boat dock has fishing, boat rental, charter trips, and cruises. The stables, with rental, instructions, and groom service, schedules children's rides and specials such as beach and picnic rides. The year-round gun club offers instruction in skeet shooting. You can also rent a bike or play a number of lawn games.

An activities director provides a full agenda that may include movies, nature talks, concerts, or other events. Parents will find babysitters and supervised programs for children. On the premises are shops and services to cover virtually every need from laundry to a chauffered car. □ **Mod to Lux** □ **G, T, W, F, R** □

Hawaii

ISLAND OF HAWAII

HILO: Naniloa Surf Hotel, *93 Banyan Dr., Hilo, HI 96720, (808) 935-0831* ☐ *Open all year* ☐ *Off Banyan Dr. in Hilo* ☐ Although a hotel has been on this site since 1938, the current Naniloa began in 1967 with the completion of a 205-room high-rise complex; two years later it reached its present size with a 185-room addition. One remnant of the original hotel is the award-winning garden, a pleasant little Eden of wandering paths and tall palms, facing Hilo Bay. The small, kidney-shaped pool is set here, right at water's edge, between the hotel's Samurai Bar and catamaran anchorage. Four dining areas range from the elegant Crown to the informal ice-cream shop. The hotel is also something of an entertainment center: the Crown has "name" acts; the Polynesian Room is a late-night disco; and the Hoomalimali Lounge features dancing and a relaxed atmosphere preferred by the local residents.

Just across Banyan Drive is a nine-hole, par-35 executive course, a double-decker driving range, and a putting green. The Hilo Municipal Course, an 18-hole, 6467-yard course, is just four miles from the hotel. For tennis players, the Naniloa Surf has two Laykold courts. Hawaii Volcanoes National Park, the black beaches of Kaimu, and a number of scenic waterfalls and orchid and anthurium gardens are a morning's drive away. ☐ **Lux** ☐ **W**

Sheraton Waiakea Village, *400 Hualani St., Hilo, HI 96720, (808) 961-3041* ☐ *Open all year* ☐ *On Hualani St., near corner of Kekuanaoa and Mililani Streets* ☐ Waiakea Village, 300 rooms in low-rise wooden bungalows, is set among waterways, waterfalls, lush vegetation, bridges, and flowers, all creating an environment out of "South Pacific." The rooms are attractively furnished with colorful fabrics and rattans, and have air conditioning, color TV, and king-size beds. Each room has a private lanai. The Kupuna Room is one of the most elegant restaurants on the island, with a view out over the lagoon. The Ohana Lounge is for a relaxing drink.

Activities include tennis (on two Laykold courts), swimming in the pool or lagoon, and golf, either at the nearby Hilo Municipal Course or the spectacular Volcano Golf Course in Hawaii Volcanoes National Park. The hotel operates a marketplace of open-air shops selling Asian treasures: tapa from Tonga, jewels from Thailand, and silks and jade from Hong Kong. ☐ **Mod**

KAILUA-KONA: Keauhou Beach Hotel, *P.O. Box 849, Kailua-Kona, HI 96740, (808) 322-3441* ☐ *Open all year* ☐ *Off SR 11 in Keauhou* ☐ Architecturally the hotel is an uninspired box, but the surrounding grounds are fascinating no less for what they were than what they are. The shoreline and tidal pools around the hotel were evidently part of an ancient Hawaiian sacred precint. Petroglyphs can be seen on one of the crusty lava outcroppings and not far off are ancient foundations of a temple and a house. Several of the tidal pools were supposedly reserved for families of the chiefs and priests of Kona. The hotel's gardens fit around all the ancient ruins without disrupting them. In fact, the hotel maintains authentic replicas of a grass hut and a hand-carved canoe to give guests an idea of island life before Captain Cook. Rooms are all modern, well furnished, and have private lanais. On the grounds are three tennis courts and pro shop, saunas, putting green, swimming pool, shuffleboard, volleyball, and badminton areas, and small sand beach. The Keauhou-Kona Country Club course is three minutes' drive away. The hotel has several dining areas and a disco, Another Foxy Lady. ☐ **Lux** ☐ **T**

Kona Village Resort, *P.O. Box 1299, Kailua-Kona, HI 96740, (808) 325-5555* ☐ *Open mid-Dec to late Nov* ☐ *6 mi N of Keahole Airport, off SR 19* ☐ The accommodations at Kona Village are thatched bungalows, and the Kona prides itself on the absence of telephones, TV, and other modern conveniences. Yet the resort's self-styled description as "primitive/plush" would more accurately be just "plush." The rooms surround a clear bay and sand beach on the Kona coast of Hawaii–a location about as close to paradise as one can get.

Set in a lush jungle at the base of the Hualalai and Mamakea volcanoes, the thatched roof "hales," as the bungalows are called, are simply but elegantly furnished. Inside, the walls and sloping roofs are a light-blond wood, complemented by bright Polynesian wall hangings, rainbow-colored bedspreads, and comfortable wicker furniture. Various accommodations from ocean-view bungalows for two to a deluxe interconnected duplex on the beach with a panoramic view from its 54-foot lanai are available. The hales are individually styled after the traditional dwellings of various South Sea islands, Samoa to Tonga. Rooms have king-size or extra-long beds.

Kona Village overwhelms guests with 12 completely different dinner menus. Dishes range from such mainland fare as rack of lamb and beef Wellington to island favorites: roast duckling lychee, fresh ono, or prawns kaupulehu. Friday night is a luau, with a pig cooked in an underground oven and carried sizzling-hot to the dining room.

Since Kona Village is dedicated to the pursuit and cultivation of pure laziness, no organized activities are scheduled. Rates, however, include all meals and recreational facilities–sailboats, outrigger canoes, snorkeling gear, tennis courts. Opportunities for surf casting or deep-sea fishing for marlin and yellowfin are abundant. The village is built on a historic site among stone platforms used by natives hundreds of years ago. Guides will take you over lava flows through the lush undergrowth to the volcanoes, petroglyphs, shrines, and shelter caves nearby. At night, torches are lighted along the beach, completing the illusion of a tropical paradise. ☐ **Lux** ☐ **W, F**

KAMUELA: Mauna Kea Beach Hotel, *Kamuela, HI 96743, (808) 882-7222* ☐ *Open all year* ☐ *12 mi from Kamuela, near Kawaihae* ☐ For a super-luxury resort, the Mauna Kea is surprisingly and pleasantly subtle. Rooms are designed for both the ascetic and the aesthetic. No bigger than a standard double and simply furnished, they are decorated with exceptionally fine pieces of Oriental and Polynesian art: robes, statues, gongs, masks, and other museum-quality *objets* adorn the walls. Rooms do have lanais and luxurious bathrooms, but the tile floors and well-fitted narrow-board paneling make them almost elegantly monkish–something an Exxon executive with an interest in Buddhism might like.

The long hotel is built into the side of a low hill on Kawaihae Bay. The building's three stories are terraced, so that the balconies on each floor are recessed from those on the floor below. In the center of the hotel are large open atriums with tall palms rising almost the full height of the building.

The hotel has nine tennis courts, large pool, horseback riding, and all the watersports you can imagine, but the sport of sports here is golf. The Robert Trent Jones course, an 18-hole pro-tester, is one of the toughest in the world. Built on a lava flow, the fairways and greens are magically watered and eternally green, but out of bounds is out of this world. Hitting balls off of lava is the golfer's version of the Last Judgment.

One drawing card that keeps islanders coming to the Mauna Kea is the food. The buffet luncheon, quite reasonable for what you get (everything), is justly famous. ☐ **Super Lux** ☐ **G, T, W, R**

PAHALA: SeaMountain at Punalu'u, *P.O. Box 85, Pahala, HI 96777, (808) 928-8301* ☐ *Open all year* ☐ *Off SR 11 at Punalu'u* ☐ These shingled, Polynesian-style condominiums, clustered around a pool, feature apartments furnished with modern, Hawaiian-style rattan and split-cane pieces. The one- and two-bedroom units have all-electric kitchens, dishwashers, and clothes washer and drier. Studios contain sleeping alcove and compact kitchenettes. All units have living room with sofa bed. All linens, kitchen utensils, and serving ware are provided.

Right across the street are four Laykold tennis courts, and a short walk away is the country club. The resort is surrounded by the 18-hole, 6492-yard golf course designed by J. Snyder. A walk toward the shore brings you to the thatch-roofed Black Sands restaurant, which overlooks a scenic lagoon. Open seven days a week, it serves continental fare.

Scenically isolated from the bustling resort towns and set amid green ranchlands, sugar-cane fields, and the black-sand beaches of the southern coast, this makes a fine choice for a get-away from-it-all vacation. ☐ **Mod to Lux** ☐ **T, G**

ISLAND OF KAUAI

HANALEI: Princeville at Hanalei, *P.O. Box 121, Hanalei, Kauai, HI 96714, (800) 367-5340* ☐ *Open all year* ☐ *On SR 56, 27 mi NW of Lihue* ☐ This is one of America's outstanding resort villages. Set among 10,000 tropical acres of cliff-lined coast, rolling pastures, river valleys, and undeveloped forest, Princeville is an unusually spacious community. The development itself covers only 1000 acres–300 are devoted to golf. Some of the the condominiums are set along bluffs above the ocean, some along the fairways. Some are slim, three-story units with broad roofs set on top like Chinese hats; others are long, wide buildings which wouldn't be out of place at a ski resort. All told, Princeville has seven different groups of lodgings, all sharing the facilities.

And what facilities they are! Robert Trent Jones, Jr. designed the 27 holes of golf, rated among the best in the world, and plans call for him to design an additional 36 holes. Each of the three nines can be played in any combination with the others. Ocean Nine is perhaps the most famous–with apparently imposing, but merely challenging, holes. One hole has a drop of 60 feet from tee to pin, over a bushy cliff; another is only 150 yards long, but all of that is over a ravine; and another has a bunker with three craggy volcanic boulders set in it, giving it the look of a Zen rock garden.

Tennis players will find seven Laykold courts, a pro staff, and pro shop. The olympic-size pool is at the clubhouses, and visitors can also trapshoot or ride horseback on the grounds. At the superb beaches surrounding the resort you can fish, skin dive, and sail. The resort organizes hay rides, trips to the undeveloped Na Pali coast via helicopter, music and dancing nightly at the clubhouse, and beach parties.

In the condominiums, each unit has a fully equipped kitchen, and living and dining room facilities. Many units have fireplaces, all have lanais. There are four restaurants on the property and coffee shop service at the clubhouse and in Princeville center. ☐ **Lux** ☐ **T, G, R, W**

KAPAA: Coco Palms, *c/o Island Holidays Resorts, P.O. Box 8519, Honolulu, HI 96815, (808) 922-6121* ☐ *Open all year* ☐ *At Wailua Beach, S of Kapaa* ☐ From the landscaping to the architecture, everything has been done to make this seaside resort a fanciful world unto itself. Nightly, native Hawaiians dressed in traditional garb light the torches around the lagoon and one native blows a conch shell to call guests to dinner. Bathtubs carved out of lava are set into mini-courtyards behind some cottages. Huge shells from deep-sea mollusks were fashioned into washbasins for the guestrooms. Everything has been done to give guests the illusion that they have dropped into an idealized island hideaway.

Guest rooms–some with satiny wallpaper, floor to ceiling mirrors, and wall to wall carpeting–have less island flavoring than the grounds and the thatched-roof buildings themselves. The Kings' and Queens' Cottages are probably the most desirable accommodations. The three dining areas–the Lagoon Dining Room, the Flame Room, and The Coconut Palace Dining Room–feature Polynesian dishes, steaks, and seafood. Hawaiian entertainment is offered every night.

Nine tennis courts, several swimming pools, and the beautiful white-sand beach appeal to the sports minded, and the resort's small zoo, library, and museum make Coco Palms a fun, even instructional, place for kids. ☐ **Lux** ☐ **T**

KEAUHOU-KONA: Kona Surf Resort, *Keauhou-Kona, Kauai, HI 96740, (808) 322-3411* ☐ *Open all year* ☐ *Off SR 11 at Keauhou Bay, 14 mi S of Keahole airport* ☐ Very white, and very space-age looking with its lodging arms stretching down to the sea, the sleek Kona surf is a striking contrast to the black and craggy volcanic rock of the headlands it sits on. The building's odd layout, something like a five-pointed star with one point missing, takes full advantage of the peninsula site by giving most guest rooms a water view. In the middle of the hotel is a large open space containing the garden courtyard. Rooms have modern split-rattan and cane furnishings, rough-sawn redwood walls, and large, step-down private lanais. The Kona has several dining areas including Pele's Court, for informal dining in a lush tropical setting, and the S.S. James Makee, for elegant dinners in a nautically styled room.

The resort has three lighted Laykold courts, with four Plexipave courts at the nearby Holua tennis complex. The Keauhou Country Club course borders the Kona, and a shuttle service to the pro shop leaves the hotel lobby every half hour. On the hotel property are two swimming pools: a large saltwater pool with adjoining men's and women's sauna baths overlooks Keauhou Bay, and a freshwater pool has its own garden setting near the golf course. ☐ **Lux** ☐ **T, G, W**

LIHUE: i Pali Kai Cottages, *P.O. Box 471, Lihue, Kauai, HI 96766, (808) 245-3023* ☐ *Open all year* ☐ *On Kalapaki Bay in Nawiliwili* ☐ In Hawaiian, "i Pali Kai" means "on the cliff by the sea," and that is exactly where these cottages are. Buttressed from below by sturdy underpinnings, the one- and two-room bungalows lean right over the cliff's edge, providing uninterrupted views of Kalapaki Bay. Each air-conditioned cottage is luxuriously furnished and has private entrance, large lanai deck, color TV, and daily maid service. Many units have fully equipped kitchenettes, but all have refrigerators, toasters, and coffee makers. Right across the bay from the Kauai Surf hotel, the i Pali Kai Cottages share its facilities (see below), and in fact, the cottages are even closer to the tennis courts and golf course than the hotel itself. ☐ **Lux** ☐ **G, T, W**

Kauai Surf Resort, *Lihue, Kauai, HI 96766, (808) 245-3631* ☐ *Open all year* ☐ *On Kalapaki Bay in Nawiliwili* ☐ On Kauai, the most remote island in the Hawaiian chain, high-rise hotels are rarities. In fact, the only one you'll find is the Kauai Surf, a 607-room mammoth done in the "ice-cube tray" style of architecture made famous in Miami Beach and, closer to home, Waikiki. In pleasant contrast to the right angles of the hotel, the grounds are a flower-filled tropical bower of bliss concocted from winding paths, a curvaceous lagoon, and Japanese garden. A palm-studded islet, connected to the sun-terrace mainland by a pair of arched causeways, sits peacefully in the middle of the beautiful, crystal-blue swimming pool. A sheet of water cascades off a "rock" shelf, really the roof of the poolside bar, onto sporting swimmers.

The hotel is on Kalapaki Beach, a long, wide strand of creamy sand. The surf is dependable, but not overwhelming here, making it a fine place to learn surfing. The hotel schedules lessons and rents boards. For sailors, Sunfisher sailboats and Hobie catamarans may be rented on the beach. A small and adventurous group of friends can take out an outrigger, too. Individualists, however, should try the agile yellow Royak kayaks in the surf. The hotel offers escorted half-day Royak paddle treks up the nearby Huleia River to view the tropical scenery, swim and picnic at a waterfall, and hear the legends of old Hawaii. Scuba equipment and windsurfing boards are also offered.

The 18-hole Kauai Surf golf course is on the hotel grounds, and a frequent shuttle service between the lobby and clubhouse is offered. Several holes on the 6808-yard course overlook the spectacular bay. Near the golf course, the hotel's racquet club has eight championship Laykold courts and two paddle tennis courts. Pro shop and pros are available for both sports.

The hotel boasts five restaurants–continental to coffee shop–and several lounge and entertainment areas. In one of these you can order a libation called Polynesian Paralysis, if you want. A program of special events, everything from coconut leaf weaving demonstrations to ukulele lessons, rounds out the already full schedule of activities. ☐ **Lux** ☐ **G, T, W**

ISLAND OF MAUI

KAANAPALI BEACH: Maui Surf Resorts, *Kaanapali Beach, Maui, HI 96761, (808) 661-4411* ☐ *Open all year* ☐ *Off SR 30, 3 mi N of Lahaina* ☐ This resort beat the shoebox look by having the long, rectangular hotel laid out on an S-curve. It's still a high rise, but it fits the lush Hawaiian landscape a little better than most. Between the hotel and Kaanapali Beach–one of the nicest strands on the islands–is a luxuriant green lawn, a huge freshwater pool, the three-court tennis club, and Surfside coffee shop and bar. Bordering the lawn and buildings are monkey pod, royal poinciana, rainbow shower, and hala trees. Right on the beach, the hotel's "surf shack" has fishing, surfing, scuba, and Hobie Cat rentals.

Across the street are two 18-hole golf courses, Kaanapali North and South;

the North course was designed by Robert Trent Jones. The golf complex also offers a complete pro shop, a driving range, club rentals, and instruction. Also across the street is a narrow-gauge steam train which tours the neighboring sugar-cane plantation.

The 12-floor hotel has 556 guest rooms, including seven suites. Each room has a private lanai and ocean view. Interiors are furnished in native koa and rattan and approximately 25 per cent of the rooms are equipped with a sofa bed in addition to the regular beds. All rooms have air conditioning, color TV and coffee makers. The Maui Surf has four restaurants, two cocktail lounges, and a 20-shop mall. ☐ **Lux** ☐ **G, W, T**

KAPALUA: Kapalua Bay Hotel, *Rte. One, 100 Bay Dr., Kapalua, Maui, HI 96761, (808) 669-5656* ☐ *Open all year* ☐ *5 mi N of Kaanapali on SR 30* ☐ The arms of two lava peninsulas gently enfold the dazzling blue waters of Kapalua Bay here on the northwestern coast of Maui. Just off the coral-protected beach, the luxurious Kapalua Bay Hotel has been woven into a palm grove. The hotel is a cluster of low-rise, peaked-roof buildings, whose arrangement follows the curve of the beach and centers on a three-story flat-roofed lobby pavilion. Laurance Rockefeller's Rockresorts corporation opened the hotel in 1978 as the focal point of a complex (which also includes condominium and summer-home leaseholds) on land that was once a ranch and is still surrounded by row upon row of pineapple plantings. Kapalua Bay is one of the most beautiful spots in the world. The drive to the hotel is lined with banyan trees and allmanda, a delicate yellow flower once reserved only for Hawaiian royalty. The air is scented by pakaluna and night-blooming jasmine, and the grounds and gardens are planted with a luxuriant variety of tropical trees and flowering plants.

Each of the 196 rooms is huge–30 by 14 feet, with 8- by 12-foot lanais–and most seem even larger because of the expanse of ocean they look out on. The rooms are divided into living and sleeping areas; each room is air-conditioned and has a refrigerator and his-and-hers bath areas (tub in one, shower in the other–take your pick). The rooms are decorated in cool pastels. Louvered mahogany doors separate the sleeping area from the baths, and a wooden-paddle fan hangs overhead in the living area as an alternative to the air-conditioner. If you feel cramped in these modest doubles, penthouse suites–40 by 22 feet–are also available.

The room rates include breakfast and dinner, and the cuisine here matches the splendor of the setting and accommodations. The Dining Room in the lobby pavilion is a spectacular two-tiered affair with a waterfall and lush tropical plantings; a waterway divides the room in half. You can begin your meal with cold banana bisque, then proceed to such specialties as capon piemonase. The Veranda, also in the lobby pavilion, is done up in the style of a plantation manor house; it specializes in broiled and roasted meats and fish. The wine cellar here features an impressive collection of California wines, with special bottlings from tiny vineyards. A stroll down the beach is the Bay Club, where the continental menu features local specialties and nouvelle cuisine. The view of the sunset may distract you from your food. And if it's one of those complicated tropical drinks you're after, you can find it at the lounge in the lobby pavilion, at the Crown Bar, or at poolside.

The 10 tennis courts are in the Tennis Garden, planted with bougainvillea and hau trees. Clubhouse, pro shop, and restaurant are nearby. The 18-hole, par-72 golf course, designed by Arnold Palmer, takes you on a tour of the grounds and finally out to sea. (To get onto the 14th green you have to drive across the ocean.) There's a swimming pool, surrounded by an open air deck, in the Bay Club. The hotel's 58-foot catamaran *Seasmoke* takes you on picnic sails and on scuba diving and snorkeling expeditions. Fishermen can arrange for deep-sea charters, which leave from the nearby town of Lahaina, through the activities desk. A three-acre shopping complex is on the Kapalua grounds.

For an extended stay in this surpassingly beautiful spot, look into the condominium rentals in the four developments scattered around the 750-acre grounds. ☐ **Lux** ☐ **G, T, W**

LAHAINA: Kaanapali Beach Hotel, *P.O. Box 637, Lahaina, Maui, 96761, (808) 661-0011* ☐ *Open all year* ☐ *Off SR 30, 3 mi N of Lahaina* ☐ The playful pool here, shaped like a whale, is the centerpiece for the hotel, which is laid out around it like a giant "C", with the open end facing the water and Kaanapali Beach. In addition to the pool, the garden-like courtyard in the big "C" contains a giant checkerboard, volleyball and shuffleboard courts, and the Shipwreck Bar, where guests are invited to broil their own steaks and drink beer. The Plantation Dining Room is for more formal meals, and the Boogie Palace is the discotheque. All rooms here have private lanai, color TV, and refrigerator. The 36-hole Royal Kaanapali Golf Course is across the street, and 11 tennis courts are at a sister hotel, the Royal Lahaina. ☐ **Mod to Lux** ☐ **G, W, T**

Royal Lahaina Resort, *P.O. Box 667, Lahaina, Maui, HI 96761, (808) 661-3611* ☐ *Open all year* ☐ *Off SR 30, 4 mi N of Lahaina* ☐ The flagship of Island Holidays Resorts on Maui, the Royal Lahaina is an extensive complex consisting of two resort hotels, six swimming pools, cottages, and 11 tennis courts (including one stadium court), all set on 50 surf-side acres on Kaanapali Beach. Accommodations range from standard double rooms to three-bedroom suites. Each room has color TV, air conditioning, and refrigerator. The Alii dining room has candlelight dining and island entertainment, Don the Beachcomber serves Polynesian specialties, and the Crown restaurant features continental cuisine. After dinner and on into the night, the Foxy Lady disco churns out the music.

For active, daytime types, the tennis ranch offers group or private lessons with videotape replay and pro shop. At the hotel and nearby Whaler's Village a wide variety of shops proffers everything from clothing to island crafts. Across the street from the resort are 36 holes of golf, at the Royal Kaanapali Golf Courses. □ **Lux** □ **T, G, W**

ISLAND OF MOLOKAI

KAUNAKAKAI: Hotel Molokai, *P.O. Box 546, Kaunakakai, Molokai, HI 96748, (808) 553-5347 Open all year* □ *2 mi E of Kaunakakai on SR 45* □ The hotel is a little village of Polynesian-style cottages clustered in a palm grove on the beach. The idea was to combine modern convenience with traditional Hawaiian design. The result is a striking and rather attractive cross between a longboat and an A-frame. Enormous saddle-shaped cedar-shake roofs extend two-thirds of the way to the ground. The decor of the 56 airy rooms is dominated by cedar walls, exposed beams, and by views of the ocean or gardens through the sliding glass doors leading to lanais. All rooms have private baths: some have air conditioning and oversized beds. The restaurant serves three meals a day and specializes in the daily fishing catch; there's nightly entertainment in the dining room, and cocktails in the bar.

You can swim in the pool, or in a nearby shallow lagoon that is especially good for snorkeling. Or you can play shuffleboard, volleyball, and horseshoe pitching on the lawn, and the resident social director posts a long list of planned amusements. The island of Molokai has three golf courses (two 9- and one 18-hole). Guests have play privileges at the two tennis courts at the Wavecrest development, about 15 minutes away. But Molokai's main appeal is to explorers and sightseers; it is probably the most unspoiled of the islands. Rent a car and head for views of the craggy northern coast from Kalaupapa Overlook, or of the rainforest and waterfalls of Waikolu Valley from Waikolu Lookout, or to the beach and rainforest at Halawa Valley. Hardy souls can hike down to the leper colony on isolated Makanalua Peninsula; you can also get there by mule, or by airplane from Molokai Airport.

On the island are remnants of the old Hawaiian religion, including the Phallic Rock, the Illiliope Heaiau (a 13th-century stone temple once used for human sacrifice), and the Kapiuwa Grove. Many of these sights are accessible by ordinary car, but when a four-wheel-drive is necessary, the hotel can arrange rentals for you. Fishing and hunting trips and sailing charters are also available. □ **Mod to Lux** □ **G, W**

MAUNALOA: Sheraton Molokai, *P.O. Box 1977, Maunaloa, Molokai, HI 96770, (808) 552-2555 or toll-free (800) 325-3535* □ *Open all year* □ *On Kepuhi Beach about 15 mi W of airport on SR 46* □ The 34 two-story buildings housing 292 rooms share the isolated beachfront with the Kalua Koi Golf Course, a 6200-yard, par-72 scenic course designed by Theodore Robinson. Guests at the Sheraton can play 18 holes a day for free, and also have free use of the hotel's four lighted tennis courts. There's swimming in the hotel pool and along the adjacent three miles of the best beach on the island.

All of the rooms have color TV, telephones, overhead fans, and private baths. Lodgings nearer the beach are larger and more expensive. Meals are served three times a day in the hotel dining room, and there's a lounge with nightly entertainment and cocktails. Though the Sheraton is complete in itself, its also a good jumping-off point for exploratory trips around Molokai. □ **Mod to Lux □ G, T**

Idaho

BOISE: Pioneer Inn, *c/o Bogus Basin Recreational Associates, 731 N 15th, Boise, ID 83702, (208) 343-1891* ☐ *Open early Dec to mid-Apr* ☐ 16 mi N of Boise on Bogus Basin Rd. ☐ Bogus Basin ski area sits between two peaks of a mountain ridge, and its trails and lifts crisscross both sides of the ridge. Six double chairlifts serve the 37 major runs, some of which are more than 8000 feet long, with vertical drops of 1800 feet. Three chairlifts and five miles of lighted trails stay open evenings. Also here are ski school, rental shop, and patrol services.

Pioneer Inn, accessible to all of Bogus's chairlifts, consists of 72 attractive, contemporary condominiums, with wood siding and glass doors onto balconies. Accommodations are usually rented on a three-night EP rate that includes lift charges. Choices consist of double rooms with sofa beds, sleeping four; small studios with kitchenettes, sleeping four; large studios with folding beds, bunks, kitchenette, and fireplace, sleeping six; and bedroom-studio combinations with two baths, kitchenette, fireplace, beds, sofa beds, and bunks, sleeping up to eight. All units have color TV and other amenities. The entertainment lodge has a candlelit restaurant, a cafeteria with wine and beer, a saloon, a snack bar, and a dance floor. A day-lodge at the other side of the basin has a cafeteria, snack bar, and day nursery. A delicatessen is also on hand, as well as two outdoor thermal whirlpools surrounded by a heated deck and sauna. Babysitters are available. ☐ **Mod to Lux** ☐ **S**

MCCALL: Shore Lodge, *P.O. Box 1006, McCall, ID, 83638, (208) 634-2244* ☐ *Open all year* ☐ *Off SR 55, 110 mi N of Boise* ☐ In the valley below Brundage Mountain ski area is Payette Lake and Shore Lodge. The modern attractive lodge takes full advantage of its lakeside location. A flower-bordered patio juts a little way out into the lake on pilings, giving guests a cool place to relax on the hottest days. In cooler weather the glass-enclosed terrace on the second floor is a warm and sunny spot to take in a view of the lake. Two docks provide a place for fishermen to tie up while they have a meal at the lodge's restaurant. Although some guests swim in the lake in summer, most prefer the heated pool. In addition, the lodge has an all-weather tennis court.

Shore Lodge is the nearest accommodation to Brundage Mountain. The ski area has more than 15 trails, with a total vertical drop of 1600 feet. During early spring, Brundage averages as much as 129 inches of powder. Also nearby is the McCall Municipal Course, with nine holes set in meadow and nine cut through forest. ☐ **Inexp** ☐ **S**

NORDMAN: Grandview Lodge and Resort, *Nordman, ID, 83848, (208) 443-2433* ☐ *Open mid-May to mid-Sept* ☐ *From SR 57 at Nordman, take Reeder Bay Rd. 3 mi E* ☐ The south wing of this lodge, a handsome peeled-log structure, is set on pilings in the lake, giving the picture windows and balconies an unobstructed view of the water. Eight bright, modern rooms are in the lodge. Upper-story rooms in the south wing have the same sweeping view of the lake and mountains as the restaurant below and have a sun terrace.

The rest of the accommodations are in 12 condominiums and cottages rented by the week. The condos and three of the cottages are right on lakeside; the cottages in the second row are 50 yards from the water. All units are modern and have fireplaces. Bedding and dishes are furnished. While all condos have dishwashers and full baths, most cottages have only shower baths and no dishwashers.

The lodge has a dock with some marina services, including boat launching and rentals. It also has a small heated pool. But its most unusual attraction is its seaplane, available for tours. ☐ **Inexp to Mod** ☐ **F, W**

PRIEST LAKE: The trout in Priest Lake are consistently among the largest in the country. The biggest kokanee (6 lb 9¾ oz) and mackinaw (57 lb 8 oz) ever taken in the United States came from Priest Lake. Dolly Varden and cutthroat also thrive here, and rainbow and eastern brook abound in the streams that flow into the lake.

Priest Lake is actually two lakes connected by a slow-moving, 2½-mile-long river called the Thorofare. Stretching 26 miles in length and five miles in width, the lakes cover 26,000 acres. The lower lake–19 miles long–is by far the larger lake.

In summer, an average water temperature of 70 degrees and the white sandy shoreline make the lake perfect for swimming and waterskiing.

During the winter, when the total snowfall is about 10 feet, Priest Lake is popular among cross-country skiers and snowmobilers. Around the lake and in the surrounding mountains are more than 100 miles of groomed snowmobile trails. Snowmobile races take place throughout the winter and in February Priest Lake is the site of the United States Pacific Coast Championship Sled Dog Races.

Hill's Resort, *Rte. 5, Priest Lake, ID 83856, (208) 443-2351* ☐ *Open all year* ☐ *Take SR 57 to Priest Lake* ☐ On the lake at Luby Bay, Hill's Lodge consists of a main lodge plus more than 30 cabins. Each of the cabins is a little different, but almost all have fully equipped kitchens, electric heat, and stone fireplaces. The most luxurious units are in the condominiums. Besides all the features of the standard cabins these have dishwashers and balconies.

Hill's has a large covered dock where guests may rent boats for fishing, trips around the lake, or waterskiing. With its many islands, Priest Lake is ideal for picnicking and exploring.

Besides boating, guests can play tennis on two Plexipave courts and golf at a nine-hole course down the road. Many Forest Service trails crisscross the area, providing hiking for beginners as well as seasoned climbers. Many types of mushrooms can be found in the forests around Priest Lake and wild huckleberries grow on the surrounding slopes.

Hill's restaurant is highly regarded in the area, especially for its baby back ribs, oysters Escondida, and huckleberry pie. ☐ **Mod** ☐ **F, W, H**

SANDPOINT: Overniter, *Schweitzer Ski Area, P.O. Box 815, Sandpoint, ID 83864, (208) 263-3331* ☐ *Open mid-Nov to mid-Apr* ☐ *Off US 2/95* ☐ The majority of the Schweitzer ski area centers on the south bowl. Here, five chairlifts and one T-bar carry skiers as high as 6389 feet from the base area 2000 feet below. Twenty-seven trails from easy slopes to daredevil inclines lace the mountainside. Over the lip of the south bowl is the secluded north bowl, where two chairlifts give access to 12 more trails. Schweitzer is only 60 miles south of the Canadian border, so it gets plenty of snow–a yearly average of 120 inches.

At the base of the south bowl are the Overniter Hotel, the rental condominiums, and the chalets. The wood-shingled Overniter is done in a rustic alpine style. Many of the rooms have private balconies. A variety of condominium and chalet lodgings are available, ranging from studios to two bedrooms. Between the hotel, condominiums, and chalets, visitors can find all types of accommodations, including suites with a fireplace in each room and a private sauna.

For dining and apres-ski relaxation, Schweitzer has a cafeteria, Beirstube, and several restaurant and lounge areas. ☐ **Mod** ☐ **S**

STANLEY: Redfish Lake Lodge, *Box 9, Stanley, ID 83278, (208) 774-3536* ☐ *Open Memorial Day to mid-Oct* ☐ *Off SR 75, 5 mi S of Stanley* ☐ For families who love the outdoors and natural beauty–but not camping–this unpretentious lodge could be fun. Set among pine trees, it offers a hodgepodge of accommodations: a rustic two-story lodge with 13 double rooms that share common restrooms and showers, 10 motel units with private bath, and 11 cabins (sleeping from two to six people) with private bath. The restaurant will prepare box lunches if you want to be out at midday.

The lodge has a bar, a marina with rental boats and boat trips, a beach on Redfish Lake, picnic and game areas. Fishing is available and the lodge can get you a horse for trail rides. The chief attraction, however, is that the lodge is on the boundary between Sawtooth National Recreation Area and Sawtooth Wilderness. It's adjacent to the exceptional Redfish Lake Visitor Center, where rangers conduct interesting programs that include lectures and demonstrations about the salmon spawning sites (Chinook salmon travel 800 miles from the Pacific Ocean to spawn here), and the beaver dams on Stanley Lake. ☐ **Inexp** ☐ **F**

SUN VALLEY: Sun Valley Resorts, *Sun Valley Co., Sun Valley, ID 83353, (800) 635-8261* ☐ *Open all year* ☐ *78 mi N of Twin Falls on US 93* ☐ Well-recognized as one of America's premiere resort areas, Sun Valley offers a full spectrum of four-season luxury vacation fun. The Sun Valley Inn and Sun Valley Lodge (located a few hundred yards apart) are set among some of the loveliest scenery Idaho has to offer. Most of the luxurious 250 double and single rooms, 13 suites, and approximately 260 self-contained condominium units have splendid views of the surrounding sage- and snow-covered knolls at the base of towering Mt. Baldy. The grounds are landscaped with sweeping lawns and winding aspen- and pine-shaded pathways. Next to the mall with all manner of shops and restaurants is a pristine little duck pond. For a resort complex with a capacity of more than 2000, Sun Valley should be commended for its success in preserving the natural features that first attracted visitors nearly a century ago.

Summer activities include swimming, sunning, and sauna. Guests may choose among three heated outdoor pool areas. For the tennis buff, 18 well-maintained courts are next to the inn, and a whopping 85 courts are elsewhere in the valley. Two- to five-day tennis clinics are offered. Computerized ball machines and video-playback are used as teaching aids. In addition to instruction in both Western and English riding, and year-round ice skating on the large outdoor rink, there are two 18-hole championship golf courses. The 6300-yard, par-69 Sun Valley course, complete with pro shop and luncheon terrace, is a stone's throw from the inn. Neighboring Elkhorn Village–just two miles away–boasts a 7200-yard, par-72 course designed by Robert Trent Jones. Completing the list of summer sports are trap and skeet shooting, whitewater float trips on the nearby Salmon River, and for the ultimate thrill, soaring among the clouds in a glider.

Sun Valley's shining reputation is perhaps even more the product of its exceptional winter sporting facilities. Whether you are a beginner testing your balance on Dollar Mountain's 127 acres of treeless slopes, or a hot-dog aerial stuntster racing down the chute on mammoth Mt. Baldy, you won't get your fill, not even in a week. Mt. Baldy, with 50 runs and some 900 acres of mostly intermediate and expert terrain, is able to handle over 16,000 skiers per hour. With ski rentals and sales, a free bus to take you to and from the slopes, and apres-ski food and entertainment in any of the several local nightspots, Sun Valley is most definitely a full-service luxury resort. ☐ **Lux** ☐ **G, T, S, R, F**

Elkhorn Village Inn, *P.O. Box 1067, Sun Valley, ID 83353, (208) 622-4511* ☐ *Open all year* ☐ *81 mi N on US 93 from Twin Falls* ☐ Located roughly a mile from the inns at Sun Valley Resort, Elkhorn Village Inn shares many of the amenities that this shoulder of Idaho's Sawtooth Range has to offer: crystal-clear air, rolling green hills, and rugged snowcapped peaks. From a distance, Elkhorn Inn, with its bell tower and loose clustering of angular stucco and shingle-style buildings, looks like a cubist-designed Utopian village. Clearly, though, the architects were concerned with providing easy access to the main facilities–olympic-size swimming pool, ice-skating rink, shops, and restaurants–without creating a rigid channeling of movement. Each of the 146 modern double rooms and the 223 one- to four-bedroom condominiums with kitchens and fireplaces are just steps away from any activity.

In summer, besides swimming and skating, you will find 18 courts for tennis tournaments and instruction, horseback riding, and even racquetball in nearby Ketchum. Golfing is on Elkhorn's 700-yard, par-72 course, designed by Robert Trent Jones. Rated as one of Idaho's five best courses, it's a challenge.

Winter really gave Elkhorn and the Sun Valley area its wide acclaim. For the advanced skier Mt. Baldy is minutes away via Elkhorn Ski Shuttle. For the beginner, the chairs and pomas of Dollar Mountain's more gentle slopes can be reached right from the Elkhorn Inn. After a tough day on the white stuff, you can enjoy a variety of menus and entertainment at one of the four restaurants. ☐ **Lux** ☐ **G, T, R, D, F**

Illinois

GALENA: Chestnut Mountain Lodge, *Galena, IL 61036, (315) 777-1320 or (312) 595-1116* ☐ *Open all year* ☐ *8 mi S of US 20 on Blackjack Rd.* ☐ From some of the rooms at this massive chalet-style lodge you can see three states—Illinois, Iowa, and Wisconsin—and the Mississippi River, their common border. For a closer look at the river, you can hop a ride on a trackless miniature train that tours the resort's 320 acres, or you can take a spectacular trip down to the river shore in the lodge's largest double-chairlift. If that's not enough river for you, Chestnut Mountain runs 90-minute cruises in its own glass-enclosed paddle-wheeler.

The hilly, wooded terrain of this palisade site is fine for exploring on foot or horseback, and the recreational facilities include a heated indoor pool and two tennis courts. But from December 15 through March 15 skiing is Chestnut Mountain's main business. Three chairlifts and seven rope tows serve 14 runs, the longest of which is 3200 feet, with a vertical drop of 465 feet. There's night skiing five nights a week. Accommodations range from dormitory space and single bedrooms all the way up to suites—ten of them—which can sleep as many as eight adults. Room rates during the ski season include lift tickets, a one-hour group lesson, and breakfast and dinner. All rooms are equipped with air conditioning and TVs; the furniture is of recent vintage and traditional style.

In addition to the rustic, knotty-pine-paneled main dining room—whose specialty is a weekend buffet—the lodge has a Bavarian-style gourmet room called the Steinhaus, and three cocktail lounges, including the slightly incongruous Banyon Tree—a little piece of Mexican village, complete with tile roofs. Rates are quoted AP during skiing season and EP the rest of the year. ☐ **Mod** ☐ **S**

ITASCA: Carson Inn/Nordic Hills, *Nordic Rd., Itasca, IL 60143, (312) 773-2750* ☐ *Open all year* ☐ *From jct US 20 and SR 53, take Rohling Rd. ½ mi N* ☐ Set on about 200 acres less than 10 miles from O'Hare airport, the Carson Inn is one of the few places in the greater Chicago area to offer a complete sports vacation. The handsome motel rooms—212 doubles and 8 suites—have walls of brick and wood paneling, and some have exposed ceiling beams. All have telephone, color TV, and private patio or balcony; a few have fireplaces or refrigerators.

Adjacent to the inn is an 18-hole golf course with pro shop and rental equipment. The resort also has five outdoor tennis courts (three lighted), eight indoor racquetball courts, six-lane bowling alley, game room, and health club with sauna, whirlpool, exercise room, and indoor and outdoor pools. Cross-country skiing is offered in winter.

The Scandinavian Room, open to the public, specializes in prime rib, surf and turf, and Dover sole. At the Vulcan's Forge guests can broil their own steaks and hamburgers, and there's also a coffee shop for quick service. □ **Lux** □ **G, T, S**

LINCOLNSHIRE: Marriott's Lincolnshire Resort, *Lincolnshire, IL 60015, (312) 634-0100* □ *Open all year* □ *From I-94 take Half-Day Rd. exit to SR 22 W, then on Milwaukee Ave. S ½ m to resort* □ This is a place for people who cannot settle on whether to take a city or a country vacation. Set on 170 acres beside the Des Plaines River, the Lincolnshire Resort has several lakes, a challenging 18-hole 6175-yard golf course, designed by Joe Fazio, six air-conditioned indoor tennis courts, an indoor-outdoor pool, and a health club. But it's only 45 minutes from the sights and the action of downtown Chicago. And the resort is only 25 minutes from the family fun of Marriott's Great America theme park.

In the warm weather guests can rent paddleboats, canoes, or Sunfish to take out onto the lake next to the main resort building, and in the winter there is skating on the ice rink and skiing.

The rooms are brightly furnished and have oversized beds, color TV, and individual climate controls.

The most dramatic of the resort's three restaurants is the eight-level King's Wharf. That has a nautical flair with pillars rigged to look like masts and dinghies hanging from the ceiling. Two of the three lounges at the resort have entertainment, and all year there is theater in the resort's own theater. □ **Lux** □ **G, T, W**

ROCKTON: Wagon Wheel Resort, *Route 75, Rockton, IL 61072, (815) 624-8711* □ *Open all year* □ *6 mi W of I-90 on SR 75 at jct SR 75 and 2* □ The Wagon Wheel opened in 1936 as a roadside restaurant with six tables, three booths, and three employees. It has grown. It is still a restaurant (though now there are four dining rooms and six lounges), but it now occupies 300 acres with 330 guest rooms, a 16-lane bowling alley, a 3200-foot airstrip, riding stables, a nine-hole golf course, a ski slope and chalet, an indoor tennis center with four courts, and two swimming pools—and, believe it or not, a non-denominational roadside church, a huge indoor skating rink, and a miniature railroad.

Guest rooms, in a long ranch-house-style lodge and matching wood-frame annexes, range from conventional single bedrooms—furnished in various styles of "old-fashioned," but with air conditioning and color TV—to what management calls "The Ultimate": a three-room suite with garden patio, sitting room with fireplace, marble-topped tables, round bed, and sunken bath.

Eclecticism is the unifying theme: one dining room, the Martha Washington, is colonial and formal; another has the proverbial wagon wheel hanging from the ceiling, and gingham tablecloths. Also hanging from the ceiling, in the Cock and Bull Lounge, is a 30-year collection of thousands of business cards. The Gay Nineties bar is tucked in an indoor wooden lean-to. Guests can buy a three-day membership in the Key Club (the Wagon Wheel's gourmet room); it's done up in an exuberant melange of styles as an opulent parlor car.

Exposed wooden trusses, a feature of the Wagon Wheel's architecture, are nowhere more impressive than in the glass-enclosed indoor swimming pool and the mammoth Ice Palace. At this "you want it—we got it" resort, you can bowl, ride, skate, ski, swim, sauna, play golf and tennis, and—for a nominal fee, and with reservations made in advance at the airstrip—you can fly. ☐ **Mod to Lux** ☐ **T, G, W**

UTICA: Starved Rock Lodge, *Schmidt Enterprises, Inc., Utica, IL 61373, (815) 667-4726* ☐ *Open all year* ☐ *From I-80 take exit 81, lodge is 3 mi S off SR 71* ☐ Legend has it that during the 1760s a group of Illini took refuge atop the 125-foot-high sandstone outcrop, now known as Starved Rock, to escape an Ottawa-Pottawatomi attack. The besieged Illini were soon starved into submission: thus the name of the rock and the state park. In 1683, the French built Fort St. Louis atop Starved Rock because of its strategic position above the Illinois River. When the French king suspended trading rights on the frontier, the fort was abandoned and rapidly fell into decay. In the 1930s the Civilian Conservation Corps built the lodge of rustic logs and stone. The lounge has a large fireplace and wood furnishings, complemented by authentic Indian handiwork. The pine-paneled dining room serves meals only to those on American Plan, and does not have an a la carte menu. During the warmer months the Pow Wow Room, in the lodge basement, offers refreshments, games, and recorded music.

West and east of the lodge are 12 log cabins connected by asphalt pathways. South of the lodge, the Teepee House has five single rooms. These lodgings are set in a wooded area where many hiking trails begin. More than 600 species of plants and numerous woodland animals, including deer, fox, raccoon, red-headed woodpeckers, and barred owls live in the 2022-acre park. A free information-interpretive center and a self-guiding nature trail with an accompanying booklet are offered. Boat launching ramps, hiking trails, and swimming are also available. ☐ **Inexp** ☐ **H, W**

Indiana

ANGOLA: Potowatomi Inn, *Box 37 A, Pokagon State Park, Angola, IN 46703, (219) 833-1077* ☐ *Open all year* ☐ *On SR 127 1½ mi S of junction I-69 and I-80* ☐ Indiana's state-administered lodges boast features that few private inns could hope to match for the price. This very attractive Old English-style inn sits amid 1200 acres of woods in Pokagon State Park. From its setting on the crown of a gentle hill, it overlooks an incredibly wide, long, and no doubt difficult-to-keep-mowed lawn that sweeps grandly down to Lake James. Besides the many sports activities–tennis (one court), basketball, volleyball, fishing, boating, bicycling, horseback riding, hayrides, cross-country skiing–the inn features a 1700-foot refrigerated toboggan run, the only one of its kind in Indiana.

The inn was constructed in 1927. The "old lobby" still has the original brick walls and an unusual floor of cement and brick. The dining room also retains its original pegged flooring. Otherwise the inn seems quite new, due to a recent remodeling. There are 55 modern rooms in the inn proper, with 16 cabin rooms and 10 motel rooms. All have air conditioning and private bath.

The inn is named for the Potowatomi tribe, who converted to Catholicism. The tribe's chief, Leopold Pokagon, for whom the park is named, built the first church in the area. In the early 1800s Chief Leopold sold three million acres of land to the government, including the land now in the state park, for three cents an acre. ☐ **Inexp** ☐ **W**

FRENCH LICK: French Lick-Sheraton Hotel and Country Club, *French Lick, IN 47432, (812) 935-9381* ☐ *Open Mar–Nov* ☐ *On SR 56* ☐ This is a mammoth resort–485 rooms, two 18-hole golf courses, 24 tennis courts, and 2600 acres. It is on the site of a salt lick and mineral springs discovered by French traders. With the Louisiana Purchase in 1803 the French abandoned the springs, but within 35 years enterprising Americans had set up a ramshackle three-story hotel and were doing land-office business selling rooms and bottles of the "miracle waters." The frame buildings burned down in 1897, but the then mayor of Indianapolis, Thomas Taggert, had them grandly rebuilt in brick. Taggert had the railroad lay a special spur and run daily trains from Chicago to and from the front entrance of the hotel. He added a golf course (and later a second) and renamed the bottled spring water "Pluto Water." His dining room, in 1917, became the first restaurant to serve tomato juice. Even though the resort became the unofficial headquarters for the Democratic party during the 1930s, the Depression and World War II did the resort in. Finally, in the 1950s the Sheraton chain bought the hotel and revived it with millions of dollars in restoration funds.

Now it is super sumptuous, with a tremendous array of facilities. In front of the hotel is a large swimming pool and children's wading pool surrounded by cabanas, and behind the building is another pool enclosed by a movable plastic dome that stays open on sunny summer days and closed in winter. Twelve indoor courts, plus 13 outdoor courts, ensure plenty of tennis. The health and spa facilities include a sauna, whirlpool, Pluto Water mineral baths, massage, steam room, universal gym, and lockers. The Country Club golf course (two miles away) is a tough par 70; the Valley course (adjacent) is a more sporty and open par 71. Besides all this, there's a skeet and trap shooting range, a riding stable, a miniature steam railroad, and a recreational center with six lanes of bowling, three pool tables, and electronic games. Skiing is available eight miles away at Paoli Peaks, which has lifts, ski instructors, snowmaking machines, and ski shop. And Pakota Lake, seven miles away, has watersports and fishing.

The resort has four dining rooms, three cocktail lounges, and one snack bar, as well as a barber shop, beauty shop, drug store, florist, and gift shop.

Each room has color TV, direct-dial telephone, and air conditioning. All rooms are rented on the American Plan. ☐ **Inexp to Mod** ☐ **T, G, R**

MADISON: Clifty Inn & Motor Lodge, *Clifty Falls State Park, P.O. Box 387, Madison, IN 47250, (812) 265-4135* ☐ *Open all year* ☐ *2 mi W of Madison on SR 62* ☐ The 1300 acres of Clifty Falls State Park cover a 400-foot-high bluff overlooking the Ohio River and the blue hills of Kentucky on the other side. Twelve miles of hiking trails meander across the wooded site following the Dean's branch tributary as it joins Clifty Creek. Along the trail are a deep, boulder-strewn canyon and two waterfalls–the idyllic Tunnel Falls and the 70-foot-high cascade of Clifty Falls.

Two additions to the motor lodge were completed in the seventies. These ultra-modern and angular buildings have pleasant wood exteriors. The color-coordinated motel rooms, like the rooms in the motor lodge, have private balconies or patio, color TV, telephone, and air conditioning. The dining room, a streamlined affair with cut-stone fireplace, hardwood floors, and cherry paneling, serves family fare, with specialties such as skillet-fried chicken and homemade salad dressings and desserts. The lodge has a playground and small pool.

The town of Madison is on the National Register of Historic Places because of the number of restored Federal, Greek Revival, and Italianate homes here.

Many are open to the public. The *Delta Queen,* one of the last of the old passenger-carrying sternwheelers, makes regular stops here. Believe it or not, Indiana produces wine, and you can stop and tour the Villa Medeo Vineyards and sample some of their products. □ **Inexp**

MARSHALL: Turkey Run Inn, *Turkey Run State Park, R.R. 1, Marshall, IN 47859, (317) 597-2211* □ *Open all year* □ *On SR 47, 25 mi SW of Crawfordsville* □ The inn is a handsome, ivy-covered stone and brick lodge set at the end of a circular drive. The inn has 52 rooms, and 21 cabins are also available. All accommodations have air conditioning and telephone.

The park got its name because thousands of wild turkeys once thrived in the forests and in the sandstone canyons of the area. The deep, rock-walled canyons are one of the park's greatest natural attractions. Sugar Creek, abounding in black bass, crappie, and rock bass, winds through these rock canyons and passes under several restored covered bridges, making it Indiana's most scenic canoeing stream. The park's 2200 acres are heavily forested with beech, tulip, walnut, and hemlock. The park naturalist conducts free hikes on the 14 miles of trails. The park also has a saddle barn, providing horses for trail rides on the 20 miles of bridle paths.

Surrounding the inn are four tennis courts, a children's playground, and a large swimming pool. Like all of Indiana's state park inns, Turkey Run has a fine restaurant offering family fare–skillet-fried chicken, smorgasbord, and so on. □ **Inexp** □ **F, R, T**

MITCHELL: Spring Mill Inn, *Spring Mill State Park, R. R. 2, P.O. Box 68, Mitchell, IN 47446, (812) 849-4081* □ *Open all year* □ *On SR 60, 3 mi E of Mitchell* □ The chief attractions of this historic park are a reconstructed pioneer village, a water-powered gristmill, and the Virgil Grissom Memorial. But nature enthusiasts will also find 1300 acres of hardwood forest, Donaldson Lake and the streams that feed it, and Donaldson and Twin caves, which may be toured by boat.

The inn is a broad, stately edifice, built of smoothed stone and fronted by thick timber columns. Rooms are pleasantly furnished and have color TV. The dining room with wood paneling and hardwood floors, is attractive, but the banquet room, with its window wall overlooking the treetops, is especially nice. The heated indoor-outdoor pool is unusual: a swimmer may go between the inside pool and the outside pool by swimming under the glass dividing wall. The saddle barn provides horses for rides on the miles of bridle trails.

Pioneer Village is composed of a number of split-log cabins that house a country store, a cobbler's shop, a tavern, a distillery, a hatmaker's shop, a weaver's, an apothecary, a post office, and several authentically furnished pioneer residences. All of the goods that the artisans and craftsmen fashion are for sale. From spring until fall you can also buy freshly ground corn meal at the gristmill. The modernistic Grissom Memorial has a Mercury capsule and one of the astronaut's space suits on display.

At Donaldson Lake and in the streams you can catch bass, bluegill, and channel catfish. Fishing boats and electric trolling motors may be rented at the lake. □ **Inexp to Mod** □ **F, R**

NASHVILLE: Abe Martin Lodge, *Brown County State Park, P.O. Box 25, Nashville, IN 47448, (812) 988-4418* □ *Open all year* □ *S of Nashville with entrances on SR 46 and SR 135* □ With 15,000 acres of wooded hills, this is the largest state park in Indiana. The forests of hardwood, pine, dogwood, and redbud trees–a blaze of color in autumn–harbor the largest herd of deer in Indiana, birdwatchers' favorites, such as the indigo bunting and the pileated woodpecker, and game birds such as quail, grouse, ducks, and geese.

The Abe Martin Lodge, constructed in 1932 of sandstone and roughhewn timber, is a rustic lodge of classic design. Two large, stone fireplaces warm the lobby, and throughout the lodge Indian relics, many recalling the famous chief Tecumseh, are on display. The lodge and the 23 adjacent cabins are named after characters created by Indiana caricaturist and humorist, Kin Hubbard. One of the dining rooms, also bears the name of one of Hubbard's figures, Grandma Pash. This intimate room with its sweeping view of the countryside, like the main dining room, serves solid family fare. Specialties of the kitchen include fried chicken, rainbow trout amandine, and homemade pies and cobblers.

The park has a stable, large swimming pool, archery area, and a wildlife exhibit of deer, elk, bear, and other animals. At night, guests may attend the park naturalist's star observatory or watch the small animals that come to the feeding stations on the flood-lit front lawn.

The town of Nashville has over 160 stores featuring hand-crafted jewelry, ceramics, and furniture. A bluegrass music festival is held in early summer in Bean Blossom, and Nashville sponsors an arts and crafts show and log cabin tour in the spring. □ **Inexp** □ **R**

SPENCER: Canyon Inn, *McCormick's Creek State Park, R.R. 1, Spencer, IN 47460, (812) 829-4881* □ *Open all year* □ *On SR 46, 2 mi E of Spencer* □ Unlike most of Indiana's state park inns, rugged wood and stone lodges inspired by Old World models, Canyon Inn with its stately white columns is reminiscent of a Southern plantation. Rooms at the inn are modern and comfortably furnished, if a bit small. Each has a color TV, telephone, and air conditioning.

The natural attractions of this 1750-acre park are subtle but compelling. Alongside the shallow pools and small waterfalls of the creek grow more than 400 different wildflowers including bloodroot, violets, and sweet William. A favorite way to see the park is by surrey or horseback. From the saddle barn,

16 miles of bridle trails wander through stands of maple, cottonwood, and giant willow. Along the trails and roads, visitors will find ravines, sink holes, caves, and gullies to explore.

The inn has four lighted courts, two pools (one olympic size), and a playground. The Park's environmental education center is staffed by two full-time naturalists who schedule free activities for interested guests.

June is the busiest month here, when two popular festivals–Pioneer Days and Outdoor Indiana–and a two-day canoe race take place in Owen County.
☐ **Inexp** ☐ **R, T**

Kentucky

The Kentucky state park system offers unusual vacation bargains in well-designed, modern accommodations at low prices. What helps keep total costs down is that most facilities are free; where fees are charged, as for golf, they are remarkably modest. Using the restaurant is optional, as these resorts are on European Plan. Lodge rooms come with two double beds, private bath, telephone, color TV, and individual climate control. Depending on the resort, the cottages vary from one to three bedrooms, sleeping two to six people; these have living room, kitchen, bath, TV, telephone, air conditioning, and come equipped with cooking utensils, tableware, and linens but no maid service. Children under 12 are free, cribs are free, and cots are a minimal charge. Pets are banned from the lodges but allowed in the cottages. Recreational directors and naturalists provide a large roster of activities, and a recreational staff supervises children in such activities as arts and crafts workshops, educational programs, hayrides, or sports lessons.

For reservations you may contact the lodges directly at the addresses given below. Or you may call Central Reservations Service, which can tell you where space is available and make reservations for any resort park: Call toll-free from Kentucky by dialing (800) 372-2961. Call toll-free from Ohio, Indiana, Illinois, Missouri, Tennessee, Alabama, Georgia, South Carolina, Virginia, or West Virginia by dialing (800) 626-2911. From other states call (502) 223-2326.

CARROLLTON: General Butler State Resort Park, *Carrollton, KY 41008, (502) 732-4384* ☐ *Lodge open all year, but some recreational facilities are seasonal* ☐ *From exit 44 on I-71, take SR 227 3 mi N* ☐ William O. Butler was a Mexican War commander, poet, politician, and a native of Carrollton. Butler Mansion, the authentically restored family home is in the park and is now open as a museum. The 800-acre park, at the confluence of the Kentucky and Ohio rivers, contains the Museum of Ohio River Lore.

General Butler Lodge, a steeply roofed, multi-angled contemporary inn, has a bright lobby with polished ceiling and floor and an immense stone fireplace. The 33 double rooms sport private patios and balconies. The 18 cottages come as efficiencies (combined bedroom/living room), one-bedroom, or three-bedrooms. The lodge has a dining room popular for its local dishes.

Also here are a 30-acre lake with fishing (for crappie, catfish, panfish, and rough fish), swimming, and boating (rentals available); a pool and wading pool reserved for lodge and cottage guests; nine-hole golf (rentals), tennis (three courts), riding, outdoor games, nature trails, and rides on a miniature railroad. ☐ **Inexp to Mod** ☐ **F, R, W**

CORBIN: Cumberland Falls State Resort Park, *Corbin, KY 40701, (606) 528-4121* ☐ *Open all year, but may close two weeks at Christmas* ☐ *19 mi SW of Corbin via US 25W and SR 90* ☐ This spectacular park is set in the wilderness of Daniel Boone National Forest. The 65-foot falls for which the park is named have a moonbow on clear nights during full moon; the only other known moonbow is in Africa.

Dupont Lodge offers 60 comfortable rooms in motel-like units of wood and stone, as well as 20 two-bedroom, two-bath suites that sleep six or more. Efficiency one-bedroom and two-bedroom modern cottages are also available. The restaurant serves three meals daily, emphasizing local dishes.

Seasonal facilities are swimming and wading pools, riding, fishing, and supervised play. There's an indoor recreation room, and a recreation director organizes entertainment and special events. This is a good choice for active families, especially if they like to hike; Natural Arch and Yahoo Falls scenic areas are in the forest within ten miles of the park, and other forest recreation areas are nearby. Children will enjoy the Levi Jackson Wilderness Road State Park, 37 miles north, with its exhibits and demonstrations of pioneer life. ☐ **Inexp to Mod** ☐ **F, R, H**

DAWSON SPRINGS: Pennyrile Forest State Resort Park, *Dawson Springs, KY 42408, (502) 797-3421* ☐ *Open Memorial Day to Oct 31* ☐ *7 mi S of Western Kentucky Parkway on SR 109* ☐ Surrounded by a 15,000-acre state forest, Pennyrile Lodge perches in heavy woods on the shore of a pretty lake. The paneled lodge rooms have private balconies or patios and two double beds. Efficiency cottages have a combination living room/bedroom, and one- and two-bedroom cottages are also available. The dining room serves Southern-style specialties, and there is a woody lounge with stone fireplace. Seasonal activities that make this a good choice for families include a pool for lodge and cottage guests, a lake beach, fishing, boat dock with rentals (no motors allowed), nine-hole golf, mini-golf, horseback riding, tennis, picnicking, and hiking. ☐ **Inexp to Mod** ☐ **W, F, R**

FALLS OF ROUGH: Rough River Dam State Resort Park, *Falls of Rough, KY 40119, (502) 257-2311* ☐ *Open all year, but some sports facilities are seasonal* ☐ *On SR 79 in Falls of Rough* ☐ Rough River Lodge offers an attractive restaurant and 40 contemporary double rooms with private balconies or patios in a two-story lodge; all the rooms have picture windows. The 15 two-bedroom housekeeping cottages sleep at least four.

The 35-mile-long, 4800-acre lake has sandy beaches, long stretches for waterskiing, and more than 60 species of fish, including largemouth and smallmouth bass, crappie, blue gill, and rough fish. Lodge and cottage guests also have a swimming pool. Besides a launching ramp, the park has a dock with open and covered slips and rental of fishing boats, rowboats, pedal boats, and pontoon boats. An excursion boat makes lake tours in summer. Other facilities include a recreation room, archery range, playgrounds, three tennis courts, nine-hole golf, mini-golf, and special events scheduled by a recreation director. ☐ **Inexp to Mod** ☐ **W, F**

GREENUP: Greenbo Lake State Resort Park, *Greenup, KY 41144, (606) 473-7324* ☐ *Open all year* ☐ *15 mi S of Greenup on SR 1* ☐ The Jesse Stuart Lodge is right on the 225-acre lake, a long, narrow body of water formed by damming up two creeks. This three-story lodge is built of native fieldstone and timbers and every article in every room—from the draperies to the faucets—is made in Kentucky. The 36 double rooms all have a private patio or balcony overlooking the lake and all have glass window-walls. The restaurant features a buffet dinner on Saturday.

In addition to the pool and wading pool for lodge guests, there is lake swimming, with a sand beach and bath house facilities; bike and boat rental; fishing; mini-golf; tennis (on two courts); hiking trails; supervised recreation for children; and special events. The park is in an area that yielded rich iron ore and supported 19 blast-furnaces; visitors can see a pig-iron furnace with an interpretive display. ☐ **Inexp to Mod** ☐ **W, F**

JAMESTOWN: Lake Cumberland State Resort Park, *Jamestown, KY 42629, (502) 343-3111* ☐ *Open all year* ☐ *14 mi SE of Jamestown on US 127* ☐ Lure Lodge sits on a palisade above Lake Cumberland, one of Kentucky's largest man-made lakes. The 48 lodge rooms, with patio and balcony, all overlook the 101-mile-long lake. The 30 cottages have one or two bedrooms; some of the two-bedroom cottages have fireplace and porch. In addition to these landlocked accommodations, houseboats may be rented.

The lodge offers a dining room serving local specialties, a gift shop, indoor game area, and during the summer, entertainment such as singing groups and square dancing. The park's pride is its interpretive nature center with displays on environment and local history. In season, a small pool is open to lodge and cottage guests, and a nearby olympic-size pool is open to the public as well. Other warm-weather activities include nine-hole golf, mini-golf, horseback riding, and two tennis courts. You may rent a bicycle, pontoon, fishing boat, or ski boat. ☐ **Inexp to Mod** ☐ **W, R**

LAND BETWEEN THE LAKES: Tennessee Valley Authority (TVA) and U.S. Army Corps of Engineers' dams on the Tennessee and Cumberland rivers created Kentucky Lake and Lake Barkley. Together the lakes make one of the largest man-made impoundments in the world. The lakes are connected by a free-flowing canal and appear on the map like the long arms of an inverted "U". Land Between The Lakes is the long (40 miles) and narrow (8 miles) peninsula that the lakes enclose. This heavily forested strip of land is being developed by the TVA as an environmental education and recreation center.

Boating, in all its forms, is the chief activity here. With a surface area of 220,000 acres, the lakes offer plenty of maneuvering room for waterskiers and sailors and many miles of shoreline for canoeists to explore. Boat rentals—from Sunfish to houseboats—are available throughout the area. The Governors Cup at Kenlake State Park, a sailboat race for all classes, is held the last two weekends in May. Another sailboat race, the Watkins Cup, is the first two weekends in October. In July, the International Outboard Championship Races bring the fastest speedboats in the nation to Lake Barkley State Park. In season, scenic cruises around the lakes are available at Kentucky Dam Village and Kenlake State Park.

The lakes abound with bass and crappie. Spring is the best season for both species since the fish, which move to shallow water to spawn, are especially hungry. Lesser catches are taken in mid-summer along old creek and river beds; but fishing there picks up again in fall. Catfish, blue gill, white bass, sauger, and rockfish are also taken year-round. There are innumerable places to rent boats and guides, and just off US 68, at the Devils Elbow Lake Access Area, is a floating fishing pier.

On the peninsula itself there's a field archery range at Golden Pond and horseback riding trails nearby at the Wranglers Camp in Lick Creek Valley. More than 400 miles of hiking trails and back roads wind across the Land Between The Lakes. Featured trails include a 26-mile historical trail, a 60-mile back-packing trail, and an 8-mile motor trail, among many others. Wildlife observers come to watch the golden and bald eagles that winter here and the herd of buffalo that roam a 200-acre pasture. Hunting, carefully managed by the TVA's resource management program, is available during fall and winter. Game includes fallow and white-tailed deer, wild turkey, and quail.

The Environmental Education Center presents programs, displays, and lectures on the history and environment of the region. Its features include a number of interpretive trails; the ruins of one of the many iron furnaces that once dotted the area; and the Empire Farm, a children's zoo and farming museum. In the southern part of the peninsula, near the Buffalo Pasture, is the Homplace-1850, a living museum modeled after the farms that flourished in the region during the 1800s.

Kenlake State Resort Park, *Hardin, KY 42048, (502) 474-2211* ☐ *Open all year* ☐ *6 mi S of jct US 68 and SR 80, 15 mi NE of Murray* ☐ Kenlake offers an attractive two-story lodge with 48 double rooms, an excellent restaurant, swimming and wading pools, and 36 kitchenette cottages with one, two and three bedrooms that sleep two to six people. The main attraction here is the water, and so the boating, fishing, and swimming facilities are tops. The year-round marina has full service for fishermen, boaters, and waterskiers. A public beach with bathhouse (open Memorial Day through Labor Day) is on the lake. Near the beach area are six tennis courts and a lighted court is near the lodge. The nine-hole golf course (one-half mile from the lodge) has rental clubs and carts. From the riding stable, trails lead around the lake near the boat dock. The Princess Excursion Line offers sightseeing cruises on Kentucky Lake in a sidewheeler that docks at Kenlake Park Marina in Aurora. ☐ **Inexp to Mod ☐ F, W, T**

Kentucky Dam Village State Resort Park, *Gilbertsville, KY 42044, (502) 362-4271* ☐ Open all year ☐ S of Gilbertsville, near jct US 62 and US 641 ☐ This state-run lodge is set on a peninsula in Kentucky Lake and has the largest docking facility in the Kentucky park system: 93 open slips, 138 covered slips, and rental houseboats, fishing boats, pontoons, rowboats, and pedal boats.

The lodge is sleek and handsome, with much glass used to give unobstructed views of the lake and woods. The 72 double rooms in the lodge have recessed private patios or balconies and picture windows. The 68 cottages have one to three bedrooms that sleep from two to six people. Some look like shingled country houses, while the executive models have cathedral ceilings and glass walls. Apart from the year-round dining room serving three meals daily, a coffee shop and beach snack bar are open in summer.

Facilities are extensive: an 18-hole championship course (with cart and club rentals) within walking distance of the lodge; horseback riding (seasonal); two tennis courts; swimming pool for lodge and cottage guests; public beach; game room; playgrounds; bicycle rental; supervised recreation; and special events. A 4000-foot paved airstrip with rental cars and fueling facilities for jet and prop aircraft is in the park, and provides transportation to the area's resort parks.

Kaintuck Territory, a re-created 1880s town with gunfights, river rides, chairlift, train rides, and other amusements is five miles south. ☐ **Inexp to Mod ☐ G, W, F,**

Lake Barkley State Resort Park, *P.O. Box 790, Cadiz, KY 42211, (502) 924-1171* ☐ *Open all year* ☐ *7 mi W of Cadiz on US 68* ☐ Designed by Edward Durell Stone, this inn takes full advantage of its setting on a peninsula on 58,000-acre Lake Barkley. Durell brought out the roughly circular shape of the peninsula by laying out the two lodging wings along a half-circle facing the water. Between the lodging wings is the restaurant, lobby, gift shop, and office complex. Large expanses of glass give the lodge an airy feeling, while timber and stone tie the building to its wooded setting. The 120 lodge rooms come with balconies. The four suites have a living room, two bedrooms with double beds, and a kitchenette. All units have fine views. Nine handsome two-bedroom executive cottages are set in the wooded hills behind their own private beach. The dining room is open all year and a coffee shop operates seasonally. The restaurant, with a lake view, specializes in catfish, country ham, fried chicken, Kentucky burgoo, and other regional dishes (for which you can get the recipes).

A recreational director and park naturalist will help you plan individual or group activities. Pedal boats are for rent near the lodge and a fully equipped marina rents fishing and skiing boats and pleasure-ride pontoons. A lakeside swimming pool is reserved for lodge and cottage guests and there's also a public beach. An 18-hole golf course has a lounge, rental clubs and carts, and instruction. The course hosts the Boots Randolph golf tournament in late September. There's also tennis (four courts, instruction), trapshooting, a canoeing program, basketball, riding, hiking, nature trails with bird feeders, and indoor games at the lodge. □ **Inexp to Mod** □ **R, F, W,**

LUCAS: Barren River Lake State Resort Park, *Lucas, KY 42156, (502) 646-2151* □ *Open all year* □ *12 mi SW of Glasgow on US 31 E* □ This 10,000-acre lake was created by the U.S. Army Corps of Engineers. The modern lodge and housekeeping cottages hug the shore of the lake, giving most units excellent water views. The attractive 51 double rooms in the lodge have private balconies or patios. The two two-bedroom executive cottages sleep at least four. In the lodge are a restaurant and recreation room. The year-round marina has 100 open and 40 covered slips, launching ramp, and fishing and pontoon boat rentals. Other recreational facilities include a nine-hole golf course, seasonal riding stable, lodge pool, bathhouse, tennis (two courts), handball, basketball, and bicycle rental. A favorite attraction nearby is Mammoth Cave National Park, 35 miles north. □ **Inexp to Mod** □ **F, W, R**

OLIVE HILL: Carter Caves State Resort Park, *Olive Hill, KY 41164, (606) 286-4411* □ *Open all year* □ *From I-64, take exit 161 6 mi NE on SR 182* □ Caveland Lodge is a good base for exploring the many natural arches, caves, and other limestone formations in this 1000-acre park. The two-story wooden lodge is built low to the ground, hugging the green hillside. It offers 28 modern rooms with balconies, 15 housekeeping cottages with one or two bedrooms, a restaurant with picture-window view of the hills, gift shop, recreation room, and swimming and wading pools.

Caves are the big thing here; some you can explore on your own. One series of caves is open for guided tours April through October; another, Cascade Caverns with an underground river, is open for tours June through Labor Day.

Right in front of the modern bathhouse on Smoky Valley Lake is a fun diving area, designed so that divers can safely plunge into the lake. A white-sand beach is next to the bathhouse, and next to the beach is a dock with pedal and rowboat rentals. In addition, there's a nine-hole golf course, horseback riding (seasonal), and hiking trails. Other specialties of the park are canoe trips and a nature center. ☐ **Inexp to Mod** ☐ **W, F, R**

PINEVILLE: Pine Mountain State Resort Park, *Pineville, KY 40977, (606) 337-3066* ☐ *Open all year* ☐ *From US 25 E in Pineville, take SR 190 W to SR 1491 and N to lodge* ☐ Set in stunning mountain scenery, in a historic area a few miles from Cumberland Gap, Evans Lodge provides 30 attractive pine-paneled double rooms with picture windows and balcony views in a modern two-story inn. In the woods are efficiency (living room/bedroom combined) and one-bedroom rustic wooden cabins for two people; these offer TV, telephone, kitchen, bath, and air conditioning and are open only April through mid-November. Two-bedroom cottages are open all year. Many cottages have screened porches. Lodge and cottage guest have a year-round restaurant with panoramic view, two seasonal swimming pools, wading pool, recreation room, and gift shop. The 2500-acre park has hiking trails, nine-hole and miniature golf, a lake, a nature center, and supervised recreation programs. The park adjoins Kentucky Ridge State Forest and scenic drives take you to the top of Pine Mountain, along the ridge, or 12 miles south to the views at Cumberland Gap National Historic Park. ☐ **Inexp to Mod** ☐ **H**

PRESTONBURG: Jenny Wiley State Resort Park, *Prestonburg, KY, 41653, (606) 886-2711* ☐ *Open all year* ☐ *2 mi NE of Prestonburg, off US 23* ☐ Jenny Wiley was a young woman captured by the Cherokees in 1787. For 11 months she was held prisoner, in an area near the present park site, but she managed to escape by crossing the Levisa Fork of the Big Sandy River and reaching Blockhouse Bottom, the first settlement in eastern Kentucky.

The park named in her honor consists of 1700 acres of forested hills set around winding, narrow Dewey Lake. May Lodge, on the shore of the lake, offers 48 rooms with balconies, a restaurant, and a recreation room. The cottages come as one-bedroom duplex units or two-bedroom cottages.

On the lake are two docks with 199 open and 40 covered slips. Guests may rent pontoon and fishing boats, or pedal boats. Fishermen will find bass, crappie, catfish, and blue gill in the lake. The park has a nine-hole course as well as a miniature golf course. In season, the lodge and community pools are open, horseback riding is available, and a chairlift carries sightseers to the top of Sugar Camp mountain. Festivals are a highlight here: the Kentucky Highlands Folk Festival, with folk music and mountain crafts exhibits, is held every Labor Day weekend, and Broadway musicals are performed regularly in the amphitheater in June, July, and August. ☐ **Inexp to Mod** ☐ **F, R, W**

SLADE: Natural Bridge State Resort Park, *Slade, KY 40376, (606) 663-2214* ☐ *Open all year* ☐ *On SR 11, 2 mi S of Mountain Pkwy at Slade* ☐ This contemporary lodge is perched on a ledge overlooking a valley forested with native hemlock. The main lodge has 35 rooms with private balconies overlooking the woods, along with six efficiency and four one-bedroom cottages. In the lodge is a dining room, gift shop, and recreational lounge, and the adjacent activities building houses a nature center. Other activities and a snack bar are at Hoedown Island, below the lodge. A large, attractive pool is nicely set beside the lake here.

The natural bridge, the park's major attraction, is a rugged, massive arch of sandstone formed by the waters of a prehistoric sea that once filled Red River Gorge. You can see the arch and gorge by canoeing in the gorge, riding the chairlift to the bridge, or taking a guided tour. Hiking trails are centered around Lover's Leap, Devil's Gulch, and Lookout Point. Seasonal stables provide guided horseback rides. Bass, bream, catfish, and other species are in Mill Creek Lake, and children can rent pedal boats near the square-dance pavilion on Hoedown Island. □ **Inexp to Mod** □ **R, H**

Louisiana

MANY: Toro Hills Hotel, *P.O. Box 1124, Many, LA 71449, (318) 586-3591* ☐ *Open all year* ☐ *On US 171, 15 mi S of Many* ☐ On one side, the long, two-story motor lodge has views of the three pools –one wading, one freeform, and one L-shaped. On the other side, it overlooks the rolling 18-hole golf course and practice range. All-weather golf carts can be rented at the pro shop, and the Rough lounge has bar service and limited menu. Right by the pro shop are two all-weather tennis courts, both lit for night play.

Each of the 123-double rooms has air conditioning and color TV, and the two suites also have kitchenettes. The Garden restaurant serves Louisiana specialties such as gumbo and jambalaya, and the Forest Club has dancing and gourmet dining.

Nearby is Hodges Gardens, with greenhouses, formal gardens, wildlife refuge, and outdoor theater. The gigantic (60 miles long) Toledo Bend Reservoir, with fishing, swimming, and boating, is only 10 miles away. ☐ **Mod** ☐ **G**

Maine

BRIDGTON: Pleasant Mountain Inn, *Bridgton, ME 04009, (207) 647-2431* ☐ *Open all year* ☐ *38 mi W of Portland on US 302* ☐ Wait around long enough after the summer waterskiing season here, and with a light change of equipment and scenery, you can continue to ski. The inn sits at the foot of Pleasant Mountain, at the edge of nine-mile-long Moose Pond, and is one of the Maine lake area's few all-season resorts.

Its 15 rooms accommodate up to 52 guests in a rustic lodge and cottages. The rooms are simply furnished; some have bunk beds in addition to two doubles, and the cottages all have fireplaces. On the premises are a restaurant and cocktail lounge; room rates include breakfast and dinner. There's a tennis court and a sandy lake beach. In summer and during the fall foliage season, the Pleasant Mountain chairlifts carry visitors to grand views of the surrounding mountains and lakes. In winter, three double chairs and three T-bars serve the Pleasant Mountain Ski Area, with a vertical drop of 1200 feet and 21 trails. ☐ **Inexp to Mod** ☐ **S, W**

Tarry-A-While Resort on Highland Lake, *Bridgton, ME 04009, (207) 647-2522* ☐ *Open mid-June to early Sept* ☐ *Take I-95 to jct US 302; follow 302 NW to resort* ☐ *Walliser Raclette mit Pellkartoffeln Silberzweibeln und Gurkins* is, unofficially, the longest menu item in the state of Maine, and is one of a number of good reasons to visit the Tarry-A-While. Hans and Barbara Jenni import a chef from Switzerland to prepare a full menu of Swiss specialties, including this dish. (It's raclette cheese, melted at the table, with onions, potatoes, and gherkins.) You can finish off your meal with a pastry–Swiss or otherwise–prepared by the resident pastry chef, and a cup of coffee the management boasts is the best you've ever tasted. Three meals a day are included in the room rates (though there's an extra charge for the Swiss menu). The 30-acre resort set amid pines on the edge of Highland Lake offers great views of the Presidential mountain range and of the lake. The grounds are visited often by deer, fox, and raccoons, as well as by bluebirds and orioles. For the energetic, there's a tennis court, and the Bridgton Highlands golf course and six more tennis courts are right across the road. The lodge will lend you a rowboat, pedal boat, or canoe to explore Highland Lake, or a bike for trips along the local country roads.

The 37 guest rooms are housed in the manor, in the Swiss-style Gasthaus, or in individual chalet units; pine paneling is everywhere, as well as pine furniture, handmade from the lodge's own trees. If you want the ultimate here, reserve the Chalet Flieder an air-conditioned, all-pine-paneled luxury unit with a screened porch. ☐ **Inexp** ☐ **W, T, G**

CARRABASSETT: Sugarloaf Inn, *Sugarloaf/U.S.A., Carrabassett, ME 04947, (207) 237-2701* ☐ *Open July through April* ☐ *Take Maine Turnpike to exit 12 (Auburn); follow SR 4 N to jct SR 27, 3 mi N of Farmington; follow 27 to resort* ☐ Located on Sugarloaf Mountain–one of the East's premier alpine skiing areas–the inn offers a novice chairlift starting right at the door of the motel-style lodge building. In addition, Sugarloaf has four more double chairs, five T-bars, and one four-person gondola. The mountain has 36 trails, and a vertical drop 2637 feet. For the cross-country skier, there are 50 miles of groomed and graded trails at the nearby Carrabassett Valley Touring Center.

The 48 guest rooms in the main building are furnished in contemporary style–ranging from single bedrooms with a double bed and set of bunks to deluxe two-level suites with five singles or two king-size beds. None of the rooms have telephone or TV, but many offer excellent views from private balconies. The most elaborate lodgings are one-, two-, or three-bedroom condominium homes rented by the inn.

Breakfast and dinner are included in the rates for the lodge units: A meal plan is available for the condominiums at an extra charge. The restaurant specializes in big breakfasts and gourmet dinners, featuring duckling a l'orange and native trout amandine.

Co-starring with the skiing in winter are a toboggan run, ice skating rink, and an indoor-outdoor pool with sauna and whirlpool. In summer, incidentally, the rates go way down, and a condominium unit is a particularly good buy for a quiet, mountain vacation. The resort has hiking trails and two all-weather tennis courts on the premises, and a golf course nearby, as well as a gondola functioning in summer for scenic rides. ☐ **Mod to Exp** ☐ **S**

GRAND LAKE STREAM: Weatherby's, *Grand Lake Stream, ME 04637, (207) 796-5558* ☐ *Open May to Oct* ☐ *Take US 1 3 mi N of Princeton; take left turn and follow road 10 mi SW to resort* ☐ Weatherby's is located right in the middle of one of Maine's finest fishing areas: 32 fishing waters are in the vicinity. This colony of 15 spacious log cottages sits at the edge of Grand Lake Stream, so you can fly-fish for salmon and trout practically at your doorstep. In the spring, there's trolling and casting on Big Lake and Grand Lake for landlocked salmon and lake trout, but the main all-season attraction is smallmouth bass. After June 21, all conventional lures and baits are legal for bass, so this is the best time to cart the whole family up for a fishing vacation. Guide service is available.

Most of the 15 one- or two-bedroom cottages have living rooms with fireplaces or Franklin stoves. Though the cabins are rustic, all have full tubs and showers, electric heaters, and screened porches. The management provides daily maid service and a supply of wood for the fireplace. The clapboard main lodge has a library, TV, and fireplace, as well as the lounge and restaurant, which specializes in home-baked pastries and Down East main courses—fish chowder and lobster stew, for example. Breakfast and dinner are included in the very reasonable room rates. ☐ **Inexp** ☐ **F**

GREENVILLE: The Cottages, *Box 1152, Greenville, ME 04441, (207) 695-2422* ☐ *Open May 1 to Oct 15* ☐ *3½ mi N of Greenville* ☐ Moosehead Lake is the largest in Maine. These nine housekeeping cottages are on the eastern shore only a few miles from Greenville—definitely in the wilderness, but not too far. Some of the cottages are on the hillside, some by the water, but all have fine views of the lake and are spaced for privacy. All linens, blankets, and kitchen utensils are furnished.

The cottages will sleep two to six people, and the screened porches and rollaway beds provide additional sleeping space. Four cottages have fireplace and all have a private bathroom with hot-water shower.

Moosehead Lake is known for its squaretail trout, lake trout, and landlocked salmon. The resort rents boats and motors to go after them. Canoes and waterskiing boats are also available. A shallow swimming cove provides a safe bathing area for children. ☐ **Inexp** ☐ **F, W**

Frost Pond, *Star Route 76, Greenville, ME 04441, messages may be left at (207) 695-2821* ☐ *Open May 1 to Nov 30* ☐ *From SR 157 in Millinocket, take Baxter State Park Rd. NW to jct Great Northern Paper Co. Rd. Follow to Ripogenus Dam, then N across dam 3 mi to resort* ☐ The cabins on this pond are 40 miles from the nearest town, isolated from the world by forests of spruce, fir, maple, birch, and beech. Seven of the cabins have no electricity; they are heated by wood, and the refrigerators, stoves, and lights run on gas. One cabin does have the conveniences that, in the woods, are luxuries: running hot and cold water, shower, and indoor toilet. All of the rustically furnished cabins contain blankets, linens, towels, and cooking and eating utensils. Six cabins are in the main waterfront area, the other two have their own docks and are farther down the shore.

Fishermen come here for the pond's brook trout and squaretails and the landlocked salmon in nearby Chesuncook Lake. Boats, canoes, and motors are available for rent, and there's a launching area for guests' own boats. The pond's gradually sloping, gravel bottom makes it fine for swimming. With advance notice, the resort can outfit wilderness canoe trips down the Allagash River and elsewhere.

The surrounding woods harbor moose, deer, and the occasional black bear along with a variety of birds from bald eagles to warblers and finches. Fifteen miles away, 200,000-acre Baxter State Park has a number of hiking and mountain climbing trails, including some on 5267-foot Mt. Katahdin, Maine's highest peak. □ **Inexp** □ **F**

JACKMAN: Briarwood Mountain Lodge, *Jackman, ME 04945, (207) 668-7756* □ *Open all year* □ *½ mi S of Jackman on US 201* □ This motel is completely modern, but the use of authentic old barn beams and siding in the dining room, lounge, lobby, and bar give it a rustic atmosphere. Rooms are furnished in a traditional decor and have color TV. Since the lodge is set on a mountainside, the picture windows in each room provide fine views of the magnificent Border Range Mountains and the surrounding birch and evergreen forests.

The motel's restaurant, open for all three meals, features prime ribs and buffet suppers. The kitchen will also make up package lunches for snowmobilers, hunters, and other sportsmen.

The Jackman area offers some the finest snowmobiling in the northeast. The 100 miles of groomed trails give access to the 1200 miles unplowed logging roads. In summer, fishermen flock to the area to angle for the landlocked salmon, brook trout, togue, and smelt that populate the numerous lakes and ponds in this corner of Maine. Jackman is on 35,000-acre Attean Lake, and Moosehead Lake is 30 miles away. Swimming is even more convenient–in the motel's indoor pool. □ **Inexp** □ **F**

KENNEBUNKPORT: The Colony, *Kennebunkport, ME 04046, (207) 967-3331* □ *Open mid-June to day after Labor Day* □ *Take Exit 2 (Wells) on Maine Turnpike; turn left and follow SR 9 to resort* □ Lobster every day and luxurious lodging by the side of the Atlantic are featured at this classic Maine resort. Most of the 139 rooms are housed in a great old Georgian mansion, which contains the wood-paneled sitting room, the elegant pillared dining room, and the cocktail lounge. Other rooms are in the Galland House annex and the Colony Motor Inn. Some of the rooms have fireplaces and some balconies; all have telephones, TV, and modern bathrooms. They are warmly and simply furnished, in traditional style, with tasteful antique touches. Three meals a day are included in the room rates. The Dining Room offers Down East specialties, and you can sample a lot of them at the lavish Sunday buffet. Luncheons are served poolside.

In addition to the heated outdoor saltwater pool, there is shuffleboard, a putting green, sailing, and boating. You can play golf at the nearby Cape Arundel Golf Club (or two others in the vicinity), which features an 18-hole, 6000-yard, Donald Ross-designed course. Nearby are eight clay/composition tennis courts, with instruction available. A full schedule of social activities–from bridge and bingo to lectures by visiting artists–includes dancing at poolside and in the Marine Room cocktail lounge. □ **Lux**

NAPLES: The Chute Homestead Cottages, *P.O. Box 127, Naples, ME 04055, (207) 693-6425* ☐ *Open June to early Sept* ☐ *From Portland, take US 302 30 mi N to jct SR 35, follow 35 1 mi N to resort* ☐ You can almost fish here without bothering to get out of bed; most of the cottages are right at the edge of Long Lake, which is filled with bass and small panfish. There are 25 units here, and all offer comfortable seclusion in various shapes and sizes; from four-bedroom cottages (equipped with fieldstone fireplaces and kitchenettes) to smaller cottages (with fireplaces) and nine chalet units.

There's a lot of roaming to be done on the trails of the resort's wooded acres. You're also entitled to free use of the canoes and rowboats, lighted tennis court, putting green, horseshoe pits, softball field, or the free-standing sauna in the woods. If you'd like to explore the 40 miles of waterway in the Sebago lake chain, you can rent motorboats and Sunfish sailboats on the premises. Or you might want to sign up for the weekly canoe trip on Crooked River.

There's no restaurant–only a snack bar–on the premises, but the Homestead holds a reasonably priced lobster-clambake on the beach every Tuesday and Friday evening. If you don't feel like charcoal broiling your own dinner on the grill outside your cottage, the village of Naples (about a mile away) has a number of good restaurants. ☐ **Mod to Lux** ☐ **W, F**

OGUNQUIT: The Cliff House, *Ogunquit, ME 03907, (207) 646-5124* ☐ *Open early May to late Oct* ☐ *From US 1 take Clark Rd. E to sign for resort. Follow 4 mi to entrance on right side of Shore Rd.* ☐ A more appropriate name would be hard to find for this summer resort. It is perched on a rugged scarp that juts into the Atlantic. Most of the promontory inclines gently down to the ocean, but on one one side the cliff has a sheer drop of 90 feet. The main inn, set back from the cliff, was built in 1872 and has the warm appeal a country inn should have. The newer motel lodges, further out on the promontory, take full advantage of the ocean view on three sides. All of the modern, carpeted rooms have color TV, and some are air-conditioned. Three efficiency apartments, right at the water's edge, are rented by the week only.

The very tastefully appointed dining room–wood paneling, white pillars and tablecloths, brick fireplace, and bentwood chairs–serves equally tasteful meals. Seafood, of course, is a specialty, and the Sunday breakfast buffet popular.

The resort has a large, outdoor, heated pool and two Plexipave tennis courts. Within walking distance is a nine-hole golf course with pro shop. ☐ **Mod to Lux**

PROUTS NECK: Black Point Inn *Prouts Neck, ME 04070, (207) 883-4311* ☐ *Open June 1 to mid-Oct* ☐ *On SR 207, 8 mi S of Portland* ☐ These craggy shores were Winslow Homer's stomping grounds as a boy. The narrow peninsula where he roamed, Prouts Neck, is now so solidly edged with private summer homes that about the only way a nonresident can enjoy it is by visiting the Black Point Inn.

A large, weather-shingled building with a mansard roof surmounted by a cupola, the inn has a long history as a traditional AP resort. Its buffet lunch with live musical entertainment is celebrated as an eminently civilized way to dine. Inside, the 80 guest rooms and suites are tastefully appointed with colonial furnishings. The grounds are nicely landscaped, and included within their 11-acre bounds is a heated outdoor pool. Guests may use the adjacent 14 tennis courts–10 clay, 4 all-weather–and the 18-hole golf course (flat seaside links) that is nearby. ☐ **Lux** ☐ **T, G**

RANGELEY: The Country Club Inn, *Box 680, Rangeley, ME 04970, (207) 864-3831* ☐ *Open mid-may to mid-Oct* ☐ *On SR 16, 2 mi NE of Rangeley* ☐ It would be hard to find a smaller place with its own golf course–the inn has only 21 rooms, one suite and one cabin, yet the 6913-yard, par-70 golf course starts and ends right outside the door. And it wouldn't be easy to find a place with a more scenic location–on a promontory overlooking Rangeley Lake and the Longfellow Mountains beyond.

The inn is an elegant version of a hunting lodge. The high-ceilinged living room has knotty-pine paneling, exposed beams and trusses, and an enormous stone fireplace with a moose head over the mantel. The individually decorated guest rooms have pine wainscoting and large picture windows to partake of the mountain views. One large room offers a sliding glass door to its own terrace, and the cabin boasts two bedrooms, kitchen-dining area, full bath, and screened-in porch.

The dining room specializes in fresh Maine seafood–particularly fresh scallops, swordfish, and lobster–and weekly steak roasts at poolside, Thursday night buffets, and Saturday night prime-rib dinners with Yorkshire pudding.

The Rangeley Lakes and streams are noted for their fine fishing for squaretail trout and landlocked salmon. With a little advance notice the inn can arrange for modestly priced seaplanes to fly you into backwoods ponds and lakes for a day of fly-fishing. The Appalachian trail runs within 10 miles of the inn, and the 4600-foot Saddleback Mountain chairlift is not much farther away. ☐ **Inexp to Mod** ☐ **G, F**

SARGENTVILLE: Oakland House, *Sargentville, ME 04673, (207) 359-8521* ☐ *Open June–Sept* ☐ *On private road off of SR 15, 12 mi S of Blue Hill* ☐ Penobscot Bay on the south and Lake Winnewaug on the north give Sargentville the feeling of an island. Oakland House, midway between beach and lake shore, has been welcoming guests since 1889, and has been run by the same family since it opened. The living room, dining room, and office are in what was once the family farmhouse, and a few rooms for guests are on the upper floors. The annex, called Acorn, has a single three-bedroom unit downstairs, and four share-a-bath guest rooms upstairs. Most of the accommodations are in Shore Oaks guest house, and in 10 cottages scattered around the resort's hundreds of acres. The cottages have fireplaces and utility rooms (refrigerator, sink, and hotplate) and are nicely angled so that guests see much more of the bay than of their neighbors.

Room rates include three family-style meals a day in the dining room; fresh lobsters, clams, and crab are specialties, along with homebaked breads and pastries. There's a weekly lobster bake on the beach. No alcoholic beverages are served and this–together with the absence of TV, telephones, and an organized social program–keeps the emphasis on peace and quiet. You may swim at the private bay beach or in the warmer lake, or play tennis (one court), badminton, or croquet. You can borrow a rowboat for either lake or bay. Management will help you arrange sailboat rental from a local boatyard, or a deep-sea expedition. ☐ **Inexp** ☐ **F, W**

SEBASCO ESTATES: Rock Gardens Inn, *Sebasco Estates, ME 04565, (207) 389-1339* ☐ *Open mid-June to late Sept* ☐ *Take I-95 to US 1; at Bath, take SR 209 11 mi S to SR 217; turn right onto 217, follow to resort* ☐ Rock Gardens Inn is a lovely family-run cottage colony perched on a rocky point of land which juts into Casco Bay south of the Sebasco Lodge area. Although there are four rooms for rent in the Main House, most of the accommodations are in nine clapboard or shingle cottages. You have your choice of a view of the bay from the ones at the edge of the point, or seclusion among the pines a little farther inland.

These are substantial buildings; even the smallest cottages have two bedrooms, and the largest has four. All have stone or brick fireplaces. Glassed-in porches, sun decks, and cathedral ceilings are among the distinguishing features of the buildings.

Rock Gardens shares the recreational facilities of Sebasco Estates with the Sebasco Lodge: swimming pool, golf course, tennis course, and bowling lawns and alleys. The room rates include three meals a day in the dining room and a weekly lobster cookout. ☐ **Inexp to Mod** ☐ **W**

Sebasco Lodge, *Sebasco Estates, ME 04565, (207) 389-1161* ☐ *Open late June to early Sept* ☐ *Take I-95 to US 1; at Bath, take SR 209 11 mi S to SR 217; turn right onto 217, follow 1 mi to resort* ☐ The choice lodgings at this beautiful 650-acre resort are in a squat, octagonal lighthouse which sits at the edge of Casco Bay, offering spectacular views of the harbor. Other accommodations are in the lovely old Main Lodge or the Early Bird house (comfortably furnished in traditional New England maple) or in one of the 18 cottage rooms (all with fireplace) tucked away amid the spruce trees. Almost all of the 94 rooms have private baths, and all are heated. None have telephones or TV.

The variety of the terrain at Sebasco encourages roaming the grounds—forest, rocky shoreline, freshwater ponds, sandy beaches, landscaped lawns and gardens. The 9-hole, par-33 golf course leads you through the woods to the edge of the sea. The olympic-size saltwater pool sits on a rocky point of its own, inches from the bay. The lodge has two all-weather tennis courts and facilities for two kinds of bowling—lawn and candlepin. You can rent a motorboat or sailboat at the lodge to cruise the bay, or pedal boat or canoe around Wah-Tuh Lake.

Breakfast and dinner are included in the room rates; especially noteworthy are the Tuesday lobster bakes on the beach, and steak and chicken broils Thursdays in the pine grove. Evening entertainment includes band concerts on the lawn, dancing, and lectures—or convivial cocktails in the Anchor Lounge. ☐ **Mod** ☐ **W**

SUNSET: Goose Cove Lodge and Cottages, *P.O. Sunset, ME 04683, (207) 348-2508* ☐ *Open early June to early Sept* ☐ *Off SR 15 in Sunset* ☐ This cottage colony, set on a scenic, sandy cove, provides a rustic retreat not far from Acadia National Park. Goose Cove's 70 acres of woodland include 2500 feet of excellent beach frontage, where hardy souls can take a quick dip in high summer. The main lodge, built of native logs, has a fieldstone fireplace and picture windows overlooking Penobscot Bay. Its dining room, open for breakfast and dinner, specializes in seafood dishes and regularly schedules lobster cookouts. Each of the seven cabins, well-spaced from each other, has a sun deck, a refrigerator, a living room with fireplace, a bedroom, and a bath. For those needing more room, there are two duplex cottages, each with two bedrooms, fireplace or wood stove, and private bath. Also available are 10 motel rooms—even these have their own fireplaces or stoves. Firewood is supplied.

With advance notice, the lodge can rent canoes, rowboats, and small sailboats. The recreation hall has a pool table and table tennis. Shuffleboard is available outdoors. Because of the diversity of its flora and fauna, the resort property was, for many years, the site of college-level course on the ecology of the Maine coast. Bird watchers will enjoy the nature trails that thread the property as well as the motor-launch trips to the Audubon Sanctuary at Mantinicus Rock. ☐ **Inexp to Mod** ☐ **W**

YARMOUTH: Homewood Inn, *P.O. Box 196, Yarmouth, ME 04096, (207) 846-3351* ☐ *Open early June to mid-Oct* ☐ *2 mi S of SR 88; follow signs to Homewood* ☐ This is a charming bayside encampment of cottages and lodge buildings complete with antique shop, general store, and laundry. You can take your pick of lodgings: In the Tides—a two-story wood-frame building on an old granite pier—all rooms and suites have sundecks and lovely views of Casco Bay. Several cabins and single and double cottages are scattered around the 20-acre site. Furnishings vary from building to building, but you are apt to find antiques, exposed beams, a stone fireplace, or some other nice individualizing touch. The excellent restaurant emphasizes seafood and is famous for its Monday night clambake.

Acres of meadowland on the inland side of the buildings are laced with paths for strolling and birdwatching. You can swim from the small sandy beach or fish from the dock, play tennis (one court on the premises, but more nearby), rent a rowboat or bicycle, or arrange for deep-sea fishing expeditions.

Portland, only 15 minutes away, offers museums, restored houses, and interesting shopping. You can also embark on a cruise of Casco Bay and the Calendar Islands, or explore the coast by car. ☐ **Mod** ☐ **W, F**

Maryland

BETHESDA: Linden Hill Hotel, *5400 Pooks Hill Rd., Bethesda, MD 20014, (301) 530-0300* ☐ *Open all year* ☐ *Wisconsin Ave. and Exit 19 on Capital Beltway* ☐ A few minutes from Washington D.C., Linden Hill Hotel provides a sensible alternative to unremitting touring. It is not only a complete resort in itself, but also an out-of-towner-oriented home base from which to see the U.S. capital. Guests can use all or part of the hotel's tourist conveniences, which include airport limousine service, tours (by special arrangement) of the capital, and a courtesy bus to Bethesda Naval Hospital, NIH, and area department stores. On the premises are a tennis club with indoor and outdoor courts, an olympic-size swimming pool with poolside cocktail service, jogging and bike trails, men's sauna and whirlpool, and for women, a health and beauty spa offering classes and private consultations, facials and massage, saunas and whirlpool baths.

Accommodations vary from single rooms to suites and apartments with kitchens. Private bath, individual temperature controls, and free coffee are standard. Many rooms have balconies. The hotel's public areas have a coffee shop, a wood-beamed, pub-style cocktail lounge, and Ristorante Firenze, one of the metropolitan area's well-known restaurants specializing in northern Italian cuisine. ☐ **Mod to Lux** ☐ **T**

OAKLAND: Will O' the Wisp Resort, *Oakland, MD 21550, (301) 387-5503* ☐ *Open all year* ☐ *On US 219* ☐ In the heart of hilly Garret County, almost surrounded by West Virginia and Pennsylvania, is Deep Creek Lake, a winding hill-bordered reservoir whose tree-lined shore is the setting for Will O'the Wisp Resort. Throughout the year Will O' the Wisp's seven-story condominium is open and active, and from May to October, four cottages expand the accommodations. The condominium houses 47 apartments with balconies facing the lake. Personalized decor, tasteful color schemes, and working fireplaces make each unit comfortable. The cottages come in varied sizes, but each has a picture window facing the lake, and an outdoor patio. The resort operates a cafe, a restaurant, and a cocktail lounge with entertainment and dancing.

Summer activity centers on watersports on the lake, and hiking and picnicking in the immediate countryside. Within an eight-mile radius are tennis, golf, and horseback riding. In winter, Wisp Ski Area, affiliated with the resort, has 12 miles of downhill trails, snowmobiles, and horse-drawn sleighs, plus a day nursery and a list of babysitters. Other facilities include a game room, handball and racquetball court, exercise room with gym, indoor swimming pool, and whirlpool bath and sauna. □ **Mod** □ **F**

Massachusetts

THE BERKSHIRES: Berkshire County has rolling hills crisscrossed with stone walls, big white colonial houses, and red barns, many of them sagging and tilting toward ruin. Tough Yankee farmers have been scratching a living out of these hills and rocky valleys for 300 years. But the area is not as provincial as it might appear. Lenox is only 120 miles from both Manhattan and Boston, and for more than 100 years citizens of those metropolitan areas have rusticated themselves here, in search of serenity and inspiration. *Moby Dick* was born here. Melville, Hawthorne, Edith Wharton, Oliver Wendell Holmes, Norman Mailer, and Norman Rockwell have all had homes in the Berkshires. Industrial barons like Andrew Carnegie built sprawling summer estates here. Sculptor Daniel Chester French, whose Stockbridge home is a major tourist attraction, said of the town: "I live here six months of the year–in heaven. The other six months I live in, well, in New York."

The town of Stockbridge is everyone's dream of an old New England town. Immaculate, white Greek-revival and Federal-style buildings line the main street. Lenox is similar, but less consistent in its layout and architecture. Both have a history as havens for the arts. In summer, the cultural calendar rivals that of many major cities. Some of the world's greatest conductors and soloists come to Tanglewood, summer home of the Boston Symphony Orchestra. The Berkshire Playhouse in Stockbridge draws top stars for its productions. In nearby West Becket, the Jacob's Pillow Dance Festival features some of the world's best dancers and companies. At the northern edge of the county is the Williamstown Summer Theatre, noted for its productions of new and classic plays.

What continues to draw artists and tourists to the Berkshires is the natural beauty of the region. Though the mountains can't compare in terms of scale or grandeur to those of the Western states, few sights compare with that of the Berkshire Hills ablaze with red, yellow, and orange foliage in late October. In winter, snowfall tends to be heavy, and the county has several good ski areas. On and off of Route 7, which bisects the county south to north, you will find historic houses, streams and farm ponds, Colonial-era graveyards, and, in the morning or early evening, grazing deer.

The Center at Foxhollow, *Rte. 7, Lenox, MA 01240, (413) 637-2000* ☐ *Open all year* ☐ *Off US 7, 251 Kemble St, 2½ mi S of Lenox* ☐ The Center at Foxhollow is something of a hybrid—the buildings and guest rooms suggest an old New England inn, but the sports facilities are those of a resort. The Manor House, once a home of the Vanderbilts, is at the end of a ½-mile long driveway shaded by 60-foot pines. The white clapboard building with black-shuttered windows looks like the typical gracious colonial house, except it is about five times as long. The Manor House served as a girl's school until 1975, so some rooms share bath. Although some rooms have fireplace and occasional antiques, the decor can be commonplace. Three cottages on the 90-acre estate have been transformed into guest houses; one was once the home of writer Edith Wharton.

If you only had a weekend you probably couldn't do everything at Foxhollow. The riding stable provides horseback rides (hourly rate) on miles of bridle paths around the center. The 4000-book library, with its old leather-bound volumes and encyclopedias, should keep you occupied indoors if it's raining. When the weather is sunny and warm you can spend time by the pool, on the two tennis courts, in front of a volleyball net, on the basketball court, or sailing a 14-foot Blue Jay on the lake. Guests have privileges on a golf course 1½ miles away. During winter, guests ice-skate on the lake and ponds and cross-country ski on the bridle paths. Ski rentals are available.

The cuisine at the restaurant aspires to gourmet status; specialties are veal Oscar and chicken Foxhollow. The resort operates on the Modified American Plan. ☐ **Mod to Lux** ☐ **R**

The Flying Cloud Inn, *New Marlboro, MA 01230, (413) 229-2113* ☐ *Open early May to late Oct and late Dec to late Mar* ☐ *From Winsted, CT 14.5 mi N on SR 183* ☐ The Flying Cloud looks as it must have shortly after the main farmhouse was completed in 1771 and still raises lambs and some crops and engages in maple sugaring. The inn sits on 200 acres of rolling farmland and forest, with a stocked pond just outside the door. Inside, the New England colonial furnishings include many genuine antiques such as the 1790 four-poster in Room 1. Each of the ten double rooms (eight with private bath) has special features and furnishings and, in keeping with the colonial ambience, you will find neither telephones nor TVs in the bedrooms.

For spring and summer amusement there are two tennis courts (one clay, one all-weather), volleyball, lawn games, plus the pond for fishing, swimming, or rowing. Four miles of trails allow ample opportunity for viewing flora and fauna, including a resident beaver family. Four golf courses are in the neighborhood. In fall the foliage is spectacular. Winter offers sledding, skating, snowshoeing, and cross-country skiing on the grounds, and downhill skiing nearby. Some sports equipment is supplied, and babysitters and sports instruction are available for children.

The kitchen specializes in traditional New England fare. The single set menu each night features breads fresh from the oven and seasonal vegetables from the inn's organic garden. Box lunches and dinners may be had on request. There is an extensive wine list and a fully stocked bar. ☐ **Mod** ☐ **S, H**

Oak n' Spruce Resort, *South Lee, MA 01260, (413) 243-3500* ☐ *Open all year* ☐ *3 mi S of Mass Pike Exit 2, off of SR 102* ☐ An informal country resort, the Oak n' Spruce is an active place with a pleasant setting–a verdant valley complete with burbling brook. The 10 buildings are set around a central expanse of lawn containing a heated outdoor swimming pool and two clay tennis courts. Within the main lodge, a rambling white structure whose half-timbered gables give it a Bavarian air, are the dining room, lounge, indoor pool and whirlpool, exercise room, saunas, and game room. Ten air-conditioned rooms are on the second floor. The remaining 50 or so guest rooms are in the lodges, long, low buildings much like motor lodges everywhere. All rooms have color TV and scenic views; some have private patios and kitchenettes. In summer, the restaurant has buffets on Saturday, chicken barbecue on Sunday, and outdoor steak fry on Thursday.

Besides the pools and tennis courts, the resort boasts a 9-hole, par-3 golf course, an 18-hole Frisbee course, an easy downhill ski slope with rope tow, a toboggan chute, and an ice skating rink. During winter snowmobile and cross-country ski rentals are available. ☐ **Mod**

Yankee Motor Lodge, *Lenox, MA 01240, (413) 499-3700* ☐ *Open all year* ☐ *On US 7/20, 2 mi S of Pittsfield* ☐ This is an example of what a motor lodge should be. The buildings, of brick and stone, are architecturally unelaborate but handsome. The 60 rooms come in five styles, from small and cozy to deluxe units with working fireplace, exposed beams, refrigerator, and breakfast bar. All have colonial decor, cable TV, telephones, and wall to wall carpeting.

With four ski areas–Brodie, Jiminy Peak, Bousquet, and Butternut Basin–within 15 miles, winter is the busy season. From December through February, the lodge's Sitzmark lounge offers complimentary beer and snacks. Although there's no restaurant here, there is one next door. In summer, the center of activity is the attractive outdoor pool, fed by a waterfall tumbling over a stone ledge. ☐ **Lux** ☐ **S**

CAPE COD: Cape Cod's beaches, pine forests, and crystal-clear ponds are extensive–the Cape's curved area would be more than 75 miles long if stretched straight. Much of the northern end has been preserved as National Seashore and is less commercially developed than the southern end. Because the northern Cape faces the Atlantic, it has higher dunes and colder water than the southern end, which is buffered by Nantucket Sound. Water on the bay side is warmer and calmer than that on the ocean side, but beaches tend to be more rocky and less white. Unless your accommodations give you access to a private beach or are within walking distance of a public beach, count on hefty parking fees to use a public beach.

The bread-and-butter industry here is fishing, and your resort hosts can advise you on where to surf cast or on the closest marina for charter or party boats or boat rentals. Sightseeing cruises and sailing trips are also available, as are golf, tennis, horseback riding, beach-buggy tours, and bicycle rental. On the cultural side, you'll find dozens of museums and historic sites rounded up in tour booklets obtainable on arrival. Several summer theaters operate; the most famous of these is at the old artists' colony in Provincetown, the Provincetown Playhouse, which staged Eugene O'Neill's plays.

The pilgrims' first landfall was not at Plymouth Rock but on Cape Cod; they are commemorated by a 250-foot granite tower in Provincetown and by exhibits at Cape Cod National Seashore. The National Seashore has several units well marked along US 6 between Eastham and Provincetown. It has, in addition to what are probably the nicest Cape beaches, a number of interpretive displays, self-guiding walking and biking trails, and special programs. At South Wellfleet is the noted Wellfleet Bay Wildlife Sanctuary, run by the Audubon Society, with self-guided nature trails, wildlife tours, and special programs. The famous Oceanographic Institution at Woods Hole is not open to the public, but the National Marine Fisheries Service operates a free aquarium at Woods Hole in summer.

Blue Rock Inn & Golf Course, *South Yarmouth, MA 02664, (617) 398-6962* ☐ *May–mid–Nov* ☐ *1 mi N of S. Yarmouth, off Highbank Rd.* ☐ This inn, which operates on European Plan, is made up of about half a dozen motor-lodge-type buildings that overlook a nicely landscaped par-3, 18-hole golf course. The modern rooms have private, recessed balconies or patios, tile baths, and color TV. The pro shop has a good selection of casual clothes as well as clubs and bags. The Blue Rock also has a restaurant, lounge, pool, and tennis court. Two full-length golf courses, Bass River and Dennis Pines, are within a five-minute drive. ☐ **Mod** ☐ **G**

The Breakwaters, *Sea Street Beach, Hyannis, MA 02601, (617) 775-6831* ☐ *Open all year* ☐ *Take SR 132 SE to Barnstable Rd., S to Old Colony Rd., SW to Gosnold St., W to Sea St.* ☐ On 1½ acres stretching back from the beach, the Breakwaters consists of nine modern houses containing 20 apartments ranging from one-room efficiencies to three-bedroom suites. All apartments have equipped kitchens and daily maid service (Monday through Saturday in summer). Rooms are furnished with quality colonial reproductions, wall to wall carpeting, and studio beds in the living room. Each unit has beach chairs, umbrella, picnic table, and grill. The cottages are adequately spaced, giving each a bit of privacy and tree-shaded lawn. Some have private decks and water views. All share the heated pool. The cottages are about a mile from Hyannis Harbor with its fishing and excursion boats. The Melody Tent nearby puts on its versions of Broadway shows. ☐ **Mod to Lux**

Cape Codder Hotel, *Falmouth, MA 02541, (617) 540-1900* ☐ *Open mid-June to Labor Day* ☐ *Take SR 28 S to W. Falmouth, turn W at Sippewissett sign and follow to resort on Sippewissett Rd.* ☐ The Cape Codder is typical of the grand Victorian seaside resort. Gray-shingled, white-trimmed, and surrounded by wide verandas, the huge hotel sits on a bluff flanked by water on three sides. In addition to the rooms in the turn-of-the-century hotel, there are contemporary rooms, each with recessed balcony or patio overlooking the sea, in a wing next to the olympic-size pool.

The Cape Codder operates on the Modified American Plan. Typical dinner entrees include baked stuffed shrimp, Cape scallops, prime rib, and seafood Newburg. The hotel has two cocktail lounges, one with large brick patio overlooking the water. Right in front of the hotel is the sandy private beach, protected by two jetties. ☐ **Mod to Lux**

Captain Gosnold Village, *P.O. Box 544, Gosnold St., Hyannis, MA 02601, (617) 775-9111* ☐ *Open May to Nov* ☐ *From US 6, take Barnstable Rd. SE to Old Colony Rd., SW to Gosnold St. and W to resort* ☐ This is a group of Cape Cod cottages closely set around a 32-room motel. The gray-shingled houses have paneled interiors, wall to wall carpeting, colonial-style furnishings, and separate kitchens with dining area. The cottages come with one, two, or three bedrooms, but one-room efficiencies with all-in-one stove, refrigerator, and sink combinations are available. The Captain Gosnold has a small heated swimming pool with sun terrace. ☐ **Mod**

Chatham Bars Inn and Cottages, *Chatham, MA 02633, (617) 945-0096* ☐ *Open May to Oct* ☐ *From US 6, take SR 137 S (exit 11) then SR 28 E 4 mi to Chatham. Resort is at end of Seaview St.* ☐ The main lodge is a handsome gambrel-roofed building laid out in a sweeping arc on a private beach on Pleasant Bay. Its 35 double and two single rooms are complemented by 25 one- to eight-bedroom cottages, each with private bath and many with porches overlooking the ocean. Adjoining the lodge are heated, non-housekeeping cottages; a little farther off are the lodge-type cottages. Rooms are furnished in contemporary reproductions of colonial pieces or in breezy, modern summer-home style.

The resort operates on the American Plan. Meals feature New England specialties, and usually two seafood dishes are offered at lunch and dinner. Baked Alaskan king crab au gratin, curried spring chicken, and broiled New York sirloin strip steak have been among the nightly dinner choices. On the resort's wide strip of sandy shoreline, the Beach House serves breakfast, luncheon, and cocktails during July and August.

On the 65-acre site tennis buffs have five all-weather Deco Turf courts at their disposal and golfers have a nine-hole, par-34 course measuring 2325 yards. The waters of Pleasant Bay are calm and warm, but those who prefer surf bathing should take the five-minute trip across the bay by hotel boat to the outer-bar beach. Bluefish and striped bass may be caught in the waters off the resort. There's a babysitting service so that parents can relax. Sailing and boat rentals are available. ☐ **Lux** ☐ **T, W**

Craigville Conference Center, *Craigville, MA 02636, (617) 775-1265* ☐ *Open all year* ☐ *exit 6 from US 6, follow signs for Craigville Beach* ☐ Although operated by the Massachusetts Conference of the United Church of Christ, this is an unusual resort open to people of any religious persuasion. Off-season the center hosts church and educational conferences and retreats (including special programs for family and other problems), but it will take individual guests if there is room. In summer it caters to vacationers, offering recreation and accommodation at very reasonable rates.

On the 60-acre grounds are Craigville Inn and Groves House, overlooking a pond and salt marsh, and the Manor, Lodge, and Seaside House on a bluff overlooking Nantucket Sound. These offer a wide variety of rooms and suites, with or without private bath, and are operated on Modified American Plan. Special rates and accommodations are made for small children and for families of four or more. In addition, two apartments and four cottages are available on a housekeeping plan.

A hearty breakfast and dinners that feature seafood are included in MAP rates. Lunch is available in the dining room on advance notice, but most people eat at the snack bar in the Beach House. There are also lounges with TV and fireplaces, where musicals, lectures, and other programs are held evenings.

In addition to a private ocean beach, which is safe for children because it has no undertow, there is tennis, volleyball, and shuffleboard on the grounds. Some sightseeing tours and boat tours are organized by the resort, and the staff will help you arrange on-your-own tours. ☐ **Inexp to Mod** ☐ **W**

East Bay Lodge, *Box N, Osterville, MA 02655, (617) 428-6961* ☐ *Open mid-June to Labor Day* ☐ This former sea captain's home is now a luxury inn with an emphasis on service. In the lounge is a ceiling-high double fireplace, and off the lounge are several small dining rooms and a large Victorian dining area. A modern wing houses 18 guest rooms. Though decorated with 18th-century reproductions, the rooms are modern, with entertaining appointments such as color TVs behind swing-away mirrors, and telephones in the showers.

Guests on MAP here are entitled to choose freely from the inn's well-regarded restaurant menu. The accent is on fish—delivered fresh daily—but chateaubriand, prime ribs, and other choices are also available. Especially famous is the Sunday evening buffet (July through September), the only time casual attire is permitted. At other times, the Dining Room observes a dress code (jackets for men; ladies may wear a formal pants suit or skirt and blouse).

The inn is on seven acres less than a mile from the beach (frequent transportation to the beach is provided) and has three tennis courts on the grounds. On request, your hosts will arrange a sail on their 41-foot sloop or an island-hopping tour by air. ☐ **Lux** ☐ **T**

The Elm Arch Inn, *Elm Arch Way, Falmouth, MA 02540, (617) 548-0133* ☐ *Open all year* ☐ *Off SR 28 (Main Street) in Falmouth* ☐ This inn, built in 1812 as a sea captain's residence, was bombarded by the British frigate *Nimrod* on January 28, 1814. One wall still bears the scar of a cannonball. All rooms here, and in the adjacent Richardson House, are decorated in early American style with quality reproductions; some rooms have fireplaces, hooked rugs, rockers, and exposed-beam ceilings. Also on the property is a cottage that sleeps five. A pleasant, screened terrace with flagstone floor overlooks the pool. Although the inn serves no food, it is in the heart of Falmouth with restaurants nearby. ☐ **Inexp to Mod**

Hargood House, *493 Commercial St., Provincetown, MA 02657, (617) 487-1324* ☐ *Open all year* ☐ *East end of Commercial St., near Howland St.* ☐ Hargood House comprises four carefully remodeled old Cape houses. Three are clustered around a lawn and garden area at the water's edge. Apartments in these three buildings have fine views of the harbor through glass-walled living rooms. Four of the five apartments in Hargood House North, across the street, have limited views of the harbor. Each of the 17 apartments is quite different from the others. For instance, one of the two studios is done in bright garden colors and has a large fireplace. The other studio has a low ceiling with exposed beams and a bed in a recessed nook, giving it the feel of a ship cabin. All of these attractive apartments, from studio to two-bedroom suites, have modern baths, and kitchens complete with china and stemware (there is no restaurant). Almost all have dishwashers.

Hargood House is close enough to walk to the center of action in Provincetown but far enough away to gain a good measure of quiet. A private beach is out front. Apartments are rented weekly during summer, but are available for two-night minimums the rest of the year. ☐ **Lux**

Iyanough Hills Motor Lodge and Golf Club, *Rte. 132, Hyannis, MA 02601, (617) 771-4804* ☐ *Open all year* ☐ *From US 6, take exit 6 (SR 132) S 1½ mi* ☐ From this 104-room, L-shaped motel on two acres, you can walk to an 18-hole golf course with pro shop and lounge and to six indoor tennis courts; both complexes offer special rates to motel guests. The resort has its own tennis court, indoor pool, saunas, and whirlpool, as well as a game room, children's playground, and shuffleboard court. All rooms have full baths, color cable TV, telephone, and climate control; some suites and kitchenettes are available. Coffee is served mornings and there's a restaurant across the street. ☐ **Inexp to Mod**

The Lighthouse Inn, *West Dennis, MA 02670, (617) 398-2244* ☐ *Open early June to mid-Sept* ☐ *Just off SR 28 in W Dennis* ☐ The Bass River Lighthouse, built in 1850, forms the core of the Lighthouse Inn. Set at the edge of the Atlantic Ocean, the Main House offers a splendid view of the water and, just outside the door, a long, jetty-sheltered private beach of white sand. On the spacious grounds, you have a heated outdoor swimming pool, two tennis courts, and facilities for shuffleboard and horseshoes. Some children's programs are conducted, and there is a recreation room. There is nightly entertainment in the Lower Deck, the casual cocktail lounge and social center. The dining room, which overlooks the water, and enforces a dress code in the evenings, specializes in fresh New England seafood.

Most accommodations are operated on MAP, but EP is also available. Singles, doubles, and suites are in the Main House or in scattered guest houses and cottages on the lawns near the beach. Farther back, in a wooded, more secluded setting, are additional rooms and housekeeping cottages. All units have private bath, and some have fireplaces, patios, or other extras. ☐ **Mod**

The Moors Motel, *Bradford St. Extension, Provincetown, MA 02657, (617) 487-1342* ☐ *Open Jan to mid-Nov* ☐ *Bradford St. Ext. and Beach Hghwy.* ☐ This weathered-shingle, low-profile motel has 32 rooms, all overlooking an unmarred expanse of moors that stretches to the sea. Streams wander through the marsh grass and many types of birds feed in the protected area. Each of the rooms has color TV and piped-in music, and during the summer (June 1 to Sept 1) continental breakfast is included in the room rate.

The restaurant and lounge are done in a hearty nautical style–ropes and pulleys hanging from massive ceiling beams, an occasional harpoon tacked to the wall, wooden kegs set around. The restaurant serves Portuguese specialties, Cape seafood, and prime rib.

The motel has an outdoor pool, and five tennis courts are available next door. Downtown is only a few blocks away. ☐ **Inexp to Mod**

Red Jacket Beach Motor Inn, *South Shore Dr., Bass River, MA 02664, (617) 398-6941* ☐ *Open May to Oct* ☐ *Exit 8 off US 6, S 2 mi to light, right 2 mi to Sea View Ave., left on Sea View* ☐ This large motor lodge sits on a sandy point on Nantucket Sound and Parkers River. The 150 guest rooms have all the usual amenities–wall to wall carpeting, color TV, and private balcony or patio. Most have a water view. But this resort offers much more than the minimum. Besides the 1000-ft. private beach and the outdoor pool, there's an indoor pool, saunas and exercise rooms with lockers, putting green, Sunfish and Sailfish rental, tennis court, restaurant, and laundromat on premises. ☐ **Mod to Lux** ☐ **W**

Red River Motel, *Rte. 28, South Harwich, MA 02661, (617) 432-1474* ☐ *Open all year* ☐ *On SR 28* ☐ The owners of this up-to-date motel, Marge and Ed Laux, are serious bird watchers who cater to birders. The outer Cape has long been famous for its impressive bird life, particularly during migration from mid-July to early November. Henry David Thoreau, William Brewster, Arthur Cleveland Bent, Henry Beston, and Ludlow Griscom are among the ornithologists and naturalists who have come to the Cape to study. The motel is close to the famed Monomoy Island Wilderness Area, Chatham (good for fall land birds), and Nauset Beach (shore birds). The waters of the Cape are generally considered good for spotting otherwise rare pelagic species.

Besides the motel rooms, all heated and having cable TV, there are four fully equipped efficiencies, each with refrigerator, stove, and sink. Canoe and Sunfish rentals are also available. ☐ **Inexp**

Salt Pond Motel, *Rte. 6, Eastham, MA 02642, (617) 255-2100* ☐ *Open June to Oct* ☐ *On US 6, ¼ mi S of National Seashore Visitor Center* ☐ Located within Cape Cod National Seashore, the Salt Pond Motel is owned by the federal government and run by a private concessionaire. The modern motor lodge overlooks Salt Pond, actually an inlet of the Atlantic with access to Nauset Bay. The sandy town beach is only 100 yards away, and rowboats are provided free of charge for fishing and for exploring the pond and bay. Great Pond, with a fine freshwater beach, is one-half mile west, and the spectacular outer beaches are just two miles east. The nearby Visitors' Center presents free indoor and outdoor lectures and guided nature walks.

In addition to the rooms in the motor lodge, there are efficiency apartments in adjacent Salt Pond House, a remodeled 19th-century building. The apartments accommodate from two to six people and all have twin beds, TV, equipped kitchenettes, baths, and daily maid service. Complimentary coffee and cake are available each morning. ☐ **Mod** ☐ **W**

DANVERS: Radisson Ferncroft, *Ferncroft Rd., Danvers, MA 01923, (617) 777-2500* ☐ *Open all year* ☐ *At jct I-95 and US 1* ☐ This long, rather boxy, brick building is set amid 400 rolling acres of woodlands. All of the 300 guest rooms and suites have telephone, color TV, air conditioning, and contemporary furnishings. In addition, the 22 executive suites have a bedroom, a parlor area, and two baths; some also have a balcony. Each of the garden court suites, surrounding the outdoor pool, has a sitting area and a bedroom, and the presidential suite, the most luxurious accommodation, boasts two bedrooms, large living room/dining room, wetbar, and exceptional view.

Ferncroft operates on European Plan and has two restaurants: the Greenery for all meals and the Running Board for dinner only. Some of the house specialties are New York strip steak, veal Oscar, sole filet Semiramis, and chicken Rochambeau.

The centerpiece of the sports facilities is the 18-hole Robert Trent Jones golf course, home of the New England PGA. A pro shop is on the premises. Ferncroft also offers four tennis courts, indoor pool, sauna, whirlpool, ski shops, and ice skating rink. ☐ **Lux** ☐ **G**

MARTHA'S VINEYARD: Only four miles off the Massachusetts coast, Martha's Vineyard was settled by Indians, and even today most of the residents of Gay Head are of Indian descent. The island was officially discovered in 1602 by Bartholomew Gosnold. Early attempts by white settlers to make a steady living off brick making and salt evaporation did not succeed, but fishing quickly took hold as the main occupation. By the 19th century, the Vineyard, together with her sister island, Nantucket, was devoted to the whale oil industry. Many of the whaling captain's houses remain in the oldest village, Edgartown, and they give the town a good part of its charm. By contrast, the equally charming village of Oak Bluffs, with its turn-of-the-century gingerbread cottages and ornate gazebos, marks the transition of the island's industry from whaling to tourism during the Victorian era.

Martha's Vineyard is roughly triangular. The maximum length and width are about 20 miles by 10 miles. Many of the beaches are open to all without restriction, including the following: Owen Park Beach, Oak Bluffs Town Beach, Joseph Sylvia State Beach, Cape Pogue Wildlife Refuge, Katama Beach, Menemsha Public Beach, Edgartown Town Beach, and Lighthouse Beach. Two public golf courses are available–Island Country Club in Oak Bluffs and the Mink Meadows Golf Course in Vineyard Haven–and there are town tennis courts in Vineyard Haven, Oak Bluffs, and Edgartown. In Edgartown (at Cooke and School Streets), the Duke County Historical Society has a museum of island memorabilia, a library, and maritime exhibits. Gay Head, at the eastern tip of the island, has spectacular, multi-colored cliffs.

Car and passenger service to the island is available year round from Woods Hole, but during the summer reservations for cars must be made well in advance; contact the Steamship Authority at (617) 548-5011. Summer passenger ferries sail from Hyannis, Falmouth Heights, and New Bedford.

Like its sister island, Nantucket, Martha's Vineyard specializes in small guest houses too numerous to be adequately covered here; for a complete list, write the Martha's Vineyard Chamber of Commerce, P. O. Box 1698, Vineyard Haven, MA 02568.

Daggett Houses, *Box 1333, Edgartown, MA 02539, (617) 627-4600* □ *Open all year* □ *On N. Water St.* □ The Daggett Houses are on a waterfront street lined with homes of whaling captains, but the main building here is even older than these noble houses, in fact it is older than the American Revolution. The snug little inn has candlelight doors, brass flintlock blunderbuss, Betty lamps, and secret stairway in the old Chimney Room. Guests take their morning meal in this room, making toast from homemade bread and pouring their own coffee and orange juice while their breakfast is being prepared. The Chimney Room overlooks the inn's expanse of lawn and garden, its broad private beach, and the harbor.

In summer, the Garden Cottage, on the water, is available as a unit or in part. It has three twin-bed rooms with baths. A few housekeeping apartments are in Daggett House II, once a whaling captain's home. ☐ **Mod to Lux**

Clark, Leland, and Clark Cottages, *Vineyard Haven, MA 02568, (617) 693-0049* ☐ *Open all year* ☐ *On Franklin St.* ☐ These 18 cottages vary in size from one-bedroom to four-bedroom. All are simple, modern frame houses with contemporary furnishings, kitchens, paneled living rooms, radios, and TVs. The four-bedroom houses have washer and drier and two bathrooms. All are provided with linens and blankets, and kitchens are complete with pots, pans, dishes, silverware, cooking utensils, toaster, iron, and ironing board. For an extra charge guests may play tennis on the Plexipave courts. A pool is also available from late May to mid-September. All cottages are rented by the week. ☐ **Mod to Lux**

Cove House, *Box 25, West Tisbury, MA 02575, (617) 693-9199* ☐ *Open mid-June to mid-Oct* ☐ *Off State Rd. S of Music St., Turn left at garage* ☐ Cove House is in the middle of the island, a five-minute walk from the village of West Tisbury. Unlike the other Vineyard towns, this agricultural and residential community remains quiet even in the height of the season.

Cove House and the adjacent Studio House are typical of Martha's Vineyard buildings–natural, weathered shingles, blue trim, New England "barn" architecture. Cove House has four apartments, each with a living room with pull-out sofa, bedroom with double bed, full bath, and kitchen. Studio House has a one-room efficiency and two one-bedroom apartments with kitchen and bath. Each apartment here is paneled and has three daybeds. All apartments have contemporary furnishings and come equipped with linens, towels, blankets, and kitchen utensils. They are within walking distance of a general store, the Post Office, and a laundromat, and four riding stables are nearby. Rentals are by the week. ☐ **Mod**

NANTUCKET: In *Moby Dick* Herman Melville called Nantucket "a mere hillock, and elbow of sand; all beach without a background." The barren beauty of this mere "elbow of sand" has always drawn people from the mainland. The Indians were well aware of this island thirty miles at sea: when white settlers arrived on the island in 1659 they found it inhabited by four tribes. The Indians gave Nantucket its name, which means "faraway land," and they created a myth to explain its origin: One night an Indian brave awoke in discomfort to find his old moccasins had filled with a bothersome amount of sand. In disgust, he threw them out to sea. The one that landed closest to shore became Martha's Vineyard, the outermost became Nantucket.

The settlers had bought the island from Governor Thomas Mayhew for the price of thirty pounds of silver and two beaver hats. They tried farming but soon found fishing to be more profitable. From the Indians, who slaughtered beached whales for meat and blubber, the settlers learned the value of the leviathans. With the capture of the first whale in 1668 and the first killing of a sperm whale at sea in 1712, Nantucket began its greatest era. By 1830 it was Massachusett's third largest port and the largest whaling port in the world.

The cobblestone streets and brick whaling captain's homes remain, giving Nantucket a 19th-century atmosphere and the sense that it is more isolated from the world than it already is. Tasteful crafts shops, antique stores, and restaurants have taken over the bottom floors of many of the homes in response to the tourist boom, but the town continues to maintain a distinctly New England grace except on the most hectic mid-summer days. Nantucket is the major town on the island, but Siasconset (pronounced and often written 'Sconset) on the eastern end is a substantial village.

The main public beach is the Jetties (N. Beach Rd. to Bathing Beach Rd.); the Children's Beach (Nantucket harbor), Surfside, Cisco, 'Sconset, and Madaket also have beaches with lifeguards. Nantucket harbor is completely sheltered and big enough for a good-sized sailboat to tack comfortably in, making it ideal for those learning how to sail.

A whaling museum as well as a windmill, a jail, a firehouse, and several old homes–Jethro Coffin House, Nathaniel Macy House, Hadwen House, 1800 House–are open for tours.

Car and passenger ferries run daily from Woods Hole. Contact the Steamship Authority, P.O. Box 284, Woods Hole, MA 02543, (617) 548-5011, for schedule and reservations. Daily passenger service is available from Hyannis in the summer. There is scheduled air service from New York, New Bedford, and Boston via Air New England or Delta Airlines.

Because Nantucket's accommodations tend to be small, only a few of the hundreds can be covered here; for a complete list, write Nantucket Island Chamber of Commerce, Nantucket, MA 02554.

The Gordon Folger Hotel and Cottages, *Easton St., Nantucket, MA 02554, (617) 228-0313* ☐ *Open early May to mid-Oct* ☐ *Between N Beach St. and Cliff Rd.* ☐ In town, two and one-half blocks from Children's Beach, the Gordon Folger is an old-fashioned resort hotel with 35 rooms and three cottages. The hotel has wide verandas, sitting and writing rooms, and a game room for children. Two of the cottages have two bedrooms, bath, and living room. The third has four bedrooms, two baths, and a fireplace. Though all have refrigerators, these are non-housekeeping cottages with daily maid service. All are heated and available in spring or fall at reduced rates. Modified American Plan is available at the Whale restaurant, adjoining the hotel. The restaurant's popular dishes include prime rib, swordfish, and Cape scallops.

The hotel's moderate rates, baby sitting service, and game room make it a wise choice for families. ☐ **Mod**

Stone Barn Inn, *North Beach St., Nantucket, MA 02554, (617) 228-0723* ☐ *Open May to Oct* ☐ *N Beach St. near Jetties Beach* ☐ This unique inn was originally constructed as an elaborate stable. It is doubtful, however, that horses ever had the chance to appreciate the structure's Tudor gables, stone walls, and marble stalls, for the building apparently became a summer residence immediately after construction. Accommodations here are in either double rooms (some share bath) rented by the day or in housekeeping apartments rented by the week (some have a two-week minimum). The inn has four apartments and two guest rooms, but five cottages on the property offer eight more apartments and three more guest rooms. The Hayloft apartment in the inn is especially nice–spacious, with two rooms, bath, and kitchenette, it overlooks the harbor and beach. The apartments in the cottages range from studios to three-bedroom, two-bath suites accommodating six people. All are completely furnished, except for beach towels, and have kitchens. ☐ **Mod to Lux**

Wauwinet House, *P.O. Box 628, Nantucket, MA 02554, (617) 228-0145* ☐ *Open Memorial Day to mid-Sept* ☐ *Take Polpis Rd. NE to end* ☐ For a secluded vacation on Nantucket, you might try Wauwinet House, nine miles away from the hustle and bustle of downtown Nantucket. On a narrow spit of land between the Head of the Harbor and the Atlantic Ocean, it is one of the few places you can stay where both rough ocean and calm harbor swimming are a few feet away. In the sprawling, handsome main building, all 24 units (singles, doubles, and suites sleeping up to six people) have a water view. Cottages with twin beds, private bath, and porch are strung out beside the main building. The restaurant specializes in vegetarian, veal, and seafood dishes, and the pine-paneled Wauwinet Casino serves cocktails. MAP is available.

There are two clay tennis courts and sailing and boat rentals on the five acres of grounds. Some fishing equipment is available, and daily fishing trips will be arranged. There is also a babysitting service. With advance notice, courtesy shuttle service is available to and from town. ☐ **Mod** ☐ **F, W**

The White Elephant, *Easton St., Nantucket, MA 02554, (617) 228-2500* ☐ *Open mid-May to Sept and mid-Sept to Nov* ☐ *Near Children's Beach* ☐ The one-, two-, or three-bedroom cottages here offer wall to wall carpeting, elegant furnishings, and a beautiful view of the harbor from private patios, as do the rooms of the main building. The sedate White Elephant is built right on the water's edge. If you are lucky enough to own a yacht or cruiser, you can dock it at the resort's private marina.

The heated swimming pool overlooks Nantucket harbor, and around it guests can sit at umbrella-covered tables, sipping cool drinks and watching the yachts, fishing boats, and ferries glide through the harbor. Sports-minded guests can be found at the resort's private tennis club or the harborside putting green.

This complex operates on European Plan, and the restaurant serves island seafood specialties as well as traditional American and continental cuisine. A trellised gazebo with hanging planters gives the dining room a garden-like feel.

For all its nearness to the active harbor, the White Elephant is quiet and peaceful, situated along a beautiful tree-lined street, surrounded by hedges, flower beds, trellised roses, and manicured lawns. □ **Lux** □ **T**

Michigan

BELLAIRE: Hilton Shanty Creek, *Bellaire, MI 49615, (616) 533-8621* ☐ *Open all year* ☐ *US 131 to Mancelona, then W on SR 88* ☐ Here's a modern resort with all the frills–recreation without deprivation. The contemporary wood-and-glass lodge is set among 1300 hilly acres between its own 18-hole golf course and ski area, with a panoramic view of the surrounding countryside, including nearby Lake Bellaire, Torch Lake, and Lake Michigan's Grand Traverse Bay. Rooms in the three-story lodge and two-story chalets are spacious, and most have balconies. The charming restaurant with beamed ceiling, fireplace, and fine view has flaming sword and a Saturday night buffet, and there is dancing and entertainment in the bar.

You can start your day with a run on the jogging course or a hike on one of the trails. Then there's the par-72, 6732-yard golf course with pro and clubhouse rentals, which hosts the Shanty Creek Open in September. All-weather tennis courts are at the racquet club, rental canoes and boats are at Shanty Creek and Lake Bellaire, and swimmers can choose between the heated outdoor pool and the private beach at Lake Bellaire. Apart from a stocked trout pond, the hotel provides full fishing services including boat charters, guides, and equipment (as well as shanties for ice fishing). Rabbit and deer hunting parties can be organized too. In winter, you can practically fall out of bed onto the slopes, which start just outside the doors of the lodge, with instruction, rentals, Kinder School, and apres-ski activities. Nordic enthusiasts have miles of marked and groomed cross-country trails. There's an ice skating rink, and snowmobile rentals can be arranged.

The hotel has saunas, well-stocked game room and electronic arcade, croquet and other outdoor game areas, and bicycles. If you'd like to hunt mushrooms in spring or go berrypicking in fall, the hotel will pack you a box lunch and point you in the right direction. Other special activities include fall foliage tours, sleigh rides, local sightseeing tours, and a number of festivals (Bavarian, Irish, Autumn). For sports not on the premises, such as horseback riding or bowling, the hotel will make arrangements for you nearby. During summer and ski season, supervised children's programs are provided for an extra fee, and babysitters are available all year. Various plans and packages are available. ☐ **Mod to Lux** ☐ **G, T, F, W, S**

BOYNE USA RESORTS, INC: This company, which developed Big Sky in Montana, has an equally ambitious project in Northern Michigan to provide complete year-round recreational facilities of superior quality. The two focal points are Boyne Mountain at Boyne Falls and, some 35 miles north, Boyne Highlands at Harbor Springs (see also Harbor Springs, below). These complexes include almost a thousand lodging units, three mountains for winter sports, three 18-hole golf courses, watersports facilities, and two airstrips (a lighted 4200-foot jet strip within walking distance of Boyne Mountain Lodge and a 6500-foot grass strip at the Highlands). Many of the facilities may be used no matter which unit you are staying in, and EP, AP, tennis, ski, golf, and other package plans are available.

Golf is big here, and the courses, designed by Robert Trent Jones and William Newcomb, are Alpine Links at Boyne Mountain and Heather Links and Moor Links at Highlands. Each complex also has a nine-holer, driving range, putting greens, and pro shop with rentals. The Golf Digest–Bob Toski Instruction School is at Highlands. Tennis is also good. Highlands has four hard courts, while the tennis center at Boyne Mountain has 22 courts (15 hard, one clay, and two grass) plus pros offering one-hour lessons to two-week clinics.

Downhill skiing is at Boyne Mountain (450-foot vertical drop, longest run a mile, eight double and quad chairlifts, Poma lift, two rope tows) Nub's Nob (425-foot vertical drop, five chairlifts plus rope tows), and at Highlands (625-foot vertical drop, longest run 1½ miles, six chairlifts, T-bar, rope tow). All have patrol, snowmaking, ski school, rentals, cafeteria, and bar; Boyne Mountain and Highlands also have nurseries, restaurants, and cross-country skiing, and the chairlifts operate in summer for sightseeing. Ice skating (rentals) and sleigh rides are also part of the winter program. Boyne Nordican, at Boyne Mountain, is especially designed for cross-country skiers; it offers 42 miles of trail for beginners, intermediates, experts, and racers, along with rentals, instruction, and guided tours.

Boyne Highlands Inn, *Harbor Springs, MI 49740, (616) 526-2171* ☐ *Open all year* ☐ *4 mi NE* ☐ More than 200 rooms and suites are offered in this motel complex, where the design is Old English. Some rooms have private balconies, patios, and refrigerators. The restaurant specializes in seafood and steak, and offers a bar, dancing, and entertainment. There is also a snack bar. Facilities include heated swimming pool, sauna, recreation room, lawn games, and bicycles. Horseback riding can be arranged nearby. ☐ **Mod** ☐ **G, T, S**

Boyne Mountain Lodge, *Boyne Falls, MI 49713, (616) 549-2441* ☐ *Open all year* ☐ *On US 131 one mile S* ☐ More than 400 accommodations are offered in the Boyne Mountain complex. The alpine-style lodge offers double rooms and suites, some with fireplaces, balconies, or kitchenette. Mountain villas and villas at Deer Lake (six miles from the lodge) all have balconies, fireplaces, and kitchens, and they vary from one to three bedrooms. Nordican Inn, the cross-country skiers' unit on Deer Lake, offers suites. Boyne Mountain restaurant, with candlelight dining, specializes in seafood and prime rib. It has a bar, patio dining, dancing, and entertainment. The Beach House restaurant, on Deer Lake, is open summers. The Nordican Inn cuisine is Norwegian and American, and its bar specialty is aquavit.

This complex offers two swimming pools, saunas, lawn games, a recreation room, and bicycles. A private beach, sailing, and boat rental are available on Deer Lake. Fishing (stocked) yields trout, coho, and bass. ☐ **Mod** ☐ **G, T, F, W, S**

COPPER HARBOR: Keweenaw Mountain Lodge, *Copper Harbor, MI 49918, (906) 289-4403* ☐ *Open early June to mid-Oct* ☐ *1½ mi S on US 41* ☐ With its all-log construction, large fireplaces, high ceilings, and manicured grounds, this 66-unit EP lodge is grand and rustic. It is just the kind of lodge you would expect to find in Michigan's rugged, beautiful Northern Penninsula. The main lodge houses a restaurant, bar and guest lounge, and two massive fireplaces. Guests stay in smaller, two-room log cottages that accommodate from one to four guests. The cottages are comfortably furnished and each has a TV, telephone, and sandstone fireplace. The main attraction is exploring the rugged shores of Lake Superior, and you can swim at a beach a half-mile away. There is a challenging nine-hole golf course on the property as well as a tennis court. ☐ **Mod**

GAYLORD: El Rancho Stevens, *Box 366, Gaylord, MI 49735, (517) 732-5090* ☐ *Open early May to late-Oct* ☐ *3 mi NE of I-75 Old 27 exit* ☐ This dude ranch offers 36 knotty-pine paneled double rooms and a 1000-acre spread for horseback riding and other open-air activity. Besides 36 riding horses, the ranch has two waterskiing boats, several canoes, sailboats, and fishing boats for use on Lake Dixon, and a tennis court. The lake is good for bass, pike, and panfish and has a sandy beach for swimming. The ranch also has a heated pool, recreation room, volleyball, baseball, and archery equipment, and a program of social activities that includes hayrides, dancing, entertainment, and movies. Sports programs for children and teenagers are available. The dining room operates for guests on MAP and is open to the public by reservation. ☐ **Inexp** ☐ **R, W**

GLEN ARBOR: The Homestead at Glen Arbor, *Glen Arbor, Michigan 49636, (616) 334-3041* ☐ *Open all year* ☐ *From Traverse City take SR 72 21 mi W to Empire, then SR 22 11 mi N* ☐ Set among manicured hardwood groves at the edge of Lake Michigan's Sleeping Bear Bay, the Homestead consists of apartment buildings and chalets strung along the water's edge. The Crystal River, which lives up to its name, flows between the dwelling units and the lake.

Seasonal facilities include a long sandy beach on the lake, where a beach club provides canoe and sailboat rental. A racquet club with tennis pros offers two-, three-, and five-day tennis clinics as well as private lessons. Also here are three platform tennis courts, bicycle rental, nursery, and supervised children's program. In winter 14 kilometers of cross-country skiing trails are maintained. The condominiums are privately owned vacation apartments ranging in size from one room with bath to four-bedroom units. All units have kitchen facilities, most have fireplaces, and nearly all have a view of the lake. From May through October a restaurant on the grounds serves breakfast, lunch, and dinner. The adjoining bar offers light meals and frequent entertainment. Light meals are also available at the beach club. ☐ **Mod** ☐ **T, W, S**

GRAYLING: Penrod's Au Sable River Resort, *100 Maple St., Grayling, MI 49738, (517) 348-3711* ☐ *Open Mar–Nov* ☐ *From US 75 take exit 254 to State St., right after crossing river* ☐ The log cabins and cottages here are strung along the banks of the Au Sable River, and Penrod's specializes in canoe trips and trout fishing. The scenic Au Sable is safe even for novice canoers, with an average depth of 18 inches over the first 25 miles. The resort will provide aluminum canoes, and plan and provide pick-up service for short and week-long float trips (with pick-up points ranging from 10–220 miles away). Guests fish and swim in the river and at Lake Margrethe, which is less than a mile from the resort. Opportunities for tennis, golf, and horseback riding are nearby. Penrod's is on the 300-mile Trans-Michigan Hiking and Riding Trail. The dozen rustic cottages have one or two bedrooms that will sleep up to six people, private showers, heat, TV, and linens. Most have kitchenettes with dishes and cooking equipment, and a restaurant is within walking distance. There is no maid service, but a playground and babysitters are available. Also on the grounds are picnic tables and grills, archery, and other game areas. ☐ **Inexp** ☐ **F, W, H**

HARBOR SPRINGS: The Colonial Inn, *Harbor Springs, MI 49740, (616) 526-2111* ☐ *Open mid-May to mid-Oct* ☐ *1 mile E on M 131* ☐ The Colonial Inn is an elegant EP resort with a strong American Colonial theme, set on meticulously landscaped grounds a short walk from Lake Michigan's Little Traverse Bay. Guests have privileges at three 18-hole courses with yardages of 6400 to 6800 (See Boyne USA Resorts, above). A tennis club two miles away has all-weather courts. The private beach and dock, just a short walk across the lawns, has boat rentals for fishing and waterskiing. The heated outdoor swimming pool serves lunch at the poolside pavilion. Double and connected rooms are available in the inn, which has been in operation since 1894, and in newer annexes. Five cottages, named after Colonial figures, are available on a daily and monthly basis. These have two to five bedrooms, equipped kitchens, fireplaces, and maid service. All units feature Colonial-style furnishings. The Candlelight Room offers gourmet dishes daily, with planked whitefish a specialty of the house. The Revere Room is a cocktail lounge patterned after Colonial Williamsburg taverns. ☐ **Mod to Lux** ☐ **G, W, F**

HARRISVILLE: Big Paw Resort, *Box 187, Harrisville, MI 48740, (517) 724-6326* ☐ *Open Apr–Oct* ☐ *18 mi N of Oscoda on US 23* ☐ Set among towering evergreens on 70 acres on Lake Huron, Big Paw's beautiful lodge and sleeping cottages are constructed inside and out of peeled and varnished logs that have a mellow golden hue. Four rooms in a motel-style unit, and two- and three-bedroom cottages are offered on MAP in summer and EP in spring and fall (there are no kitchens). The cottages face the mile-long private beach and are screened from each other by trees; seclusion is further assured by private lanes to the beach and dining room. The combination of privacy, log walls, pleasant furnishings, picture windows, and the large stone fireplace makes each cabin almost idyllic. The dining lodge serves tasty family-style breakfasts and dinners. Many of the vegetables are grown on the premises, and steak and fresh lake fish are regular menu items. Sundays feature a beach cookout.

On the white-sand beach are sailboats, motorboats, and rowboats to take out on the lake, as well as a charter fishing boat. On the flower-bordered lawn between the beach and the lodge are shuffleboard courts, putting green, playground, and croquet area. Big Paw also has hiking and a lighted tennis court, and can get you babysitters or arrange nearby golf privileges. This is a pleasant and well-kept resort. ☐ **Inexp to Mod** ☐ **W, F**

HOUGHTON LAKE: The Woodbine Villa, *R1, Box 333, Houghton Lake, MI 48629, (517) 422-5349* ☐ *Open all year* ☐ *From US 27, take SR 55 exit E to West Shore N* ☐ Each of these nine rustic log cabins, on a private Houghton Lake beach, has kitchen, TV, screened porch, accommodates two to six people, and comes with a boat. The chief activities are swimming, fishing, and waterskiing. On the grounds are a fish-cleaning house, picnic tables and grills, sauna, lawn games, and a playground. There is no restaurant. Off season is noted for hunting, skiing, snowmobiling, and ice fishing. ☐ **Mod** ☐ **W, F**

INTERLOCHEN: Interlochen Center for the Arts, *Interlochen, MI 49643* ☐ *(616) 276-9221* ☐ *Open during academic sessions and from June–Aug* ☐ *From Traverse City S on US 31 to SR 137 S* ☐ This center, founded in 1928 as a music retreat for talented students, has grown to an internationally renowned center for all the performing arts. High school and junior-high students from all over the world come to the National Music Camp (late June to late August) and the Arts Academy facilities (September to May), which are set in pine forest. Although not a true resort, the center welcomes a limited number of guests who wish to partake of the unique spirit of this rural artistic center. Guests are housed in small lodges scattered in the woods, and all rooms have private bath. Rates usually include three cafeteria meals daily, but some EP units are also available. The 1200-acre campus adjoins Interlochen State Park. The freshwater lakes have sandy swimming beaches, good fishing, sailboat and canoe rentals, and playgrounds. In winter, there is ice skating and cross-country skiing, and downhill skiing is within driving distance. The center offers a remarkable opportunity to hear and see promising students and guest musicians at work in a beautiful and contemplative setting. The summer season offers over 350 concerts and performances. ☐ **Mod** ☐ **W, F**

IRON MOUNTAIN: Duval's Lakeside Resort, *Star Rte. #1, Box 63, Iron Mountain, MI 49801, (906) 774-1689* ☐ *Open all year* ☐ *From Iron City N on US 2, turn left at sign and follow signs 4½ mi N* ☐ Duval's is a rustic 25-acre hideaway in pine forest. The seven housekeeping cabins, rented by the week, are set at the edge of Harding Park Lake. Cabins are paneled, have two bedrooms and porches, and the kitchens come with cooking and eating facilities (there is no restaurant). The lake has walleye and northern pike year round, and perch, blue gill, bass, and crappies in season. It offers access to miles of inland waterways for canoeing and boating (rentals available), and has a sandy beach for swimmers. Set up primarily for the nature lover who wants a beautiful and secluded setting, Duval's offers relatively easy access to ski areas, golf courses, and shopping. ☐ **Mod** ☐ **W, F**

Pine Mountain Lodge, *Pine Mountain Road, Rte. 2, Iron Mountain, MI 49801, (906) 774-2747* ☐ *Open all year* ☐ *½ mile N on US 2 then follow signs* ☐ This three-story, Bavarian-style, 40-room lodge sits at the base of the Pine Mountain ski area. Some of the motel-style rooms have balconies. In addition to the dining room and cocktail lounge at the lodge, a cafe with terrace is at the slopes. Rates are EP from April to late November, and MAP in winter. Noted for its world-class ski jumps of 30, 50, and 110 meters, Pine Mountain has snowmaking, 14 slopes, three double chairlifts, three rope tows, rental shop, and a ski school. Cross-country enthusiasts will find 10 miles of maintained trails. The lodge has an indoor pool, wading pool, a tennis court, a recreation room, and a babysitter list. Lake and river fishing is nearby, and golfers can play at the 18-hole Pine Grove Country Club. ☐ **Mod** ☐ **S**

ISLE ROYALE NATIONAL PARK: Isle Royale, in Lake Superior, is a wilderness preserve of one large island (about 45 miles long and 7 miles wide) and some 200 small islands, all densely wooded with evergreens and hardwoods. The main island, 15 miles offshore, is regularly serviced in summer by ferries from Houghton, Grand Portage, and Copper Harbor, and by seaplane from Houghton. These services land you at Rock Harbor at the north end of the island. The island has no roads but is laced with 160 miles of hiking trails, along which are more than 30 campgrounds. Without side trips, count on four or five days if you want to hike from one end of the island to the other. However, you need not hike the whole distance: Rock Harbor Lodge provides water-taxi service to some campgrounds, and at others the cruiser *Voyageur* will drop you off or pick you up. You can reach the smaller islands with a rented boat or by water taxi.

Wildlife includes moose, red fox, beaver, red squirrels, and 200 species of birds. The famous wolves prey on the moose herds and keep them at a level the island can support. Freshwater streams and lakes are good for fishing, but are not recommended for swimming because of leeches. No bicycles or other wheeled vehicles are allowed. Insects reach their peak in May and June, but repellent is handy at all times. For more information, write: Isle Royale National Park, 87 N Ripley St., Houghton, MI 49931.

Rock Harbor Lodge, *National Park Concessions, Inc., P.O. Box 405, Houghton, MI 49931, (906) 482-2890, June to Sept; rest of year c/o Mammoth Cave, Kentucky 42259, (502) 773-2191* ☐ *Lodge open late June through Labor Day; housekeeping units open late May to late Sept* ☐ Offering the only accommodations in the park, Rock Harbor has lodge rooms on AP and 20 housekeeping units on EP. The lodge buildings, set in a string along Rock Harbor, offer small, basic, uncarpeted double rooms with private bath and heat; the saving grace is that each has large windows overlooking the water and islands. Housekeeping units are one-story cabins set in the woods near the marina store and lodge. These are also basic, with electric stove, refrigerator, utensils, and china, table and chairs in the studio-like living area, private bath with dressing room, and one double bed and two bunk beds with linen and blankets. The lodge dining room is open to EP as well as MAP guests. The lodge also has a snack bar and gift shop.

The focus of activity is the full-service marina with 450 feet of dock space. It offers rowboat and motorboat rental, but canoes should be rented in advance through the carrier you use to reach the island. The marina has water-taxi service, charter boat service, and cabin cruisers for trolling the reefs for lake trout. The marina store sells fishing supplies in addition to staple foods and sundries, and has a laundry. A number of sightseeing trips are offered. *Voyageur* circumnavigates the island three times a week. Water taxis will drop you off and pick you up at the points you would like to explore on your own. Guided scenic cruises of a couple of hours to a full day stop off at interesting sites on the islands. The National Park Service maintains self-guiding nature trails at the lodge and conducts a number of guided hikes and other interpretive programs. ☐ **Inexp** ☐ **F, H, W**

MACATAWA: Point West, *Macatawa, MI 49434, (616) 335-5894* ☐ *Open all year* ☐ *From US 31, take 32nd St. 6 mi W and follow signs* ☐ These are sleek, modern buildings of wood and stone, a contrast to the older summer cottages that occupy the remainder of this peninsula. Most guests are housed in the single-story motor lodge, laid out in a "U" around a grassy court with a swimming pool at its center. Some of these rooms have a kitchenette, but vacationers who want full housekeeping facilities should try the one- to three-bedroom apartments overlooking Lake Macatawa. These have laundry facilities, private patio, and dining area. For the most space and privacy, there's the cottage, which has a large living room with fireplace, a kitchen, a screened porch, and two baths. All units have color TV and maid service. Babysitters are available. One of the Point West's attractions is its restaurant. The handsome dining room, dominated by a massive stone fireplace, has a wide-open view of Lake Macatawa. The menu includes both steakhouse and continental fare–prime rib and chicken Kiev are specialties. Everything, from breads and pastries to soups and salad dressings, is made on the premises. The cocktail lounge has dancing and live entertainment.

Perched as it is between mild-mannered Lake Macatawa and the vast and often unruly Lake Michigan, Point West is especially appealing to fishermen, sailors, and other watersports enthusiasts. The resort has a private beach, and several marinas are close by. Five tennis courts are on the property, and guests are automatically enrolled as members of the Holland Tennis and Racquetball Club, which has eight indoor tennis and four racquetball courts, sauna, whirlpools, and exercise facilities. On weekends in winter, Point West provides cross-country skiing instruction. ☐ **Mod to Lux** ☐ **T, W, F**

MACKINAC ISLAND: Because of its strategic position, this island played an important part in the Northwest fur trade. When the fur trade declined, it began to develop as a resort. Wealthy Southern planters brought their families here for vacation before the Civil War; industrialists and socialites discovered it in the Gilded Era. Today, Mackinac is known as the "Bermuda of the North." The natural beauty that first inspired the Indians to revere this island has been largely preserved in the 2000-acre state park, which has a visitor center. Mackinac is rich in reminders of its exciting history, including Fort Mackinac, built by the British in 1780, one of John Jacob Astor's furtrading posts, and other historic buildings. Contributing to the special charm is the absence of automobiles. You leave your car in Mackinac City or St. Ignace and take the ferry. On the island, you can get about by horse and buggy, bicycle, horseback, or afoot. On the island almost every kind of summer recreation is available, including golf, tennis, and fishing. Special events include the lilac festival in early June and the annual Chicago and Port Huron Yacht races in July.

The Chippewa Hotel, *Mackinac Island, MI 49757, (906) 847-3341* ☐ *Open mid-May to late Sept* ☐ This older hotel is a budgeter's choice, offering single and double rooms EP or MAP. A long, white four-story inn, it looks out over Lake Huron and the Mackinac Island Yacht Club on one side and a quiet horse-and-buggy boulevard on the other. The swimming pool and Cabana Club deck are on a terrace just above the water. The Emerald Dining Room and an airy and informal cafe share the lake view, and there is a cocktail lounge. The hotel will help you arrange babysitters, riding, and golf, and rents bicycles. ☐ **Mod**

Grand Hotel, *Mackinac Island, MI 49757, (906) 847-3331, May to Oct; 222 Wisconsin Ave., Lake Forest, IL 60045 (312) 234-6540, Oct to Apr* ☐ *Open mid-May through late Oct* ☐ Grand is a good name for this venerable 500-acre resort. The gracious, white Greek Revival structure, with an impressive, three-story colonnaded porch, sits high on a bluff overlooking the straits of Mackinac. The building is so much a symbol of the elegance and opulence of another era that Hollywood has more than once used the Grand as a backdrop—Esther Williams performed a memorable water ballet in the hotel's serpentine pool. The Grand first rolled out its red carpet in 1887, with a guest list that read like a social register of the Midwest. Since then many presidents and celebrities have signed the guest book.

Today the Grand claims to be the largest summer hotel in the world. You are met by a carriage driven by a red-coated coachman in beaver top hat and pulled by plumed horses. Flowers bank the main entrance. The 800-foot-long-front porch, lined with 5000 geraniums, looks across lawns with lounging chairs and the flower-bordered pool to the beach. The 300-foot-long main dining room overlooks the water and serves the three meals included in the AP rates. There is a special breakfast room for late sleepers who miss the regular breakfast. Outdoor buffets and picnic baskets are available, too. The sumptuous Main Lobby, with comfortable conversational arrangements, is the setting for the traditional (and complimentary) afternoon tea in silver service and the after-dinner demitasse concert hour. In addition to other lounges and drinking nooks, there is nightly dancing in the Terrace Room. Men must, of course, wear jacket and tie, and women may not wear slacks in public areas after six. In lieu of tipping, 18% service charge is levied.

As in all older hotels, rooms and suites vary. Yours may have high ceilings, huge windows, lovely views, or especially appealing decor, but in any case, it will be comfortable.

Apart from beach and pool swimming, the Grand has a nine-hole golf course with pro rental shop, carts and caddies, and putting green, as well as four tennis courts with pro, rentals, and instruction. Nearby is a stable, and bridle paths weave over the grounds for 40 miles. In addition to the usual lawn games, a baseball diamond, and indoor game rooms, unique features are surrey tours and Palm Beach-style rolling chairs for trips to the village. A social director organizes movies, bridge parties, and other events. A playground and babysitters are available. ☐ **Lux** ☐ **G, T, R**

Hotel Iroquois, *Mackinac Island, MI 49757, (906) 847-3321* ☐ *Open mid-May to late Oct* ☐ This three-story Victorian inn is a good choice for those who prefer intimacy. Because it sits on a point of land in the straits of Mackinac, most rooms have fine views (all have private bath). The airy restaurant, noted especially for fresh whitefish, offers lunch and dinner indoors and outdoors; the cocktail bar also has a water view. The inn has sun deck, bicycle rental, room service, and babysitting service. ☐ **Lux**

MANCELONA: Schuss Mountain, *Mancelona, MI 49659, (616) 587-9162* ☐ *Open all year* ☐ *6 mi W of US 131 off SR 88* ☐ If Jean-Claude Killy and Walt Disney got together to create a resort in Northern Michigan, the result would probably be Schuss Mountain. With its fairy-tale, alpine-style lodge, enchanted forest–and, reportedly, its own language, "currency," and king–this is not your run-of-the-mill resort. Emphasis is on the ski season. The mountain has 13 slopes above the "Village Square," three chairlifts, a T-bar, ski school, and rentals. Also on the premises are an 18-hole golf course, heated swimming pool, and six tennis courts with a staff of pros. Supervised children's programs and babysitters are available, and a social director organizes cookouts and other activities for adults too. Schuss Mountain has entertainment in the nightclub with music by the "Schussycats." The Ivanhoff restaurant has seafood and ribs.

Rooms accommodating up to six people are in the Village Square; all have bath and color TV, and children under 16 sleep and ski free. Suites and condominium units sleep up to 12 people and have equipped kitchen; some have sun decks and fireplaces; maid service is available; children under 12 sleep and ski free. Chalets scattered in the 1500-acre Enchanted Forest have two to four bedrooms, equipped kitchen, and views; some have fireplaces and maid service is available; charges here include children. Both American and Schuss currency are accepted. ☐ **Mod** ☐ **S, G, T**

RICHLAND: Gull Lake View Golf Club, *7417 M-89, Richland, MI 49083, (616) 731-4148* ☐ *Open Apr–Nov* ☐ *½ mi E on SR 89* ☐ The name sums it up pretty nicely. The resort is adjacent to the lake from which it takes its name, but the draw is 36 holes of good golf. The older West course is 6262 yards. The front nine are wide-open with no water, while the back nine are wooded and have six water holes. The East course is 5712 yards but more demanding–greens are smaller, the fairways are lined with trees, rolling, and well trapped, and water comes into play on 11 holes. Accommodations are in modern two- and three-story villas along the fairways. These offer one or two bedrooms and bath, plus living room, kitchen, and private balconies or patios. The clubhouse has a seasonal restaurant, lounge for drinks and sandwiches, and complete pro shop. Package plans include greens fees. ☐ **Mod to Lux** ☐ **G**

THOMPSONVILLE: Crystal Mountain, *Dept S-790, Thompsonville, MI 49683, (616) 378-2911* ☐ *Open all year* ☐ *From US 31 take County 602 7 mi E* ☐ This modern lodge of stone and wood offers motel-style rooms, two- to four-room chalets units, and condominiums featuring private whirlpools–all available on a variety of package plans for golf, skiing, family holidays, or canoe river trips. Features include a dining room, cafeteria, cocktail lounges, live entertainment, lawn games, playground, outdoor pool, and special events.

The ski season begins at Thanksgiving. Vertical drop is 375 feet, and the area has snowmaking, ski shop with rentals and lessons, double chairlifts, a Poma lift, six rope tows, and children's programs. Cross-country skiing enthusiasts will find 5½ miles of groomed trails connecting to a trail system on adjacent state land. The 18-hole golf course winds through rolling meadows and forest, and carts and rentals are available. Also in summer there is canoeing, and nearby Crystal Lake has swimming, fishing, and boat rental. ☐ **Mod** ☐ **S, G**

TRAVERSE CITY: Park Place Motor Inn/Ranch Rudolph, *Traverse City, MI 49684, (616) 947-9529* ☐ *Open all year* ☐ *300 E State St., Traverse City, for Inn; Ranch 8 mi S on Garfield Ave., then follow signs 6 mi* ☐ The well-appointed 150-room Park Place Motor Inn in downtown Traverse City, complete with pool, restaurants, and the usual conveniences, has a unique feature: Guests may use the facilities of Ranch Rudolph, a secluded, 200-acre ranch 14 miles out of town and surrounded by 10,000 acres of state forest. Or, if you prefer, you can stay at the ranch and use the Park Place facilities.

The ranch, on a bend in the Boardman River, has a rustic lodge with restaurant and bar. This is flanked by one-story extensions containing 16 comfortable double rooms; these are side by side, so all rooms face the river. Between the lodge and the river are two stocked trout ponds, a large outdoor swimming pool (open seasonally), picnic tables and grills, a playground, and areas for outdoor games and archery. The ranch operates on EP (family rates available), and whether you stay here or at the Park Place motel, you pay the same separate fee for each ranch activity. In summer, you can rent a canoe or take a guided canoe trip, or rent horses for trail rides, instruction, or pack trips. Other activities include backpacking, barbecues, and hayrides, and the ranch will arrange for golf, tennis, charter fishing, hang gliding, and racquetball in the vicinity. In fall, there is hunting. Winter activities include cross-country skiing, snowshoeing, and snowmobiling (equipment, rentals, and lessons) as well as sleigh rides. Most exciting of all are dogsled excursions and "skijouring," which is cross-country skiing behind Siberian Huskies. Parents will find supervised play and babysitters available. ☐ **Mod** ☐ **F, W, R, S**

WAKEFIELD: Indianhead, *Indianhead Mountain, Wakefield MI 49968, (906) 229-5181* ☐ *Open all year* ☐ *Just off US 2, 10 mi E of Ironwood* ☐ Indianhead operates on EP, MAP, and package plans and offers standard rooms sleeping up to six, deluxe rooms sleeping up to four, and economy rooms sleeping up to four with baths across the hall. Also available are single- and multiple-unit chalets with one to four bedrooms, equipped kitchens, and linens; most have fireplaces and some have saunas. The Chamonix provides fine dining, plus less formal restaurants and cafeterias; also here are a tavern and a Gay Nineties saloon.

The skiing season opens in late November, offering pro shop, ski school, Kinderschule, NASTAR racing programs, four chairlifts, two T-bars, a Poma lift, patrol, snowmaking, and cross-country trails. Some evenings free supervised activities for children are provided, and babysitters are always available. On the grounds are an outdoor swimming pool, saunas, tennis courts with pro instruction, and a nine-hole, par-three golf course. You can rent a bicycle or go hiking. The lodge also will point you in proper directions for fishing and other activities. ☐ **Mod to Lux** ☐ **S, T, G**

WHITEHALL: Lakeside Inn, *5700 North Scenic Dr., Whitehall, MI 49461, (616) 893-8315* ☐ *Open May to mid-Oct* ☐ *1½ mi S on South Shore Dr.* ☐ With its well-groomed lawns and gardens and modest two-story profile, the Lakeside Inn has the feeling of a sprawling family home. Set on the edge of White Lake, just a few miles from Lake Michigan, this is an attractive, informal family resort run on EP. The lake is just across the lawn from the glassed-in dining room. The inn maintains a dock for sailing, fishing, and waterskiing, and swimmers can choose between the lake and the heated outdoor pool. There is a tennis court on the grounds and, next door, an 18-hole golf course—one of six in the area. Rooms feature knotty-pine paneling and contemporary furnishings. Cottages with kitchens and without maid service are also available. Specialties in the dining room include shrimp and scallops broiled with lemon butter. ☐ **Inexp** ☐ **W, G**

Minnesota

ALEXANDRIA: Radisson Arrowwood, *P.O. Box 634, Alexandria, MN 56308, (612) 762-1124* □ *Open all year* □ *Take County 82 W to County 22 and resort* □ Facing Lake Darling, the front of this lodge is a steeply angled slope broken only by the balconies set back in their private recessed nooks. The novel architecture of this modern lodge is a sign of its uniqueness. Set on 450 acres among eastern Minnesota's galaxy of lakes, this is a sophisticated resort in a very rural area. The rooms—all with color TV, climate control, telephone—are furnished in contemporary style (some, notably those in the cathedral-ceilinged royal suites, in a very high style). Most rooms have private balconies.

The resort has an 18-hole, 5074-yard golf course with clubhouse, four Laykold tennis courts, a stable of 20 horses, bicycle rental, lake and pool swimming, and a marina with canoe, fishing boat, pontoon, ski boat, sailboat, and pedal boat rentals. The lake and nearby lakes harbor pan fish and northern and walleye pike. Use of the tennis courts, indoor swimming pool, whirlpool, and sauna is free to guests. In winter there is ice fishing, snowmobiling (suits and boots may be rented), and cross-country skiing (rental equipment available). In addition to the restaurant and bars, you can have a box lunch or use the seasonal snack bar (EP, AP, MAP, and packages are available). Services include an activities director, sports instructors, and children's programs in summer. □ **Mod** □ **T, G, W, F, S**

BAUDETTE: Sportsman's Lodge, *Lake of the Woods, Baudette, MN 56623, (218) 634-1342* □ *Open all year* □ *12 mi N on SR 172* □ This barnlike modern building is located on the Rainy River where it meets Lake of the Woods. It offers 30 modern guest rooms, two cocktail lounges (one with entertainment), sauna baths, indoor pool, and game room. The large, fully carpeted main dining room serves good steaks, walleye pike, and chicken dishes, and offers salad bar with homemade bread. For family groups desiring more privacy, the lodge has ten modern housekeeping cabins that will sleep two to ten people. These come with stoves, refrigerators, bath, and linens. In summer there's swimming, and fishing for walleye and northern pike, bass, mackerel, saugers, and perch. At docks and launching ramps immediately in front of the lodge, boat and motor rentals, guides, bait, fish cleaning services, and freezer lockers are available. During hunting season, the game includes deer, ducks, and geese. Snowmobiling and ice fishing are popular winter sports. □ **Mod** □ **F, W**

This cabin in Lost River State Park in Mathias, West Virginia, has peace, quiet, and backwoods beauty. Lost River, with inexpensive lodgings, hiking, and horseback riding, is part of the state's park system that includes historic sites, mineral baths, rivers, ski areas, golf, tennis, and even resorts with gourmet dining.

Rancho Viejo sits at the southern tip of the continental United States in Brownsville, Texas, on a plot of land that once belonged to Spain. The oversized golf tee is a water tower overlooking 36 holes of golf and eight tennis courts. At the end of a day of play guests will find the boat waiting to take them on an evening cocktail cruise.

Hotel du Nord is one of the simple resorts in Wisconsin's Door County, on the peninsula which juts into Lake Michigan. All the Door County resorts in the book are open in summer for swimming, boating, and fishing at the lake. Many stay open in winter for cross-country skiing, snowmobiling, and skating.

The Doral Country Club in Miami, Florida, offers golf, golf, golf, and more golf. Four courses in all ring the elegant hotel, and, for those who want to cool off between rounds, a shuttle bus connects this with its sister hotel on the nearby beach.

This skier and his dog are off into the wintertime quiet of the Vermont countryside around the town of Woodstock, near the Woodstock Inn. At many ski resorts, especially in New England, cross-country skiing facilities are as extensive as those for alpine skiing.

The field these cowpokes are crossing happens to be in Hawaii at Princeville at Hanalei on Kauai. If you want a riding vacation you can head west, of course, but you can also go east to a New York ranch or south to another in North Carolina.

Some resorts offer a choice between the ultramodern and the old-time. At the Boca Raton Hotel and Club in Boca Raton, Florida, guests can stay in the space-age-style tower, eat in the cathedral-ceilinged dining room in the original hotel, and play on the four golf courses or 40 tennis courts at their disposal. *(Photo courtesy of National Resort Photographers)*

The spectacular Rainbow Bridge and the surrounding Arizona desert are two of the natural wonders outside the doors of resorts described here. *(Photo courtesy of U.S. National Parks Service)*

No place in the 50th state, Hawaii, is far from the winds, the surf, and the sands. Behind the palms here are the peaked roofs of Kapalua Bay Hotel on Maui. The hotel is a member of the Rockresort chain.

This stately manse, The Greenbrier in White Sulphur Springs, West Virginia, is one of the old-time resorts that blends the best of the past with the pleasant amenities of the present. Pictured here in the springtime with a yard full of flowers, the resort is open all year with horseback riding and hiking through the Blue Ridge Mountains, golf, tennis, and baths in the springs for the weary.

The Desert Inn and Country Club in Las Vegas, Nevada, offers a host of diversions from the gambling tables. Those in search of less conventional accommodations need not look too hard. One Las Vegas hotel has a circus midway and another is modeled after an ancient Roman retreat.

Outside these dining-room windows at the Country Club Inn in Rangeley, Maine, the leaves are beginning to turn to their spectacular fall colors. This is a country resort with golf, hiking, and boating, and the inn can even arrange to fly guests to backwoods lakes for private fly-fishing.

Kentucky State Parks, many of which are in the Land Between the Lakes created by Tennessee Valley Authority dams, offer some of the best family-vacation bargains in the country. At Barkley Lodge, above, near Cadiz, every room has a lake view.

The red-jacketed coachman in the beaver hat is from the Grand Hotel on Mackinac Island, Michigan, an island where motorized vehicles are forbidden. Guests and residents must travel by carriage, wagon, bicycle, or foot to protect the island quiet.

This lone skier is floating down through the powder snow on "Drainpipe" at Copper Mountain in the Colorado Rocky Mountains, which, along with their western neighbors in Utah, have some of the best skiing in the world.

For anyone interested in wildlife and majestic scenery Alaska's wilderness lodges and camps are hard to beat. Here, the view across Nugget Pond from Camp Denali reaches up 20,320 feet to the top of Mt. McKinley, the highest peak in North America.

BEMIDJI: Birchmont Ruttger, *Bemidji, MN, 56601, (218) 751-1630* ☐ *Open June to late Sept* ☐ *7 mi N on Old Hwy 71* ☐ This well-established lakeside resort has been under the same management for almost half a century. In addition to rooms and suites in the main lodge, Ruttgers' has 28 one- to four-bedroom cottages, most of them with fireplaces, knotty-pine paneling, and screened porches; six have full kitchens. Among the cottages, the six villas are the deluxe units. These large, well-furnished cabins at the east end of the resort are set in a semicircle overlooking Lake Bemidji and their own swimming beach. All units, both cottages and lodge rooms, are fully carpeted and have daily maid service.

Ruttgers' has more than ¼-mile of sandy beach on the lake and plenty of rowboats, canoes, pedal boats, sailboats, and ski boats. On the grounds are a heated pool, a kiddie pool, playground, recreation room, and four Laykold tennis courts. Less than a mile away is the Bemidji Town and Country Club, with a fine 18-hole golf course. Castle Highlands, another 18-hole course, is seven miles north. Golf packages with both courses are available. A full children's program is scheduled in summer.

The attractive lakeside dining room has a menu featuring homemade soups, pastries, and garden vegetables. One favorite dish here is Minnesota walleye pike, caught in Lake Bemidji or nearby. ☐ **Mod** ☐ **W, F, T, G**

Kohl's Last Resort, *Box 247, Bemidji, MN 56601, (218) 586-2251* ☐ *Open all year* ☐ *Take County 15 N to County 22 E, follow to resort* ☐ Kohl's 18 cabins come in a variety of styles, from one bedroom to four-bedroom duplexes. All have kitchen and private bath, and many have screened porches, handsome knotty-pine paneling, and fireplace. Linens, maid service, and use of an aluminum boat come with each cabin. Kohl's takes great pride in its food. Tuesdays feature steak broils; Thursdays charcoal and hickory barbecued ribs and lemon pie; and Sundays a big brunch with baked ham, bacon, scrambled eggs, lumberjack pancakes, fruits, juices, and more.

Big Turtle Lake is five miles long and two miles wide, plenty of room to do some secluded fishing for walleye and bass. Most of the eight lakes in the Turtle River chain are accessible without portage. The resort has a sandy beach and all types of boat rentals. A nearby ranch provides horseback riding. In fall the resort is popular with hunters and in winter with snowmobilers and cross-country skiers. Hundreds of miles of marked snowmobiling and cross-country trails surround the resort, and Kohl's has a number of snowmobiles to rent. AP is available fall, winter, and spring, and there are honeymoon and ski packages. ☐ **Mod** ☐ **F, W, S**

BRAINERD: Grand View Lodge, *Rte. 6B, Box 22, Brainerd, MN 56401, (218) 963-2234* ☐ *Open May–Oct* ☐ *Take SR 371 14 mi N, then County 77 1 mi E* ☐ One thing that sets this resort apart is its commitment to tennis. The facilities include seven Laykold courts, one cement court, two hitting walls, ball machine, and pro instruction—all free to guests except for the ball machine and instruction. Right next to the courts is a sporty par-three golf course. The golf-tennis pro shop has rental equipment, carts, and sports clothes.

The lodge is a handsome log building with single and double guest rooms with large windows, carpeting, and climate control; some have porches. Most guests, however, are housed in the 50 cottages set among the pines, each with picture-window views of the lake. The cottages have from one to six bedrooms accommodating two to fourteen people. All have carpeting, refrigerators, and heat. Some have kitchenettes, air conditioning, fireplaces, or sun decks. All units have maid service and most are rented on MAP, although some seasonal EP accommodations are available here and at the lodge a mile away. A specialty in the main dining room is beer-batter walleye pike; Sunday smorgasbord and outdoor barbecues are held seasonally. In addition to the indoor bar (with dancing and entertainment) you may have cocktails on redwood decks under the pines in balmy weather.

At the wide sandy beach, you may go waterskiing or rent sailboats, canoes, water bikes, speedboats with driver, fishing boats, or Kayats (popular little red kayaks). Fishing for walleye, northern, bass, and sunfish is good because of stocking, and guides are available. Also here are a heated, semi-enclosed pool, sauna, and volleyball area. Throughout the summer, a children's program includes sports, nature hikes, cookouts, movies, and boat rides, and babysitters and supervised meals are available. The adult social program includes dances, bingo, movies, and card parties. ☐ **Mod to Lux** ☐ **T, W, F**

Madden's in Minnesota, *Box 387, Gull Lake, Brainerd, MN 56401, (218) 829-2811* ☐ *Open all year* ☐ *From Brainerd, N on SR 371 for 4 mi, W on County 77 for 8 mi* ☐ Occupying about 750 acres on a spit of land on Gull Lake, Madden's calls itself *four* resorts even though most facilities are shared. It offers a wide range of recreation and almost 300 air-conditioned rooms and cottages. Most units have lake views, and many have small refrigerators, living rooms, and balconies. The four distinct resorts are: *Madden Lodge,* a traditional inn building, with a lounge with a stone fireplace, a glass-walled dining room overlooking the lake, and an outdoor deck for beverage sipping. In addition to bedrooms and suites in the main lodge, one-, two-, and four-bedroom cottages and apartments are strung out along the lakefront. *Madden Inn,* a beautiful old hunting lodge connected to the low-slung Golf Club, offers modern bedrooms and suites (some with balcony) in the inn; one- and two-bedroom cottages with wetbar and patio, overlooking the golf course; and luxury Sunrise Villas, chalet-style units with wetbar, color TV, lakeside balcony, and perhaps an extra such as a fireplace or loft. The Prime Rib Room and the Sirloin Room restaurants are here, as well as a coffee shop and a lounge on the 19th hole. *The Voyageur* consists of motel-like studios and apartments overlooking the lake; some units have kitchenettes, and a grocery is nearby. This resort does not accept children under six. The fourth resort, *Pine*

Portage, specializes in groups; it has a main lodge with dining room overlooking the beach and houses guests in log cabins nestled among the pines.

The Golf Club has 45 holes made up of two 18-hole championship courses and a par-3 executive course; equipment and carts may be rented. Six of the eight tennis courts are Laykold, and a clubhouse provides rentals. Also here are three outdoor and three indoor pools, saunas and whirlpools, bicycle rentals, game rooms, outdoor game areas, playgrounds, trapshooting, and a shopping center. In the center of the complex is Lumbertown, U.S.A., a reconstructed 19th-century logging village. The Lumbertown Dining Room, popular with families, serves all-you-can-eat chicken or walleyed pike dinners.

Three sandy beaches and marinas are accessible to all four resorts, and Madden's has a fleet of rowboats, pontoon boats, canoes, kayaks, paddleboats, sailboats, and runabouts for waterskiing. Bait and guides are available for fishermen. Walleye pike, northerns, and bluegills are caught. Parents on MAP at Madden Lodge will find a free supervised children's program in summer; the service is available to guests of the other resorts for a fee. Adult activities include dances, movies, and cookouts.

Madden's is proud of its food and features native products including Minnesota butter, wild rice, chokeberries, pike, and local vegetables.

The rate schedule here is complex. Depending on the type of accommodations and type of plan (EP, AP, MAP, and special plans are offered), certain activities are free and others may involve a charge. Thus an avid golfer might do best in a plan offering unlimited golf, food enthusiasts might prefer the choice of five restaurants open to EP guests, parents might want to consider the children's program at Madden Lodge, etc.–so a careful study of the many options is advised. Beaches, pools, rowboats, shuffleboard, and maid service are included in all plans at all resorts. ☐ **Inexp to Mod** ☐ **T, G, W, F**

GRAND RAPIDS: Quadna, *Hill City, MN 55748, (218) 697-2303* ☐ *Open all year* ☐ *1 mi S on US 169* ☐ A good family place on 1400 acres, Quadna has a lodge, with plenty of roughhewn wood inside and out, containing the lakeview dining room, cocktail lounge, indoor pool, sauna, and recreation room. Flanking this core are two-story lodging wings offering 48 rooms overlooking five-mile-long Hill Lake. Another 48 rooms are in the nearby and cheaper Quadna motel. And 130 luxurious town houses have studio to four-bedroom units. All units have modern amenities such as color TV. Accommodations are rented EP or under a variety of family and sports package plans.

Quadna's sports facilities are shared by all units. Included in the rates are use of two outdoor and one indoor pool, sauna, lake beach, hiking trails, playground, and areas for games such as bocci, croquet, and badminton. Other sports, unless included in a package, involve extra but moderate fees. On the grounds are a nine-hole golf course with rentals, mini-golf, four indoor and two outdoor tennis courts, and bicycles, pedal boat, canoe, and pontoon boat rentals. Anglers can also rent fishing boats to go after Hill Lake's walleyes, northerns, sunnies, and crappies. Just down the road is Quadna ski area, with rentals, instruction, and snowmaking. The rope tow and T-bars give skiers access to 14 trails of varying difficulty. An apres-ski chalet is at the base of the mountain. A shuttle-bus ferries skiers to the lodging area at Quanda. □ **Mod to Lux** □ **T, S, W, F**

Sugar Hills Resort, *Box 369, Grand Rapids, MN 55744, (218) 326-3473* □ *Open Nov–Apr and May–Sept* □ *Off US 169, 12 mi S of Grand Rapids* □ Sugar Hills has four places to stay: At the lodge or cabins on the lake, at the inn at the base of the ski slopes, or at one of the town houses set in the woods behind the lodge. The rustically furnished cabins are the most basic, and the air-conditioned one- and two-bedroom town houses (the only units with kitchens) are the most luxurious. Most of the other rooms have knotty-pine paneling and contemporary or rustic furniture. The lakeside Sugar Lodge restaurant specializes in trout and barbecued ribs, and there is a bar. Units may be rented MAP, EP, or on family, skiing, or other plans.

The lake has the walleye, crappie, and panfish common to Minnesota lakes. Fishermen may rent boats and motors, and for recreational sailors there are sailboats, pedal boats, canoes, pontoons, and waterskiing runabouts. Sailing lessons–from one-hour introductions to three-day cruises through the Apostle Islands–are also offered. Artisans demonstrating crafts encourage participation. The resort has two tennis courts, a nine-hole golf course, putting greens, an indoor pool, sauna, a children's program, and special events such as horse-drawn hay wagons. The resort also has its own airfield. Come winter, Sugar Hills opens up its 23-trail, 9-lift (2 chairs, 3 T-bars, 4 rope tows) ski area. Ski rentals, night skiing, instruction, and Kinderski school are available. The resort also has more than 80 km of cross-country ski trails. □ **Mod to Lux** □ **W, F, S**

INTERNATIONAL FALLS: Northernaire Floating Lodges, *Box 510, Island View Rte., International Falls, MN 56649, (218) 286-5221* □ *Open mid-May to mid-Oct* □ *At jct US 71 and US 53, take SR E to County Rd. 94. Follow 1 mi to lodge* □ All the rooms here have water views–Northernaire rents completely furnished houseboats by the week. They come in six models, from the 36-foot Wanderer that sleeps two to the 64-foot Executive that sleeps eight and has two baths, sauna, hi-fi system, and ice cube maker. Guests are given a free dockside orientation, charts, and a set of tools. With that and a week's supply of food they're ready to explore enormous Rainy Lake, winding across more than 50 miles of the United States and Canada. Trip planning assistance, guides, and other services are provided. Rainy Lake is famous for walleye and smallmouth bass. Northerns are common all over the lake, and crappies average two pounds. Northernaire rents small boats and motors that can be carried on the houseboats. □ **Lux** □ **F, W**

Thunderbird Lodge, *Box 407, Island View Rte., International Falls, MN 56649, (218) 286-3151* ☐ *Open Dec through Oct* ☐ *10 mi E on SR 11* ☐ Fishermen and boating enthusiasts will appreciate the extensive dockage and marina services of this lodge, set on the shores of Rainy Lake. Besides fishing, waterskiing, sailing, and canoeing rentals, guides can be hired to direct you to Rainy Lake's many muskie, walleye, smallmouth, and crappie. Fly-in fishing trips to remote areas can be arranged. You can swim in the lake or heated pool, or go berrying. Hunting is in October and November. In winter, cross-country skiing, snowmobiling, skating, and ice fishing are the big sports. Thunderbird has modern, air-conditioned, and heated lodge rooms (with color TV and direct-dial telephone) and pine-nestled cabins (with cook-top stove, refrigerator, and sink–no dishes or utensils provided). Some units have private docks, and most have beautiful lake views. The lodge has a lounge and bar with a fireplace and a lakeside dining room serving steakhouse fare. ☐ **Inexp to Mod** ☐ **F, W**

LAKE ITASCA: Itasca State Park, *Lake Itasca, MN 56460, (218) 266-3654* ☐ *Open late May to early Oct* ☐ *Near jct US 71 and SR 200, 28 mi N of Park Rapids* ☐ Itasca Lake is where the Mississippi begins. It is such a small stream at this point it can easily be crossed by the stepping stones provided just for that purpose. The 30,000-acre forested park surrounding the lake offers Douglas Lodge with a restaurant, a snack bar, and all types of accommodations (on EP), one- and two-bedroom cabins, lodge rooms and suites, motel units, even "clubhouses"–two-story, 10-bedroom log cabins. None of these units is equipped for housekeeping, but all come with linens and maid service; some have fireplaces or screened porches. In addition, six rustic housekeeping cabins at the campground provide sleeping room for four people, electric heat, and kitchenettes with cooking and eating utensils.

The headquarters area, about three miles from the lodge and one mile from the campground, has boat rentals and swimming beach. Naturalists stationed in the park during tourist season present free programs on natural history and conduct guided hikes. The park has a 17-mile scenic drive, 28 miles of foot trails, 12 miles of snowmobile trails, and 21 miles of cross-country trails. Also here are Indian mounds plus the University of Minnesota arboretum and exhibits of native flora and fauna. ☐ **Inexp** ☐ **F, W, S**

PARK RAPIDS: Brookside Resort, *Itasca Star Rte., Park Rapids, MN 56470, (218) 732-4093* ☐ *Memorial Day weekend to Labor Day weekend* ☐ *Take US 71 11 mi N, County 41 2 mi W, and County 50 1½ mi S* ☐ This lakeside resort for gregarious families has the usual boating, fishing, and swimming plus guided horseback riding, a nine-hole golf course with rentals, and two tennis courts. Guests are housed in two- to four-bedroom units, ranging from mobile homes on a brook to log cabins and A-frame cottages along the lake. Most cabins have knotty-pine paneling and carpeting or braided rugs. Each has a kitchenette equipped with cooking and eating utensils; linens may be rented if you do not bring your own. A 14-foot aluminum boat comes with each cabin, and a nearby dock is shared with one other cabin. The recreation lodge has a small indoor-outdoor pool with sun deck, sauna, snack bar (there is no restaurant) and game tables. A supervised morning program is conducted for children aged two through eight in a fenced-in play area; scavenger hunts, free pony rides, or other activities are scheduled weekly, and babysitters are available. Besides fishing boats, the Brookside furnishes canoes, sailboats, and waterbikes free of charge. You may also rent a large pontoon or ski boat. Two Inlets Lake has walleyes, northern, crappie, bass, perch, sunfish, and bullhead. ☐ **Inexp** ☐ **R, W, F**

Val Chatel, *Park Rapids, MN 56470, (218) 266-3306* ☐ *Open all year* ☐ *On Lake George Rd. (County 4), 16 mi NE of Park Rapids* ☐ Imaginatively and tastefully decorated to suggest a European alpine inn, this motor lodge on 450 acres offers a lake and a modest ski area. Three rope tows and one triple chairlift run up the 270-foot-high hill behind the lodge. Some of the runs are lighted and have snowmaking; there is a ski school, patrol, and cafeteria. An additional eight kilometers of cross-country trails thread through the pines. In summer, activity centers on the lake, where you swim or fish for northern pike, largemouth and smallmouth bass, sunfish, or crappie. Fishing equipment is available, and you may engage a guide for Rainy Lake trips. Also here are sailing dinghies, canoes, and water bikes.

In the spacious two-level dining room with a circular balcony, steaks and barbecue ribs are featured, with Sunday smorgasbord a highlight in summer. Adjacent is a bar. Val Chatel, operated on EP, AP, and MAP, and offers eight lodge rooms and five cabins. Decor is enhanced by the attention to detail characteristic of owner-operated establishments, and your accommodations may have elaborate woodwork, handsome patterned wallpaper, or an arched doorway. All units have wall-to-wall carpeting, modern baths, and color TV; cabins have private patios and efficiency cooking units (no cooking utensils supplied; disposable dish packs may be bought). ☐ **Mod** ☐ **S, F, W**

SUPERIOR NATIONAL FOREST: This five-million acre wilderness of thousands of lakes, streams, muskeg, and conifer forests abounding in wildlife is the historic Voyageur country. The northern third of the forest has been designated for preservation purposes as the Boundary Waters Canoe Area. On the Canadian side, this wilderness is buffered from civilization by Quetico Provincial Park. Packing into the BWCA requires a Forest Service permit issued on a quota basis at ranger stations at Cook, Ely, Grand Marais, and other outfitting centers, but similar landscape can be freely enjoyed at resorts in the remaining vast expanses of Superior National Forest. East of the forest, on Lake Superior, is Grand Portage National Monument. At Ely is the Voyageur Visitor

Center, where the Forest Service provides exhibits and interpretive history and natural history programs in season. The forest has several scenic auto routes such as the Gunflint Trail, as well as interpretive and nature walking trails.

Burntside Lodge, *Box 4215, Ely, MN 55731, (218) 365-3894* □ *Open mid-May–Sept* □ *Off County 88, 6 mi W of Ely* □ This family resort offers 25 log cabins located on or near the shore. Some of the summer cabins are on islands just offshore. Most have kitchenettes, knotty-pine paneling, and pleasant, solid furnishings. The largest cabin has three rooms and handmade tree furniture. The EP rates include maid service and use of the tennis court. MAP is popular because it includes waterskiing and use of a boat or canoe. The restaurant specializes in walleye pike and lake trout, of course, but also offers steaks and salad bar. The picturesque, 7000-acre lake has more than 150 islands and plenty walleye, smallmouth bass, lake trout, northern, and crappie. □ **Mod** □ **F, W**

Gunflint Lodge: *Grand Marais, MN 55604, (218) 338-2294* □ *Open mid-May to mid-Oct* □ *43 mi NW of Grand Marais, on the Gunflint Trail* □ If you take out a fishing boat here you'll probably end up going to another country: Gunflint Lake, only about a mile wide, is invisibly divided by the United States–Canadian border. Right from the opening of the season (mid-May) the lake is good for walleye and lake trout, and smallmouth bass start hitting in the second week of June. Guided canoe trips into the BWCA are outfitted by the lodge. Fly-ins can also be arranged.

Most of the lodge's 14 handsome modern cabins overlook the lake and are sited for privacy. They come in a variety of styles—A-frame, log, and so on—and sleep 2 to 14 people. Most have electric heat and electric blankets; some have sun decks and kitchenettes (AP is usual but can be modified). Maid service and linens are provided. The lodge's big lounge has knotty-pine paneling, polished-wood floors, exposed beams, large stone fireplace, a library, and picture-window lake views. Breakfasts are hearty; lunch may be indoors or out, or you may have a box lunch and thermos. The Sunday smorgasbord with sauerbraten and fruit breads is a highlight in summer, as are the frequent outdoor barbecues. Beer and setups are available but guests bring their own wine or liquor. An 8% service charge replaces tipping.

For an outdoors family, this is a fine choice. Complete services for fishing include boat rental, licenses, gas, and bait. Canoes, waterskiing boats, a sand swimming beach, playground, outdoor game areas, tennis, archery, and a sauna are also available. In winter the lodge opens up 80 miles of groomed trails to cross-country skiers. Gunflint Lodge, run by the Kerfoot family for more than half a century, is noted for its naturalist programs. These 30-odd activities offered six days a week include evening animal watches, bird and wildflower identification, overnight canoe or hiking trips, berrying, and wilderness survival skills. And one naturalist specializes in similar programs geared to children. □ **Mod** □ **F, S, W**

Ludlow's Island, *Cook, MN 55723, (218) 666-5407* □ *Open May to Oct* □ *Take County 24 3 mi NE to County 540 E to resort* □ Most of this resort is on an island–not a large island, but enough of a separate world to be something special. Nine cottages, sleeping two to eight or nine people, and the lodge are set around the perimeter of the four-acre island, with plenty of space and trees between. All cabins have knotty-pine or cedar interiors, carpeting throughout, and tasteful furnishings. Kitchen equipment and linens are furnished. There is no restaurant. Some units have fireplaces. On the mainland are three similar cabins, parking area, and grocery store. One very small island nearby has a campsite for those who want an overnight stay on their own island. The resort furnishes canoes, rowboats, sailboats, some fishing and boating equipment, and transportation to and from the mainland without charge. Fishing guides are available. Motors, waterskiing boats, and pontoon boats may be rented. The island has a sandy swimming beach, a sauna, outdoor game areas, an exercise-hiking trail, and a nature trail. Golf and horseback riding are available nearby. Vermillion Lake is actually a chain of lakes 25 miles long, with plenty of walleye, northern, bass, crappies, and pan fish. □ **Mod to Lux** □ **W, F**

Lutsen Resort, *Lutsen, MN 55612, (218) 663-7212* □ *Open all year* □ *Off US 61* □ On 3000 acres of Lake Superior shoreline in a less remote area of Superior National Forest, Lutsen's offers lakeside vacations in summer and ski vacations in winter. The big, three-story main building looks like many a hunting lodge from the 1930s and is furnished with rustic Scandinavian pieces. It has a large lobby and dining room with fireplace, bar and lounge, game room, heated pool, and sauna. The front rooms here have a view of Lake Superior; other rooms have a view of the central courtyard. Rates are MAP. Just off the resort's nine-hole golf course, on a hill overlooking the lake and main lodge, is Cliff House (rooms here, and the villas and apartments following, are on EP). More luxurious accommodations are at the Sea Villas, about three miles south of the main lodge. These one- and two-bedroom villas have furnished kitchens, sliding glass doors, decks, fireplaces, and sleep as many as 7 to 12 people. Finally, the All Season apartments, just south of the villas, have furnished efficiency units with complete kitchens, fireplaces, and large picture windows overlooking the lake. These sleep between two and ten people.

In summer, the resort offers swimming, golf, tennis, playground, trout fishing in stocked streams, hiking, and a 2000-foot alpine slide. In winter, the three chairlifts, Poma, and T-bar start up at the Lutsen ski area, two miles from the main lodge. It has 19 trails on three slopes and a 630-foot vertical drop, the largest in Minnesota. The chalet has cafeteria service, ski rental shop, and ski school. □ **Mod** □ **S, T, G, W, F**

Nelson's Resort, *Crane Lake, MN 55725, (218) 993-2295* □ *Open mid-May–Sept* □ *25 mi NE of Orr on SR 23-24* □ These 27 red-roofed cabins are set beneath the pines on a sheltered cove of Crane Lake. Built of spruce, pine, and cedar logs with stone foundations and fireplaces, they are rustic on the outside but quite up-to-date within. All have private bath; some have fireplace, kitchenette, or porch. Cabins are rented on AP or MAP. The restaurant's smorgasbord and standing rib are very popular with local residents. Canoes, fishing boats, waterskiing boats, guides, and all other marina rentals and services can be had at the resort's ample dock area. Crane Lake, at the western end of Superior National Forest, gives water access, without portage, as far north as Rainy Lake. ☐ **Mod** ☐ **F, W**

Nor'Wester Lodge, *Gunflint Trail, Grand Marais, MN 55604, (218) 388-2252* ☐ *Open May–March* ☐ *On Gunflint Trail, 30 mi NW of Grand Marais* ☐ Constructed in the 1930s of peeled pine and spruce logs, the older cottages and the lodge, with its hunting trophies, fireplace, and rustic furniture, have the look of classic wilderness cabins—sort of like the set for Sergeant Preston of the Mounties. Of more recent construction are the two-level lakeshore villas, frame units with four bedrooms, two baths, wall-to-wall carpeting, and sliding glass doors overlooking the water. The resort is on Poplar Lake, a fine spot for walleye, bass, northern, or lake trout fishing. Six canoe trails start at the lake, and the lodge is just as ready to outfit a complete canoe trip as it is to rent a rowboat for an afternoon's paddle. All cabins are equipped for light housekeeping, but there is a dining room that serves home-style cooking. ☐ **Inexp** ☐ **W, F**

Mississippi

BAY ST. LOUIS: *Fairway Villas, Diamondhead Yacht and Country Club, P.O. Box 22, Bay St. Louis, MS 39520, (601) 255-1421* ☐ *On I-10* ☐ The rental units at Diamondhead resort community are in handsome attached villa units with wood siding, balconies or porches, and sliding glass doors. Accommodations range from studios to multi-bedroom suites, all with air conditioning and some with kitchens, overlooking the fairways. The country club–a structure inspired by the soaring rooflines of Polynesian architecture–is a short walk away. It has two elegant dining rooms, two cocktail lounges, and pro shop. The tennis center has eight championship clay courts, pro shop, and pro staff. For riding enthusiasts, the stable provides riding ring, tack room, escorted rides, and five miles of bridle paths. The fisherman will find a full-service marina with bait and tackle shop and rentals. The Yacht Club dining room has a wide menu of seafood, steaks, and other fare, and overlooks the docks. Ten swimming pools and recreation centers are scattered on the lovely 6400 acres. EP, MAP, and package plans are available. ☐ **Lux** ☐ **T, G, F, W**

BILOXI: The Broadwater Beach, *Biloxi, MS, (601) 388-2211* ☐ *Open all year* ☐ *On US 90* ☐ An outstanding facility at this resort is the amazing marina. The man-made harbor is almost a perfect square, lined on three sides by covered berths. Each berth is equipped with fresh water, electricity, telephone, and TV hookup. Besides the usual fuel, service, and fishing facilities, the marina has a restaurant with appropriately nautical atmosphere. Free transportation to nearby shopping areas is provided. Also quite extensive are the golf and tennis facilities. All told, the resort has 45 holes of golf, two 18-hole courses and a lighted nine-hole course. Of the championship courses, the Sea course has thick woods and narrow fairways and the Sun course, a bit tougher, has hazards on 14 holes. The seven Rubico tennis courts, complemented by a pro shop, are set behind the main building near the lodging wings, cottages, and one of the two pools.

Most of the lanai rooms, many facing the garden courtyard, have wall-to-wall carpeting, beamed ceilings, paneling or exposed brick walls, and separate dressing rooms. Several styles of cottages surround the courtyard. The executive cottages are the largest, with two bedrooms, a living room with kitchenette area, and three baths. Each of the three main rooms may be rented individually as a bedroom or in combination with other rooms in the cottage. The single cottages have bedroom, living room, kitchenette, and bath, while the garden cottages have bedroom, dinette-kitchen, dressing room, and bath. In the main hotel building are several dining areas–from the elegant Royal Terrace to the informal Patio coffee shop–cocktail lounges, shops, and lobby. ☐ **Lux ☐ T, G, F, W**

Missouri

GRAVOIS MILLS: Millstone Lodge, *Box 515, Rte. 1, Gravois Mills, MO 65037, (314) 372-5111* □ *Open Mar–Nov* □ *At the Lake of the Ozarks, 25 mi N of Camdenton on SR 5 then 6 mi E on State Road "O"* □ Elevated on the grassy knoll of one of the lake's many wooded promontories is the unpretentious and pleasant Millstone Lodge. With 48 double rooms, 7 singles, and 10 suites (up to 3-bedroom units; some with kitchenettes), it's a family-style place that operates at a relaxed pace. The informal restaurant overlooking the lake specializes in trout and barbecue ribs, and the cocktail lounge has entertainment. The resort operates on EP. The lake is famous for its hungry bass and crappie, and the full-service marina offers everything from canoes to waterskiing. Also on the 30-acre grounds are two asphalt tennis courts, an outdoor pool, a playground, a recreation room, and a general store. The activities director can arrange for babysitters. □ **Mod** □ **W, F, T**

KIMBERLING CITY: Bittersweet Resort, *Kimberling City, MO 65686, (417) 739-4492* □ *Open all year* □ *From Kimberling City, 5 mi from SR 13 on Lake Road 13-35* □ Scattered under old trees on six acres on the shore of Table Rock Lake are a dozen housekeeping cottages. Each cottage is different–some are frame, some are brick and timber, some have fireplaces–but all have cathedral ceilings with exposed beams, paneled walls, TV, climate control, kitchenette, cooking equipment, tableware, and linens. Outside each cottage are fireplaces, grills, and picnic tables. On the grounds are a large shaded pool, a playground, shuffleboard, horseshoes, and volleyball and basketball courts. Your host is an avid bass fisherman who will rent you boats and motors or arrange for a guide. This is a homey establishment where coffee and donuts are served at a get-acquainted session on Monday morning and where comedy films are shown by the pool twice weekly. □ **Inexp** □ **F, W**

LAKE OZARK: Lodge of The Four Seasons, *Lake Road HH, Lake Ozark, MO 65049, (314) 365-2381* □ *Open all year* □ *1 mi S of Bagnell Dam, 3 mi W of US 54* □ Perched on a rise just a few yards from the lush shoreline of Lake Ozark are the imposing structures of the Four Seasons Lodge and resort. The luxurious lodge is a sprawling complex of various architectural styles and motifs surrounded by multi-leveled terraces, patios, manicured lawns, and a curious collection of Japanese-style garden sculpture. Many of the lodge's 288 double rooms look out upon the large freeform outdoor pool and sun deck. All rooms and each of the 12 plush suites are fully air-conditioned and have TV. Some sport fireplaces or private balconies or patios.

Aside from the gift shops, boutiques, and beauty salon that lie right off of the main lobby, there are several establishments for dining and cocktails; these range from the elegant Toledo Room to a snack bar to the Fifth Season nightclub that offers live entertainment. Catch a movie at the Four Seasons cinema and then bowl a few lanes at their in-house alley. If that isn't enough, there is a sumptuous spa area, apart from the cloverleaf outdoor pool, for indoor swimming, saunaing, and whirlpooling. If outdoor sports are more your bag, tee off at the nearby Four Seasons Country Club, where Robert Trent Jones has created one of mid-America's finer 18-hole championship courses. Or, for a day on the lake, sail, ski, or take a sightseeing trip on the *Season's Queen* party Boat at the lodge's marina. Or you might prefer horseback riding, trapshooting, bicycling, or archery. Special for parents are two supervised youth programs–for ages 4 to 7 and 8 to 14; these operate in summer 9 am to 5 pm, and the daily fee includes lunch and all activities. The Four Seasons operates on EP, AP, MAP, and package plans. ☐ **Lux** ☐ **W, G, T, R**

LESTERVILLE: Wilderness Lodge, *Box 87, Lesterville, MO 63654, (314) 637-2295* ☐ *Open all year* ☐ *1½ mi E on SR 21, then 2 mi S on turnoff* ☐ These handsome cottages are set on the banks of the Black River. Many of the one-, two-, and three-bedroom units have a refrigerator, a stone fireplace, a rustic interior, and a balcony overlooking the river. Recreation includes swimming, fishing, canoeing, float trips, trail rides, tennis, platform tennis, square dances, hiking, fishing, hayrides, and lawn games. A social director is on hand to schedule evening entertainment, and an outdoor pavilion is available for dances and dinners. The lodge runs on MAP spring through fall. The dining room, open to the public with advance notice, serves old-fashioned, family-style meals. ☐ **Inexp to Mod** ☐ **R, F, W**

OSAGE BEACH: Antlers Resort, *P.O. Box 118, Osage Beach, MO 65065, (314) 348-2278* ☐ *Open all year* ☐ *From Lake Ozark, 1 mi N off US 54-52* ☐ The emphasis at this small five-acre fishing resort is on simplicity. The nine two-story frame cottages, rented on EP, accommodate up to four persons with adequate comfort. Seven more modern apartments have two bedrooms and a sleeper couch in the living room. All units have air conditioning, heat, TV, and kitchen facilities. What the Antlers Resort may lack in glamor, it makes up for in closeness to the pleasures of Lake Ozark. If bass isn't your bag, you can rent pontoon boats, paddleboats, and motorboats. Near the heated pool are a snack bar, recreation room, and playground. ☐ **Mod** ☐ **F, W**

Malibu Beach Resort, *P.O. Osage Beach, MO 65065, (314) 448-2233* ☐ *Open all year* ☐ *US 54 to Grand Glaize Bridge, then SE on Lake Road 54-49* *for ½ mi* □ This water-surrounded resort, occupying eight wooded acres on a peninsula in Lake of the Ozarks, has a slogan, "It's a real resort for real people." For, with 15 charming cottages, nicely spaced and secluded by trees, the accent is on simple and quiet comfort. Each cottage is different, but sited for a good lake view. Each has special features–perhaps cathedral-beamed ceiling, fireplace, or lakeside terrace. The cottages range from cozy nests for two to spacious family units sleeping up to six, and may be constructed of stone, brick, or wood. All have air conditioning, TV, bath, kitchen, and come with cooking equipment, tableware, and linens. Lovingly decorated and well-maintained, these are an exceptional buy. There is no restaurant, but on the grounds are a boat dock with rentals, a dock with deck for outside fishing and a heated interior with a fishing well, a sandy swimming beach with a diving raft and slide, a playground, and picnic areas. □ **Inexp** □ **W, F**

Marriott's Tan-Tar-A Resort, *Osage Beach, MO 65065, (314) 348-3131* □ *Open all year* □ *On county KK, 2 mi W of jct US 54* □ This giant deluxe complex has 700 rooms and suites and covers 70 acres on one mile of private shoreline. From the lake, the jumble of brown and white gables and balconies plus broad horizontal lines of large lodges makes the complex look something like a Tudor village interpreted by Frank Lloyd Wright. The well-appointed units are scattered in buildings ranging from an eight-story lodge to individual villas. Many units have ocean views, some have kitchenettes, and about half have refrigerators. The resort operates on EP, AP, MAP, and package plans, with a variety of restaurants ranging from casual to elegant to a family corner to lakeside dining to a Bavarian room. There are also several cocktail lounges, two discos, and many shops.

Action is the keynote here, and water lovers will find not only a swimming beach on the lake but also three outdoor pools, an indoor pool, a health spa, and saunas. On the lake you can fish; rent boats by the hour, day, or week; take waterski lessons; go sailing; take a cruise; or take a seaplane ride. Stretching back from the water are golf courses offering 27 holes. For tennis buffs there are six lighted outdoor courts and three air-conditioned indoor courts, plus a pro. You can also horseback ride on miles of shaded trails through the woods. In winter there is skiing, with instructors, a lift, and a day lodge, as well as an indoor ice rink. In case it rains, you'll find an eight-lane bowling alley, a billiard parlor, and other indoor game areas. Parents will be pleased by the nursery, babysitters, and supervised children's programs. □ **Lux** □ **W, T, F, R, G**

SHELL KNOB: Bass Haven, *Star Rte. 1, Box 286 M, Shell Knob, MO 65747, (417) 858-6401* □ *Open all year* □ *I-44 Mt. Vernon exit, then 40 mi S on SR 39, ½ mi S of Central Crossing Bridge* □ With a fine view of Table Rock Lake, here is an unpretentious stop for the avid bass and crappie fisherman which invites you to "come in and set a spell." Accommodations are simple–one- to three-bedroom motel units and six- to eight-person housekeeping cabins–all with TV, heat, and air conditioning. Most have good lake views. No food is served, but five restaurants are within 1½ miles. You can rent rowboats and motorboats, or sail or waterski, at the resort's covered dock (which you reach via a self-operated electric cable car). Also here are a swimming pool, a recreation room, and horseshoes. Bass Haven has a distinctively folksy flavor and obliging hosts who will pick you up at the airport, get you a babysitter, or even invite you to share their Thanksgiving dinner. □ **Inexp** □ **W, F**

Montana

BIGFORK: Flathead Lake Ranch, *Bigfork, MT 59911, (406) 837-4391* ☐ *Open about mid-May to mid-Sept* ☐ *2 mi SE of Kalispell on SR 35* ☐ This 2000-acre working ranch—still run by the founding Averill family—at the northeast corner of Flathead Lake has been popular for almost half a century because it offers the enticements of a watersports resort, the excitement of a dude ranch, and, away from the lakeshore, where the property borders Flathead National Forest, the seclusion of the backcountry. The two-story main lodge is as Western as all get-out—logs inside and out, a large stone fireplace in the dining room, and trophies and a birchbark canoe for decoration. The building looks out across lawn toward the swimming pool, boat dock, and placid lakeshore. This and a similar lodge, plus 14 modern log cabins with picture windows, house the guest rooms; many rooms have lake views; cribs and cots are provided free. The AP rates include all meals, activities, and on-premises facilities, making this a good buy for active families. Lunch is often served at lakeside picnic tables, and there are steak fries in the backcountry. Facilities include tennis courts, sailboats, canoes, fishing boats, an indoor recreation hall, and horseback riding. The daily schedule might include waterskiing, song fests, boat trips, or calf-roping and other wrangler demonstrations. Counselors offer special activities for children, and the wranglers give riding instruction. Available at extra cost are whitewater river trips, pack trips, fly-in wilderness fishing, and tours of Glacier National Park. ☐ **Mod to Lux** ☐ **R, F, W, T**

BIG SKY: Big Sky, *P.O. Box 1, Big Sky, MT 59716, (406) 995-4211; outside MT, call toll-free (800)-548-4486* ☐ *Open all year* ☐ *45 mi SW of Bozeman off US 191* ☐ Chet Huntley, the late newscaster and prime mover behind the creation of Big Sky Montana, once said "Just as important as working for the good life, is finding a place to enjoy it." And without a shadow of a doubt, he did just that. Encompassed within the roughly ten square miles of staggering alpine scenery is a complete community of lodges, condominiums, and campgrounds with a full spectrum of four-season recreation. The 3-story, 204-unit Huntley Lodge is a huge, widespread complex at the base of 11,200-foot Lone Mountain, offering double rooms and suites with amenities such as color TV and picture windows. In addition, more than 500 exceedingly attractive condominiums are available; these include studios and one-, two-, and three-bedroom units, some with lofts, kitchenettes, private balconies, or other extras. There is also an 88-room, economy-rate mountain lodge with a cafeteria and pub. Most units are rented on EP, but MAP is available at Huntley Lodge and some condominiums.

Huntley Lodge, attractively decorated with Indian portraits and Navajo rugs, offers a main dining room with a spectacular view and changing gourmet menu, a steakhouse that also has a fine view, and an intimate Fondue Stube. The bar contains a handsome 19th-century bar of carved oak found by Chet Huntley in an old Montana saloon. In summer, there are also outdoor buffets. At nearby Mountain Mall, an indoor shopping center, are additional eating and drinking facilities, including a cafeteria. At Meadow Village, a sprawling condominium complex, is the Yellow Mule, specializing in steaks and Sunday smorgasbord. Entertainment in the various lounges varies from Austrian yodelers to dance and rock bands.

In addition to indoor and outdoor pools, and a health club with saunas and whirlpools at Huntley Lodge, there are many fine sports facilities (most involve extra charges): An 18-hole golf course designed by Arnold Palmer is 6725 yards, par 72; the pro shop provides rentals and other services. The Guest Ranch, built in 1927, offers horseback riding, instruction, guided trail rides, trips to fishing lakes, sightseeing pack trips, a restaurant, and a swimming pool. Also available is tennis on Laykold courts, archery, trapshooting, courts for basketball and other games, fishing, walking trails, and a seven-mile hiking trail from Huntley Lodge into the Spanish Peaks Primitive Area. The Lone Mountain Sports Shop in the Mountain Mall has rental equipment and can arrange guided backpacking and hiking trips, guided fishing tours, and float trips. A full schedule of activities includes bridge tournaments, craft and cooking demonstrations, bus trips to rodeos or Yellowstone National Park or other points of interest, and a fly-fishing school. There is a day-care center in addition to special day and evening activities for children and teenagers that include hayrides, movies, and cookouts.

Two mountains, receiving an average snowfall of 400 inches, offer 35 miles of trails, 2531-foot vertical drop, a gondola, a triple chair, three double chairs, and a ski school. Free winter evening activities include ice skating, horse-drawn sleigh rides, chestnut roasts, and ski movies. There are also 35 miles of cross-country skiing trails, and the winter tours through Yellowstone National Park may be made by bus or by snowmobile. □ **Mod to Lux** □ **S, R, W, F, T, G**

ENNIS: El Western Resort, *P.O. Box 487, Ennis, MT 59729, (406) 682-4217* □ *Open early Apr. to early Dec; a few units may be open all year* □ *¾ mi S on US 287* □ This 150-acre spread on the Madison River offers comfortable EP lodgings with beautiful views across the valley to snowcapped mountains. Ringing a landscaped little pond, snug, knotty-pine-paneled log cabins are set among spruces. Single and double rooms, suites, and two-room efficiencies with kitchenette all come with carpeting, color TV, and ranch-style furnishings. Deluxe housekeeping units include two-bedroom cottages with fireplace, a two-bedroom ranch house with an exceptional view, and an "executive" house that sleeps up to 15 people. The restaurant is open from 7 am to 9 pm. Madison River trout fishing is excellent, and the resort will arrange for guides and float trips. Your hosts can also advise you on big-game hunting, the upland bird season, and pack trips. Ennis, surrounded by Beaverhead National Forest, is a good base for family touring. Within the forest are recreational sites at lakes, rockhounding, ghost towns and old mining camps, numerous hot springs, and a 150-mile scenic tour road through the Gravelly Range. Virginia City is only 14 miles west. About 40 miles south is the Madison River Canyon Earthquake Area, where a visitor center provides interpretive programs and exhibits. The valley and forests are also good for birdwatching and wildlife observation. □ **Mod** □ **F, W**

PRAY: Chico Hot Springs, *Pray, MT 59065, (406) 333-4411* □ *Open all year* □ *26 mi S of Livingston on US 89, then follow the signs E from Emigrant* □ The large hot springs here, long known to Indians and painted by Charles Russell, became a spa in the late 1800s. Rooms in the old three-story inn, topped with gables and surrounded by a Western-style covered boardwalk, come either with or without private bath. These and the newer motel units are moderately priced. In addition, there are luxury one- or two-bedroom modern chalets with fireplaces, kitchens, and private decks and saunas; these rent by the day or the week and sleep up to ten people. The lodge, run on EP, has a restaurant decorated with century-old barn wood, woodcarving, and 19th-century mementos. The Chico Saloon often has live music, dancing, or other entertainment.

The features, of course, are the two pools fed by the hot springs, where dressing rooms and a snack bar are provided, but there is also trout fishing in the private lake, hiking, and seasonal cross-country skiing. You can rent horses by the hour, for short trips, or for overnight pack trips. The resort will also arrange for float trips on the Yellowstone, fishing, pack trips, whitewater floats, or tours of Yellowstone National Park (about 35 miles south). Chico Hot Springs, surrounded by Gallatin National Forest, is set in the beautiful Absaroka Range, two miles from the Yellowstone River and near lake recreational sites. □ **Mod to Lux** □ **W, F, R, S**

Nevada

INCLINE VILLAGE: One of the most scenic coves on beautiful Lake Tahoe is Crystal Bay, where rugged 9000-foot mountains drop precipitously to the shore. Nestled here is Incline Village, a planned community with year-round recreational facilities. Ski Incline is a full-service ski area with six lifts servicing 21 runs from late November to early April. The ski school offers programs for all levels, for racers, and for children, and the shop has rentals and supplies. The base lodge and the lodge on the slopes have restaurants, cocktail lounges, and sun decks with magnificent views of Lake Tahoe. Cross-country skiing and rentals are also available. Similar facilities are also available 12 miles north at Mt. Rose. Golf Incline offers two 18-hole Robert Trent Jones courses, one a 7000-yard championship course, and the other less challenging. Tennis is available at Tahoe Racquet Club. Ponderosa Ranch, a re-created frontier village inspired by the *Bonanza* television series, offers horseback riding, trail rides, stagecoach rides, and other services. In winter, the ranch rents snowmobiles. Lake Tahoe State Park has a beautiful swimming beach on the lake plus fishing and boating.

Most accommodations here are luxury condominium units with access to private beaches as well as the above facilities; for information, write Golf and Ski Incline, P.O. Drawer AL, Incline Village, NV 89450.

Hyatt Lake Tahoe, *P.O. Box 3239, Country Club Drive and Lakeshore Blvd., Incline Village, NV 89450, (702) 831-1111 or toll-free (800) 228-9900* □ *Open all year* □ In a beautiful wooded setting on the north shore of Lake Tahoe, the Hyatt offers convenient access to Incline facilities and has its own 24-hour casino. Most of the 461 rooms and 1-, 2-, and 3-bedroom suites–in an 11-story main tower and the adjoining Aspen Terrace–have good views. The 24 Lakeside Suites, four to a building, on the hotel's private beach, feature private balconies or decks, living rooms with fireplaces, and wetbars with refrigerators. On the 15-acre grounds are an outdoor heated pool, health spa, two tennis courts, game room, and shops. Canoes, sailboats, pedal boats, and other boats may be rented at the beach, and drivers are available for water skiing. Restaurants include the rotisserie, featuring spit-roasted duckling, a buffet, a 24-hour coffee shop, and a delicatessen. There are three cocktail lounges and entertainment. Babysitting and other services are provided. □ **Mod** □ **W, G, T, R, S**

Lake Tahoe Resort, *Box 3566, Incline Village, NV 89450, (702) 831-0360* □ *Open all year* □ Offering luxury accommodations and formal and informal dining, this resort is the site of the Tahoe Racquet Club, which offers 27 courts and complete instruction facilities with the latest teaching aids. In addition, there is a special health and exercise program. Children are especially welcome, with programs for kids over age three and teenagers. □ **Lux** □ **T, G, R, W, S**

LAS VEGAS: Some Las Vegas hotels provide recreational facilities rivaling any in the United States, but the idea is to keep you perfectly happy *indoors.* So each hotel is a little self-contained city, with shopping arcades, a variety of restaurants, complete services, and, of course, casinos. Each hotel exists in its own private time zone; in the artificial daylight of the casinos it is whatever hour you imagine it to be. Most rooms have blackout drapes, so that you can be independent of the sun, which tends to rise about the time you're ready to turn in.

The hotels will try anything to establish their uniqueness–each offers the biggest or most of something, and many are designed around the themes their names announce: the Frontier, the Sahara, Caesar's Palace. But there are features common to all. Most of them are enormous: around 1000 rooms is standard. Accommodations vary from elegant to garish, and deluxe units may feature wetbars, sun rooms, lanais, or split levels. Plumbing is especially inventive–you may have a telephone in the bath, or a Roman tub, a steam bath, a whirlpool, or your very own swimming pool. The restaurants and bars at each hotel are as impressive and eccentric as the management can manage. Among the bewildering array there's usually a gourmet room where the captain will personally set your food on fire; a 24-hour restaurant-coffee shop; and a couple of steakhouses or other specialty dining areas. The big-name shows for which Las Vegas is famous take place in the hotel showrooms, and reservations are essential.

Since children are not permitted on the gaming floors, families should seek places that have facilities for their amusement. Most hotels have electronic games and pinball machines, as well as babysitter service, but those with outdoor facilities and planned children's programs are best (see suggestions below).

A Las Vegas vacation can be a remarkably good deal. Many hotels offer packages. Room configurations are so various that you can sometimes house a family in a lavish suite at less per person than at an off-Strip motel. Meals can also be a bargain; the gourmet rooms may be cheaper than equivalent restaurants elsewhere, and the buffets are especially worth checking out.

If you manage to break free of the one-armed bandits, Las Vegas has some sumptuous sightseeing. Watersports at Lake Mead National Recreation Area are 40 miles away, hiking in Red Rock Canyon is spectacular, ghost towns are nearby, and the road through magnificent Valley of Fire State Park winds through a surreal world of blazing red sandstone—nature's answer to the eerie neon glow of the Las Vegas Strip.

Among the best-known hotels are:

Aladdin, (702) 736-0111: Oriental theme.

Caesar's Palace, (702) 731-7110 or toll-free (800) 634-6661: Roman decadence; 12 tennis courts (four indoor, with pro).

Dunes Hotel and Country Club, (702) 737-4110 or toll-free (800) 634-6971: Excellent 18-hole golf course, two swimming pools.

Flamingo Hilton, (702) 733-3111.

Frontier, (702) 734-0110 or toll-free (800) 634-6966: Wild West motif; four tennis courts.

MGM Grand, (702) 739-4111 or toll-free (800)634-3434: 10 Plexipave tennis courts.

Sahara, (702) 737-2111 or toll-free (800) 634-6666: Good 18-hole golf course.

Sands, (702) 733-5000 or toll free (800) 634-6901: Four tennis courts.

Tropicana Hotel and Country Club, (702) 739-2222 or toll-free (800) 634-6212: 10 Plexipave tennis courts (eight indoor).

Circus Circus Hotel/Casino, *P.O. Box 14967, 2880 Las Vegas Blvd., Las Vegas, NV 89114, (702) 734-0410 or toll-free (800) 634-3450* □ A 120-foot neon clown named Lucky announces what management boasts is the world's largest permanent circus, complete with its own midway, located one level above the casino floor. Circus Circus is the Las Vegas version of a family resort; since the circus operates 13 hours a day, children can spend their time playing carnival games, eating cotton candy and hot dogs, and watching trapeze artists and wild animal trainers, confident that their parents are keeping themselves sufficiently amused downstairs. □ **Mod**

Desert Inn and Country Club, *3145 Las Vegas Blvd. S, Las Vegas, NV 89109, (702) 735-1122 or toll-free (800) 639-6906* □ This lavish, ultramodern resort city of glass towers and pyramids occupies 200 acres. Lodgings come in an endless variety of shapes, sizes, and prices—conventional luxury rooms, two-story fun suites with decor evoked by their names (Istanbul, Cervantes, Michelangelo, etc.), and the enormous penthouse Presidential Suite with two marble fireplaces, an onyx whirlpool, and satin-covered walls. Notable here, apart from the usual enormous outdoor pool and lots of whirlpools, is the two-level, bronze-and-glass clubhouse, complete with grill room and a pro shop. Here the par-72 golf course (about 7100 yards from the championship tees) is tournament quality; a profusion of trees and traps makes this desert course especially challenging. Also, surrounded by fairways, the tennis center offers five lighted Grasstex courts with a full-time teaching pro. □ **Lux** □ **G, T**

Las Vegas Hilton, *Avenue of the Hiltons, Box 15087, Las Vegas, NV 89114, (702) 732-5111* □ Children will like this hotel—huge even by Las Vegas standards (2783 rooms and suites)—because each floor is decorated to a different theme (Spanish, French, Oriental, etc.). Also, the split-level Lanai Suites on the third floor open onto a 10-acre rooftop park with six lighted tennis courts and a giant heated outdoor pool. Guests have golf privileges at the Las Vegas Country Club, whose par-72, 6700-yard course is one of the best in the area. The dozen dining areas inlcude the Beihana Village complex and the Imperial Crown Room where the table service is of 14-karat gold. Best of all for families is the Youth Hotel. This is a full-time day camp offering a variety of supervised activities for children between the ages of three and eighteen. There's also dormitory space available for a limited number of young visitors. ☐ **Mod to Lux** ☐ **T, G**

RENO: Harrah's, *Box 10, Reno, NV 89504, (702) 786-3232* ☐ *Open all year* ☐ *Center St. at 2nd* ☐ Besides 325 guest rooms and suites, this enormous high rise, one of the landmarks of the Reno skyline, contains a casino, nightclubs, restaurants, heated pool, and health club. And of course, there's the famous Harrah's automobile collection, over 1000 beautifully restored and maintained antique vehicles, including planes, boats, and motorcycles. There's even a Pony Express Museum here, a real kid-pleaser. The hotel's three restaurants provide plenty of dinner choices: the Steak House has continental food in addition to steaks and chops, the 24-hour Garden Room offers Chinese food, and Seafare has roasts and seafood.

Besides gambling, Reno has much to offer, including several interesting cultural stops, such as the Nevada Historical Society Museum and the Fleischmann Atmospherium-Planetarium. Reno also has good indoor and outdoor swimming pools and four worthwhile golf courses open to the public: Washoe, Sierra Sage, Brookside, and Lake Ridge, a Robert Trent Jones course. ☐ **Lux** ☐ **G**

New Hampshire

BRETTON WOODS: Mt. Washington Hotel, Motor Inn, and Townhouses, *Bretton Woods, NH 03575, (603) 278-1000* ☐ *Hotel open June to mid-Oct; motor lodge and townhouses open May–Nov and Dec–Apr* ☐ *Off US 302 in Bretton Woods* ☐ Mt. Washington Hotel is one of those rambling, white, old-fashioned resorts where the fashionable rich and near-rich used to spend their summers. In keeping with the hotel's age, its 200-plus rooms do not have TVs or radios, and some share baths. But the grounds are spacious here, and there's much to do.

The centerpiece among the sports facilities is the 18-hole, par-72 golf course designed by Donald Ross. The rolling course has pro shop with rentals and other essentials. Tennis players will find 12 clay courts, and swimmers have both indoor and outdoor pools at their disposal. Horseback riding is available, and fishermen will discover trout in the nearby Ammonoosuc River.

The hotel operates on MAP, but the small restaurant, Stickney's, can provide sandwiches, soups, and salads in between meals. In skiing season, when the unwinterized hotel closes, the motor lodge and town houses come into their own. The two-story motor lodge overlooks both Mt. Washington and the hotel. The 57 contemporary rooms all have color TV and modern bath. The lodge's restaurant serves both continental breakfasts and dinners. The town houses, just above the Bretton Woods' bunny slope, are modern one- to four-bedroom condominums, many privately owned but available to rent. The motor lodge and town houses share the use of the hotel's sports facilities.

The Bretton Woods ski area, within walking distance of the hotel, motor lodge, and condominiums, has two double chairlifts and T-bar, ski school, snow making, rentals, base lodge, nursery, and 50 miles of cross-country skiing trails. ☐ **Mod to Lux** ☐ **G, T, S, R, F**

DIXVILLE NOTCH: The Balsams: *Dixville Notch, NH 03576, (603) 255-3400* ☐ *Open late May to mid-Oct and late Dec to mid-March* ☐ *11 mi E of Crawford on SR 26* ☐ Tucked in a valley between the sheer rock walls of Dixville Notch, the Balsams resembles a Swiss spa. This hotel is a rambling castle prickling with turrets and towers. A state unto itself, the Balsams is set on 1500 forested acres, maintains its own independent telephone company, and produces much of its electric power. The hotel faces Lake Gloriette, a private, stocked trout lake. No license is required to fish here, but you are limited to a catch of five fish or five pounds per day. At the waterfront are the heated, olympic-size pool, surrounded by a sun deck, canoes and rowboat rentals, and three clay plus three Plexipave tennis courts.

Adjacent to the hotel are a putting green and nine-hole executive course, the Coashaukee. A little farther off is the Panorama course, an 18-holer designed by Donald Ross and named for its spectacular mountaintop view of New Hampshire, Vermont, and Canada. Unlimited use of both golf courses is complimentary to resort guests. The Wilderness ski area, only a short distance from the hotel and accessible by regular shuttle, makes the Balsams a year-round resort. The 12 trails, up to two miles in length, are served by a 4000-foot double chairlift and two T-bars. At the base is the lodge, with stone fireplace, sun deck, cafeteria, sales, rentals, and repairs.

The resort operates on AP in summer, MAP in winter. The dinner menu offers a nightly choice of about eight entrees, with dishes such as coquilles St. Jacques and veal Marsala. The buffet luncheon, a daily extravagance, includes beer, stuffed Atlantic salmon, fried Maine shrimp, and a dozen salads. Evenings bring plenty to do—dancing to the Balsam Orchestra in the ballroom or to a combo in the Wilderness Room, movies in the theater, pool in the Billiard Room, or cards in the Game Room. ☐ **Lux** ☐ **T, G, W, S, F**

FRANCONIA NOTCH: Mittersill: *Franconia Notch, NH 03580, (603) 823-5511* ☐ *Open all year* ☐ *From US 3, 1 mi N of Franconia Notch, take SR 18 NW 1 mi* ☐ Most resorts done in a self-consciously "Swiss" or "alpine" architecture are an embarrassment to behold, but Mittersill is a wonderful exception. The inn, a multi-level building painted brown and white, has a classic look complemented by the other buildings at the ski area. Even the chalets built by private owners blend with the resort's Tyrolean theme. The 85 modern rooms have wall-to-wall carpeting, TV, telephone, and private bath. The public rooms are also nice; the dining room and cocktail lounges have exposed beams and paneling. The inn also has a game room and nursery. Chalets are rented through the inn for three-night minimums; these sleep up to 12 people and have kitchen facilities; most have fireplaces. Rates are EP.

On the 165 acres are clay tennis courts, a large heated pool, and a putting green, but the big sport here is skiing. A 4000-foot double chairlift and 2700-foot-T-bar carry skiers up to trails that interconnect with renowned and challenging Cannon Mountain slopes. Services include rental shop, ski school, patrol, snowmaking, trail grooming, lounge, and cafeteria. For cross-country skiers, there are groomed trails and rentals. Near the inn is an ice rink. ☐ **Mod** ☐ **S, T**

JAFFREY: Woodbound Inn, *P.O. Jaffrey, NH 03452, (603) 532-8341* ☐ *Open mid-May to mid-Oct and Dec 26 to mid-Mar* ☐ *S of Jaffrey off US 202, on Contoocock Lake* ☐ The Woodbound, which operates on AP, is somewhere between an inn and a resort. The main lodge is a three-story colonial building set among tall pine trees. Inside, rooms have an old-fashioned decor–four-poster beds, rocking chairs, patterned wallpaper, and antique prints prevail. Each of the pine-paneled cottages, which can accommodate two to eight people, has a living room, bedroom, private bath, fireplace, and picture windows. The living room in the main lodge, furnished with comfortable armchairs and rockers, boasts the original 1819 fireplace and mantel. The dining room, bright and cheerful, features traditional homemade New England fare. All pastry is baked in the Woodbound's ovens, and fresh vegetables and salad ingredients come from nearby gardens. Features are the Sunday buffet suppers and the weekly cookout. The resort has its own ski slope, perfect for novice and intermediate skis. It has two tows–one 500 feet long, the other 600 long–plus rentals and instruction. For cross-country skiers, 18 miles of trails begin at the doorstep. In summer, swimming, waterskiing, rowing, and canoeing are available off the private beach that slopes down to two-mile-long Lake Contoocock. A par-three golf course, a tennis court, and lawn games also keep guests busy. Supervised programs for children and a babysitter service are provided, and square dances, movies, and golf tournaments are scheduled frequently. □ **Mod** □ **W, S**

LINCOLN: The Village of Loon Mountain, *P.O. Box 508, Lincoln, NH 03251, (603) 745-8157* □ *Open all year* □ *3 mi W of Lincoln on SR 112 (Kancamagus Hwy.)* □ This development, in White Mountain National Forest, has been praised by architects and planners for being exceptionally well designed: The roads climb the steep hills at gentle angles to keep them passable even in icy conditions. Roofs are covered with Teflon-coated steel so that snow slides off. Large windows face south. Streams and most of the trees on the site were preserved. Homes are modern, angular, and high-ceilinged, with exteriors sheathed in weathered-gray pine siding to give them a New England look. Apartments–from studios to four-bedrooms with loft and two baths–are furnished in contemporary style. All have kitchens and fireplaces (except for studios). Most styles are available for periods of two nights to four months. The village has ten tennis courts, health club facilities (sauna, whirlpool, indoor pool, exercise room), outdoor pool, recreation room, restaurant, and grocery store. The Loon Mountain Ski Area, across the highway, has 18 trails and a 185-foot vertical drop, and is serviced by a four-passenger gondola and three double chairs. □ **Mod to Lux** □ **T, S**

MIRROR LAKE: Pick Point Lodge and Cottages, *P.O. Mirror Lake, NH 03853, (603) 569-1338* □ *Open early May to late Oct.* □ *6 mi N of Wolfeboro, off SR 109 on Tuftboro Neck* □ This small, informal resort with one-half mile of shore frontage on Lake Winnipesaukee offers attractive housekeeping cottages and lodge rooms in a woods setting. Seventy-five acres of the property are set aside as a wildlife refuge, with feeding stations for deer and wild ducks, more than 60 bird houses, and three miles of trails. The main lodge, a handsome white building with spacious front porch, sits among trees on a knoll not far from the water's edge. The game room, TV and writing room, and library are here. There are four large fireplaces in the building and the interiors have exceptionally nice woodwork and ceiling beams. All of the motel-style lodge rooms have direct-dial telephones and color TV. The lodge provides only buffet breakfasts and tray lunches on advance request, but there are restaurants in the area. The cottages are widely spaced for privacy, and most are within a hundred feet of the lake. All have heat, equipped kitchens, maid service, linens, porches, and picture windows. The three- and four-bedroom cottages have two baths, and almost all cottages have fireplaces.

Pick Point has a sandy, gently sloping beach, and sailboats, rowboats, and canoes may be rented at the dock. Use of the clay tennis court, and the cookouts scheduled throughout the week, are free. □ **Lux** □ **W, F, H**

NORTHWOOD: Lake Shore Farm, *Jenness Pond Rd., Northwood, NH 03261, (603) 942-5921* □ *Open all year* □ *From Concord take US 4/202 15 miles E to jct SR 107; resort is off 107 on Jenness Pond Rd.* □ Over a span of seven generations Lake Shore Farm was transformed from a five-room farmhouse into a 39-room resort lodge. Although no farming has been done here for quite a while, the 150 acres of fields and forests and the barn and outbuildings are a constant reminder of the farming heritage. The lodge buildings overlook Jenness Pond, which is easily big enough to be considered a lake. A huge expanse of lawn stretches down to the sandy private beach and the dock, where you may take out a canoe or rowboat for the asking. Around the recreation hall (with pool, Ping-Pong tables, and a fireplace) are croquet, horseshoe, badminton, volleyball, and basketball areas. In the field on the other side of the lodge buildings are two acrylic tennis courts and a duck pond. In winter, Lake Shore caters to snowmobilers. Accommodations vary from share-bath rooms in the old house to contemporary rooms with private bath in the annex, all on AP. The menus are largely made up of old family recipes prepared with local produce and dairy products. Meals are served family-style. □ **Inexp** □ **W**

PITTSBURG: The Glen, *First Connecticut Lake, Pittsburg, NH 03592* □ *Open mid-May to mid-Oct* □ *7 mi NE of Pittsburg on US 3, take private road 1 mi SE to resort* □ The Glen, tucked away from it all on the Canadian border in a town not on many maps, features an informal country atmosphere. The main lodge building was constructed in 1900 as the manor use of a private estate. Wood-paneled inside, with wood floors and an immense stone fireplace in the living room, the Glen still has the charm of a turn-of-the-century woodland retreat. The dining room, surrounded by a porch on three sides, also has a fireplace and pine paneling. Rooms in the lodge have 19th-century furnishings and a rustic feel. The cottages, with one to three bedrooms, have living rooms, kitchenettes, shower-baths, and large porches, most of which have a view of the lake. All units are rented AP, and food at the Glen is home-cooked and served family-style. Friday nights are buffet nights, and cookouts and salad bars are a regular feature. The Glen is set on 200 acres of birch and poplar woodland on First Connecticut Lake, which has a swimming beach. Plenty of rental boats are available to go after the landlocked salmon, lake trout, rainbows, browns, and squaretails in the lake or its tributary streams. ☐ **Inexp** ☐ **F, W**

PORTSMOUTH: Wentworth-by-the-Sea, *Portsmouth, NH 03801, (603) 436-3100* ☐ *Open May to late Oct* ☐ *1 mi SE of jct US 1 and SR 1B, on SR 1B* ☐ Set on a spit of land overlooking Portsmouth harbor and its wooded islands, this 19th-century resort hotel seems to ramble on interminably. Besides the 250 rooms, furnished in contemporary style, the Wentworth also has New England cottages, some with ocean views and fireplaces. A traditional New England resort run on AP, the Wentworth is perhaps best known for its clambakes, regularly scheduled in summer. The Wentworth is also known for its golf. Its 18-hole course, designed by Geoffrey Cornish, runs along the ocean, and five of the approaches pass over water. In the gardens around the hotel are a putting course and par-three course. Tennis players also do well here. They have four clay courts and three all-weather courts at their disposal and easy access to eight covered courts nearby.

By the Ship Theater—a 550-seat, boat-shaped hall—are the heated olympic-size pool and children's pool. Fishing charters are available at the long pier which runs into the bay from the theater. In July and August the resort has an extensive children's program which includes games, arts and crafts, and picnics. There is also a separate children's dining room. ☐ **Lux** ☐ **T, G, F**

SUNAPEE: Dexter's Inn and Tennis Club, *Stagecoach Rd., Sunapee, NH 03782* ☐ *Open mid-june to mid-Oct* ☐ *From SR 11 W of Sunapee, take Winn Hill Rd. S 1½ mile to inn* ☐ It could be argued that three tennis courts don't make a tennis club, but Dexter's is small enough (a maximum of about 36 guests) to preserve a good ratio of guests to courts. The main house, built in 1801, is attractively furnished in colonial style with some antiques. The living room has a grand piano, fireplace, and a library of more than 500 books. Besides the MAP rooms at the inn, Dexter's has a suite in the Sugar House, a restored log cabin. Housekeeping suites and apartments are in the "Place at the Inn"—a modernistic structure in the woods a quarter mile from the inn. These range from a two-room suite with kitchenette to a three-story apartment with two baths, full kitchen with dishwasher, sleeping loft, and fireplace. All units are fully equipped with china, cooking utensils, and linens. The grounds are beautifully landscaped with flower-lined walkways and stone walls. Besides the three Plexicushion tennis courts, Dexter's has a swimming pool and a large barn with game tables and such. Trails through the 200-acre estate, preserved as a bird and wildlife sanctuary, are used for cross-country skiing in winter.

Nearby Mt. Sunapee is a full-service ski area (five double chairlifts, four slopes, and 16 trails). The lifts also operate in summer to provide sightseers with panoramic views from the mountaintop. ☐ **Inexp to Mod** ☐ **T, S**

WATERVILLE VALLEY: Waterville Valley is the name of this White Mountain ski area, but the main slopes and trails are on 4000-foot Mt. Tecumseh. Eight lifts (two triple chairs, four double chairs, one T-bar, and one J-bar) carry skiers over 2000 feet to the top. Skiing is guaranteed here from Thanksgiving to mid-April: snowmaking covers at least one major run on seven out of the nine lifts. At the bottom of the slopes are base lodge, cafeteria, nursery, and ski school. The ski school here takes great pride in teaching beginners and children, and package vacations are available that include unlimited lessons at no extra charge. The valley's second, much smaller, ski area is Snow's Mountain. Open weekends and holidays, the area is serviced by one chairlift. At the base are lodge, cafeteria, restaurant, cocktail lounge, and nursery. The Ski Touring Center has 55 kilometers of graded cross-country trails, many of which are groomed and patrolled daily.

Lodging plans with unlimited use of lifts, unlimited lessons, and daily meals at any of three local restaurants are offered by nine lodges in the area. Also summer vacations, with lessons on the Tennis Center's 18 clay courts and privileges on the 9-hole golf course, are available. For more information contact the Waterville Valley Resort Association, Waterville Valley, NH 03223, (603) 236-8686. For condominium units, many with kitchens, you can also write Waterville Valley Gateway, Upper Mad River Rd., Campton, NH 03223, (603) 726-3724.

Snowy Owl Inn, *Waterville Valley, NH 03223, (603) 236-8383* ☐ *Open Thanksgiving to Apr and Memorial Day to Columbus Day* ☐ *Off SR 49, on Snow's Brook Rd.* ☐ This is roughhewn elegance. From the wood barn-board siding of the exterior to the foot-thick mantel of the three-story fieldstone fireplace inside, the Snowy Owl's woodsy design displays extraordinarily good taste. This is a clean-cut and sturdy place, but there are some frills, such as an 1890 grand piano and Indian rugs, that give the lodge a warm and personal air. Besides doubles, loft rooms and rooms with sitting areas are among the 37 suites. Room 401, with large circular window and skylights is frequently used as a honeymoon suite. The lodge has a heated pool but no restaurant. Guests are invited to participate in activities which might include free cocktail party, torchlight parade, feature movies, ice skating, Swiss fondue, or kid's disco party. ☐ **Mod to Lux**

WHITEFIELD: Mountain View House, *Whitefield, NH 03598, (603) 837-2511* ☐ *Open late May to mid-Oct* ☐ *3 mi N off US 3* ☐ This yellow clapboard Federal-style giant is one of New England's finest resorts. Although it can accommodate 300 guests in 154 single and double rooms and six suites, and it offers complete recreational facilities on a 350-acre wooded mountain site, the feeling is intimate. Staying here is like being at an unusually well-managed inn. Or perhaps more like visiting the enormous country house of rich and hospitable friends. Mountain View House has been run by the Dodge family for four generations, and they *do* manage the place as though it were a home. The meals–three a day are included in the room rates–are all home-cooked. The steaks, chops, and fresh seafood are all of the highest quality. Featured also are fresh breads and pastries and elaborate homemade desserts. Lunch is a sumptuous buffet served on the flagstone terrace of the large heated pool. The golf course is a beauty: only nine holes but perfectly maintained and gently rolling, with views almost everywhere of the White Mountains. There are two sets of tees, so you can play the course as 18 holes. The Plexicushion tennis courts are grouped near the Sports House, which is the central clubhouse for all activities.

The guest rooms are in the main building. Each is individually decorated in traditional style, with floral-print wallpaper, bedspreads, and drapes. All rooms have wall-to-wall carpeting and telephones, and all but a few suite-like connecting rooms have a private bath; most have lovely mountain views. The main building also contains the large dining room, an elegant coffered-ceiling guest lounge with fireplace–decorated with antiques and colorful arrangements of fresh flowers–and various other nice places for just sitting around, including a well-supplied library. And if you want to relax a bit more actively, there are strollable gardens, easy forest trails nearby, and good streams for trout fishing. ☐ **Mod to Lux** ☐ **G, T, F**

Spalding Inn Club, *Whitefield, NH 03598, (603) 837-2572* ☐ *Open late May to late Oct* ☐ *2 mi N of Whitefield on US 3* ☐ One of New England's classic country resorts, this gracious inn has been owned and managed by the Spalding family for more than a half century. The club and cottages, all painted brown with white trim, are set amid well-manicured lawns and gardens. One landscaped section of lawn, overlooked by the dining room, contains nine holes of par-three golf. There are also four clay tennis courts near the heated pool and the sports house. The most passionately pursued activity here is the venerable and distinguished sport of lawn bowling. This is not bocci or nine pins, later variations of the game, but a sport far closer to that practiced by the English and French kings–a game so aristocratic that commoners were forbidden to play. Like all things associated with the English, this once-popular sport lost favor during the Revolution. George Washington had his own green plowed under. The object of the game seems simple: to roll "bowls" as close as possible to a small white target ball ("jack"). The catch is that the "bowls" are not perfectly round; they curve left or right as they roll. Though the greens here are quite professional (U.S. championships have been played here), even rank amateurs are welcome to learn the game.

Spalding Inn Club, run on AP, is also famous for its food, which might include poached fresh Atlantic salmon hollandaise, roast Cornish game hen with wild rice stuffing, or omelet with sour cream and caviar. Many fruits and vegetables are purchased locally. All breads and pastries are made on the premises. The 70 rooms in the main building are handsomely furnished with colonial reproductions. The half-dozen cottages have a sitting room with fireplace, service bar, and one or more connecting bedrooms.

Within a few minutes' drive are four golf courses and several trout streams. Lectures, concerts, bingo, and card parties are regularly scheduled for the evenings. There's a well-stocked library, too. ☐ **Mod to Lux** ☐ **T**

WOLFEBORO: Point Breeze Resort, *P.O. Box 420, Wolfeboro, NH 03894, (603) 569-3211* ☐ *Open late-May to Sept* ☐ *From SR 28 in Wolfeboro, take road to Lake Wentworth's south shore* ☐ This cottage resort is on Lake Wentworth–five miles long, three miles wide, and studded with 17 islands. The string of white cottages is separated from the water by a broad lawn and a sandy beach. Each cottage has knotty-pine paneling, private bath, and screened-in porch. The social hall, with fireplace and wraparound porch, is on a point overlooking the lake. At the boat dock, you may go waterskiing or rent fishing boats, motors, rowboats, Sunfish, or canoes. The lake is known for its black bass fishing, which is especially good in spring. Two Plexicushion and one clay tennis court are on the grounds, plus facilities for badminton, horseshoes, shuffleboard, croquet, and Ping-Pong. ☐ **Inexp** ☐ **W, F**

WOODSTOCK: Jack O'Lantern, *Woodstock, NH 03293, (603) 745-8121* ☐ *Open mid-May to mid-Oct* ☐ *At exit 30 on I-93* ☐ This well-established family resort boasts its own 9-hole, par-35 golf course, two tennis courts, golf and tennis pro shop, fishing on the adjacent Pemigewasset River, playground, and swimming pool. Guests are housed in 50 cottages and in 15-room motor lodge. Rooms have contemporary furniture, wall-to-wall carpeting, and TV; some have air conditioning, private patio, wood paneling, or exposed beams. Cottages, with one to three rooms, are set in a pine grove adjacent to the golf course and surrounding the heated swimming pool. The Cabana Club, overlooking the pool, features movies and other entertainment on summer evenings. The resort offers MAP during summer. The terrace dining room, open for three meals, has a dinner menu of eight entrees, changed nightly. The Continental Room, for more sumptuous dining, has dancing and entertainment. ☐ **Mod**

New Jersey

MCAFEE: Playboy Resort & Country Club at Great Gorge, *McAfee, NJ 07428, (201) 827-6000* ☐ *Open all year* ☐ *Off SR 23* ☐ If you want a bit of razzle-dazzle with your relaxation, Playboy Great Gorge offers a seven-story hotel and resort complex with 650 double rooms and a score of suites, and top nightclub entertainment and lounge acts of the caliber found in New York City (which is only 1½ hours away). As you might expect, many waitresses wear the famous Bunny costumes, yet you need not have a Playboy Key to come here, and this is not an adults-only resort. Children are catered to with a supervised day camp, special playground, and babysitting service.

Great Gorge first became known because of the Vernon Valley ski center that adjoins the 800 acres of hill country belonging to the Playboy. In winter, you can spend days on the slopes or, conditions permitting, go cross-country skiing. From March to November, the Country Club side of Playboy functions with a 27-hole golf course, pro shop plus instructors, cocktail lounge-restaurant, and lockers. Designed by George Fazio in consultation with Doug Sanders, the course is actually three nine-holers that are playable in combinations of varying difficulty. Also here are indoor and outdoor tennis and swimming, horseback riding, paddleboats, a nature trail, a health club, an electric-game room, several boutiques, a photo studio, and a unisex hair salon. The hotel also runs participatory activities, including TV-game-show-type games plus yoga, disco-dancing, and jogging classes.

The rooms are decorated in a modern vein that matches the cubistic architecture of the hotel's facade as well as the design of lounge areas. Many rooms feature a wall of wood paneling and a wall of windows leading to a private balcony. Plush carpeting, color TV, individual temperature controls are standard, as is room service. Suites are equipped with wetbars. For dining you may choose among four restaurants, a cafe, and a delicatessen. EP, AP, MAP, and package plans are available. ☐ **Mod to Lux** ☐ **T, G, S**

New Mexico

ELEPHANT BUTTE: Elephant Butte Resort Inn, *P.O. Box E, Elephant Butte, NM 87935, (505) 744-5431* ☐ *Open all year* ☐ *E of Truth or Consequences on SR 52* ☐ Elephant Butte is a strange area, named for an ancient volcano core that has eroded to resemble a pachyderm. This Sonoran desert landscape has weird volcanic necks and purplish ash-beds where dinosaur fossils have been found and rockhounding is popular. In this stark yet colorful landscape, man-made Elephant Butte Lake has become a fisherman's favorite, with catfish, crappie, bass, and walleye and northern pike. Overlooking the lake is the inn, a two-story, air-conditioned motel at Elephant Butte State Park. The motel has 10 single rooms, 35 doubles, and a suite, plus restaurant, lounge, game room, outdoor pool, and tennis courts. South is Elephant Butte Golf and Country Club, an 18-hole course with a pro and rentals. On the lake is a full-service marina with boat and tackle rental, waterskiing, and other services. The park also has lake swimming and picnic grounds. In Truth or Consequences, five miles away, are Geronimo Springs Museum and the mineral springs used for cures since Indian times. ☐ **Inexp** ☐ **G, F**

MESCALERO: Inn of the Mountain Gods, *Mescalero, NM 88340, (505) 257-5141* ☐ *Open all year* ☐ *Off US 70* ☐ Surrounded by peaks over 9000 feet in Lincoln National Forest, and hosted by the Mescalero Apache Tribe, this is one of America's most attractive and enjoyable resorts. The lakeside inn is ultramodern in design yet, through the use of natural woods, in complete harmony with the rugged timberland of the Sierra Blanca foothills. Three sides of this unusual structure are all roof; the fourth side, overlooking the swimming pool, Lake Mescalero, and an impressive mountain vista, is a wall of windows. Inside, a wood-beamed cathedral ceiling, plenty of natural light, and a three-story copper-sheathed fireplace in the lobby create a warm atmosphere. Facilities include a coffee shop, restaurant (dress code after 6 pm), and two cozy cocktail lounges as well as a large lounge with live entertainment and dancing nightly. The 134 air-conditioned, spacious bedrooms have double beds, color TV, telephones, full baths, and room service. Some have private balconies and fireplaces.

On the grounds are six lighted tennis courts with pro, and the Ted Robinson-designed 18-hole golf course with pro, putting green, and driving range. The stable offers lessons, trail rides, and pack trips. Fishing dock and boat rental are available on 130-acre, spring-fed Lake Mescalero, plentifully stocked with trout and great for sailing and canoeing. There's also skeet and trapshooting, and hunting for deer and elk. In winter, ski packages are offered for the Sierra Blanca Ski Area, where some of the state's best downhill trails are located. □ **Mod to Lux** □ **T, G, R, W, F**

RUIDOSO: High Country Lodge, Inc. *P.O. Box 137, Alto, NM 88312, (505) 336-4321* □ *Open all year* □ *5 mi N of Ruidoso via SR 37* □ On seven mountain acres, this cluster of 32 efficiency cabins and a lodge overlooks Alto Lake and is surrounded by the pinetree country of Lincoln National Forest. Each of the pine-paneled, homey cabins sleeps up to six adults and has a bedroom, living-dining area, fully equipped electric kitchen, brick fireplace, outdoor grill, picnic tables, and private porch with view of the Sierra Blanca mountains. Linens are provided, and maid service is available. The lodge restaurant opens in late afternoon for dinner, and specialties include deep-fried frog legs, stuffed mushrooms, and mountain trout. The cocktail lounge has weekend entertainment. Also in the lodge are an indoor swimming pool, sauna, whirlpool, and recreation room. Outdoors are a tennis court, horseshoe court, and playground. On Alto Lake, which is regularly stocked with trout and bass, is a private beach. The lodge rents bicycles and fishing tackle, and Eagle Creek runs through the property. Jogging and nature trails lead through the forest. The lodge can arrange horseback riding. Being the closest hostelry to the Sierra Blanca and Ruidoso ski areas, High Country offers ski packages and rents sleighs in winter. □ **Inexp** □ **F, S**

SANTE FE: The Bishop's Lodge, *Bishop's Lodge Rd., Santa Fe, NM 87501, (505) 983-6377* □ *Open mid-March to early Nov* □ *3 mi off US 84/285 via SR 22* □ In the Sangre de Cristo foothills, this was once the retreat of Archbishop Lamy, immortalized in Willa Cather's novel *Death Comes for the Archbishop.* The chapel he built more than a century ago still stands in a hilltop garden. The buildings consist of a pueblo-style main lodge, two villas built by Joseph Pulitzer for his daughters, and newer wings. Although all 70 units are decorated in Southwestern colors and furnishings, the rooms and suites vary. Some lodgings have beamed ceilings, fireplaces, balconies, or regional furnishings. In addition to dining room meals and buffets, you may lunch on a terrace, at poolside, or enjoy a steakfry. The El Charro cocktail lounge has Spanish background music and a copper-hooded fireplace.

The swimming pool is set in a garden below the main lodge. On the grounds are tennis courts with a shop and pro, and a health club with sauna and whirlpool. Some 60 horses are available for trail rides across the 1100-acre grounds, and wranglers offer lessons, breakfast rides, and luncheon rides. For skeet and trapshooting you can rent guns and get instruction. The sports director will arrange for you to play golf at two nearby courses or point out the best trout streams. There are also the usual croquet and badminton courts and recreation room. Children get special attention here. They have a wading pool, a stocked fishing pond, paddleboats, pony rides and special wrangler attention, sitters, and their own cheerful dining room. In summer, at no extra cost, there is an all-day activities program for ages four through twelve. Teenagers have their own club. Rates are AP in July and August, EP the rest of the time, and family rates are available. ☐ **Mod to Lux** ☐ **R, T**

Rancho Encantado, *Rte. 4, Box 57 C, Santa Fe, NM 87501, (505) 982-3537* ☐ *Open early May to early Nov* ☐ *8 mi N of Santa Fe; off US 285 at Tesuque* ☐ The charm of this elegant little guest ranch, nestled in the Sangre de Cristo foothills, has drawn celebrities from around the world. Low adobe buildings, hidden courtyards, gardens, and beautiful vistas have made many guests feel this 168-acre retreat is truly enchanted. Public rooms and guest quarters are spectacular, with viga ceilings, whitewashed walls, hand-painted tiles, Indian rugs, and art objects of the Southwest. In the main lodge are seven guest rooms which, though large, are the cheapest accommodations. The other units, prized for their woodburning abobe fireplaces, consist of large cottage bedrooms, kitchenette studios, two-room casita suites with notable views and refrigerators, and gorgeous luxury suites (with refrigerators) in Casa Pinon.

Rates here are EP, but few guests pass up the opportunity to eat in the dignified dining room renowned for its special tournedos, sour cream enchiladas, and other interesting twists on Mexican cuisine. Saturday's feature a seafood buffet and Sundays there is brunch. Also in the main lodge are a library, comfortable lounge and bar. You can also enjoy drinks and snacks in a pretty patio. On the grounds are a lovely pool, tennis with a pro, archery, rifle and trapshooting, a recreation room, and a playground. Trail rides, cookouts, and riding instruction are available, and the ranch will arrange for golf nearby. ☐ **Lux** ☐ **R, T**

TAOS: Abominable Snowmansion Ski Hostel, *Box 3271, Taos, NM 87571, (505) 776-8298* ☐ Nine miles from the Taos Ski Valley slopes, this is the cheapest, closest place to stay. It has dormitory accommodation, a fireplace, and a game room. ☐ **Inexp** ☐ **S**

Starfire Lodge, *Angel Fire, NM 87718, (505) 377-2301* ☐ *Open all year* ☐ *26 mi E of Taos on SR 38* ☐ Angel Fire is a condominium development with year-round recreation. From November until April, Angel Fire ski area is activated. Three double chairlifts service 18 miles of trails (most rated intermediate and beginner), and the area has snowmaking, ski school, rental shop, ice skating, snowmobiling, and ski touring. In balmier weather the country club offers nine holes of golf, tennis, and a pool. Angel Fire Lake is stocked with trout, and you can rent paddleboats or sailboats (no motors allowed). Stables offer horse rental and guided rides. Also available are hot-air ballooning and gliding.

Starfire Lodge is a condominium hotel built in stepped design that gives every unit a balcony. Available here are double rooms, studios with fold-out couches, or kitchenette apartments with living/dining area, two bedrooms, and three baths. All units are spacious and decorated in attractive colors. The lodge has a heated pool, laundry center, and shuttle service to the ski slopes and country club for dining and entertainment. Other accommodations are in condominium units named Snowbird (studios and kitchenette), Wren (one to three bedrooms plus living/dining area and equipped kitchen), and Angel Fire Common (one to three bedrooms, equipped kitchens); all of these units have fireplaces and are available on package plans. ☐ **Mod** ☐ **S, R, F, T**

Tennis Ranch of Taos, *P.O. Box BBB, Taos, NM 87571, (505) 776-2211* ☐ *Open all year* ☐ This condominium hotel offers 70 studios and one- or two-bedroom apartments; some have kitchenettes and balconies, and all have fireplaces. They share a restaurant with continental cuisine, a cocktail lounge, sauna, and whirlpool. The summer highlight is tennis, with lighted courts and a pro. In winter, you are only 12 miles from Taos Ski Valley, to which the ranch provides shuttle service. The shuttle is included in ski packages, and tennis plans and AP are also available. ☐ **Mod** ☐ **T, S**

TAOS SKI VALLEY: Whatever the season, the contrast between the sagebrush-speckled desert and the forested 9000-foot Sangre de Cristo peaks in Kit Carson National Forest is abrupt and awe-inspiring. Sacred to the Taos Indians, these mountains and their atmosphere of enchantment make a lasting impression. Wisely, the developers of Taos Ski Valley cut as few as possible of the blue spruce and aspen to make way for the 61 ski slopes, six double chairlifts, and two cable lifts. Downhill skiing is superb. The longest run is more than five miles and the vertical drop is 2612 feet. This is a full-service facility, with snowmaking, patrol, rentals, and cross-country skiing trails. The ski school is excellent, offering special programs, NASTAR clinics and races, an all-day Kinderkäfig for ages 3 through 11, and guided touring of spectacular Wheeler Wilderness Area. The ski season runs from mid-November until after Easter, but some bowls have snow until June.

Development has been restricted by a concerned group of Europeans and Americans who have seen other ski areas ruined. At the foot of the lifts there are only five small, quaint lodges; three condominium developments complete the lodging at Taos Ski Valley. These offer a number of package plans in skiing season. For people not booked here, reserving space in the dining rooms is difficult, and this discourages overcrowding by day visitors. After the snow melts, Taos Ski Valley keeps on going as a mountain retreat. Kit Carson National Forest waxes surprisingly verdant, offering hiking trails and trout streams galore, and horseback riding is available. The town of Taos, with the Millicent Rogers Museum of Art and Taos Pueblo, are worth visiting, and the literary-minded make a pilgrimage to the D.H. Lawrence Ranch and Shrine, where Lawrence spent the last years of his life.

For general information, contact Taos Ski Valley, Inc., Taos Ski Valley, NM 87571, (505) 776-2291. Or write to the individual accommodations listed below, all at Taos Ski Valley, NM 87571.

Hondo Lodge, *(505) 776-2277* ☐ *Open mid mid-Apr* ☐ Some of the 25 paneled double rooms in this small lodge are spacious enough for a family of four; all have bath and three have a balcony. EP and meal plans are available. The attractive dining room incorporates the oldest structure in the valley—a mountain warming-hut left over from the area's copper boom. The American menu features game on Sunday night, an 18-foot salad bar, and a wine list. The adjacent bar has a TV set and entertainment that ranges from a dance band to ski movies. Downstairs is La Bodega, a combination boutique-cafe. The lodge has a sauna and whirlpool. ☐ **Mod**

Hotel Edelweiss, *(505) 776-2301* ☐ *Open mid-Nov to mid-Apr; Memorial Day to Labor Day weekends* ☐ Former French ski champion, Bernard "Dadou" Mayer and his Austrian-born wife, Ilse, run this Swiss-chalet style lodge, offering 16 double rooms all with private bath and some with balconies. Most guests stay on AP in ski season, but four kitchenettes are available, and EP is offered in summer. Morning meals at Hotel Edelweiss are what *Gourmet* magazine has called "breakfasts of gargantuan legend"—including crepes, omelets and fresh croissants. In winter, luncheon and dinner are at the next-door Hotel St. Bernard, owned by Mr. Mayer's older brother, Jean. In summer, the Edelweiss serves both breakfast and dinner, and a box lunch is available. The lodge offers a sauna and whirlpool, tennis court, playground, and babysitters. Guests may swim at the St. Bernard's pool. ☐ **Mod to Lux**

Hotel St. Bernard, *(505) 776-2251* ☐ This small friendly lodge is run by Jean Mayer, a former ski champion, instructor at the ski school, and brother of the owner of the Edelweiss. This chalet ringed with balconies offers facilities and charm equivalent to the Edelweiss, has a swimming pool, and shares the Edelweiss tennis court. Especially noteworthy here is the cuisine, prepared by a French chef. The Rathskellar lounge has entertainment and dancing nightly. ☐ **Mod to Lux**

Innsbruck Lodge, *(505) 776-2313* ☐ *Open mid-Nov to mid-Apr* ☐ Owned by Austrians, this two-story lodge is an Old World pension run on AP. Your hostess, Therese Voller, doubles as chef, and the host, Franz Voller, teaches at the Taos Ski School. The 16 wood-paneled double rooms sleep up to four persons, and two small dormitories accommodate skiers on a modest budget. Copious family-style meals combine Austrian treats such as chicken paprikasch and American dishes, and a choice of domestic and imported wines or beers. The lodge has a cocktail lounge with music for dancing, a game room, and a whirlpool. □ **Inexp to Mod**

Kandahar Condominiums, *(505) 776-2226* □ *Open all year* □ This cluster of long, motel-like condominiums is stepped down the mountainside, affording views from the balconies. The duplex apartments can be rented as separate units or together. Upstairs is a large studio with equipped kitchen, bath, and fireplace that sleeps two. Downstairs are two bedrooms (sleeping up to six people) and a bath. □ **Mod**

Rio Hondo Condominiums, *(505) 776-2646* □ *Open all year* □ A tributary of the nearby Rio Grande, the Rio Hondo is a trout-filled mountain stream alongside of which rise the Rio Hondo condominiums. Built of exposed wood and rough timber, this ultramodern complex of semi-attached buildings offers apartments with two, three, and four bedrooms–the largest sleeping up to nine people. Each has a fireplace, equipped kitchen, and private balcony. Guests share a recreation lounge, sauna, whirlpool, and laundry. □ **Lux**

Sierra del Sol Condominiums, *(505) 776-2981* □ *Open all year* □ These 32 two-story condominium units have one or two bedrooms, equipped kitchens, fireplaces, and balconies. Guests share a recreation lounge, sauna, and laundry. □ **Mod**

Thunderbird Lodge and Chalets, *(505) 776-2280* □ *Open all year* □ The main lodge here is a three-story inn with 24 carpeted bedrooms, each spacious enough for a family of five. Nearby are two chalets, each offering five double bedrooms and three two-bedroom suites. The dining room is under the command of a French chef who prepares three fine meals daily. Saturday features a sumptuous buffet. The wine cellar boasts an outstanding stock of domestic and imported wines. Jazz or folk music is heard nightly in the Twining Tavern, with roughhewn beams and a blazing fire. On the musicians' night off, skiers turn amateur entertainers and take over the stage. In summer and autumn, this lodge becomes a cultural center offering workshops in painting, language, and other specialized subjects. □ **Mod to Lux**

VADITO: Sipapu Lodge Ski Area, *Rte. Box 29, Vadito, NM 87579, (505) 587-2240* □ *Open all year* □ *25 mi SE of Taos via SR 3* □ A lovely spot in the Sangre de Cristo Mountains is Tres Ritos Canyon, the location of Sipapu Ski Area. From the large alpine lodge, some of the vistas of steep slopes covered with pine, spruce, and aspen seem almost like scenes in Switzerland. From mid-November until April, the mountain rising directly behind the lodge offers 3 Poma lifts, 16 patrolled trails (3 novice, 10 intermediate, and 3 expert), a ski school, and a ski shop with rentals. In warmer months, life at Sipapu Lodge encompasses archery, fishing in the pond and trout stream, hiking, and in autumn, hunting. Sipapu Lodge contains a cafeteria and dining room featuring Mexican specialties, and a grocery. On the third floor, which can be reached from an outside stairway and balcony, accommodation in 12-bed, carpeted dormitories is less than the price of a lift ticket. Beyond the lodge are 13 adobe housekeeping cottages with apartments sleeping from two to eight people. These heated quarters contain kitchenettes and bathrooms, and the larger ones come with a stone fireplace and living-dining area. Mobile-home units are also offered. □ **Inexp** □ **S, F, H**

New York

THE ADIRONDACKS: Alpine Village, *Box 672, Lake George Village, N.Y. 12845, (518) 668-2193* ☐ *Open all year* ☐ *1 mi N on SR 9N* ☐ On landscaped grounds on Lake George, this resort offers 43 MAP double rooms in an inn, guest lodge, and log cabins. In summer there is swimming on the private sand beach, canoeing, boating, and waterskiing. In winter there is cross-country skiing, ice skating, and sleigh riding. ☐ **Mod** ☐ **W, S, F**

Big Moose Inn, *Eagle Bay, N.Y. 13331, (315) 357-2042* ☐ *Open all year* ☐ *6 mi N of Sr 28* ☐ On Big Moose Lake, site of Theodore Dreiser's **American Tragedy,** this rustic inn, built in 1903 offers 18 rooms, half of which share a bath. The restaurant is known for its fine food; dinners are by candlelight and in good weather, there is outdoor dining. There is a bar plus a lounge with TV. You can swim at the inn or rent boats and water skis at the nearby marina. In winter there is cross-country skiing and snowmobiling. ☐ **Inexp** ☐ **W, F, S**

Black Mountain Ski Lodge and Motel, Inc., *North Creek, NY 12853, (518) 251-2800* ☐ *Open all year* ☐ *On SR 8 5 mi S of North Creek* ☐ Four miles from Gore Mountain Ski Center, this U-shape, two-story Black Forest-style motel has 25 double rooms, a restaurant (specialties include sauerbraten, beef strogonoff, and veal paprikasch, and beer and wine are served) and a small swimming pool. Gore Mountain has five double chairlifts, ski school, rentals, nursery, bar and cafeteria, a skating rink, and snowshoe and cross-country trails. Additional cross-country trails are at the Ski Touring Center in North River. ☐ **Mod** ☐ **S**

Covewood Lodge, *P.O. Eagle Bay, NY 13331, (315) 357-9744* ☐ *Open all year* ☐ *Off SR 28 at Eagle Bay, then 5 mi N on Big Moose Rd.* ☐ Set on the shores of beautiful Big Moose Lake, this rustic inn offers 11 rooms and 18 housekeeping cottages in summer; a dozen of the cottages are winterized and open all year. There is a restaurant, and a TV set in the lounge. Although tennis, a recreation room, and outdoor game areas are available, most activity centers on the lake, where there is a fishing dock and boat rental. The canoe trail on the Fulton Chain of Lakes originates at Old Forge and runs 125 miles north. Nearby is the Old Forge ski area. ☐ **Mod** ☐ **W, F**

Hidden Valley Guest Ranch, *Lake Luzerne, NY 12846, (518) 696-2431* ☐ *Open mid-May to mid-Mar* ☐ *5½ mi N of Lake Luzerne off SR 9N* ☐ On lovely wooded hills on Lake Vanare, this is a Western-style ranch offering 85 MAP rooms in more than a dozen low lodges. Many rooms are paneled, and some have Western decor and fireplaces. The three dining rooms have fireplaces, and in addition to a weekly smorgasbord and Sunday buffet breakfasts, cookouts are featured. There are six lounges for cocktails, TV, or indoor games, and dancing and entertainment are featured nightly.

The year-round emphasis here is riding, with instruction, picnic rides, and surrey rides. In winter the horses are shod for winter riding and sleigh rides are offered. In addition to the large outdoor pool there is a pretty sand beach on the lake, where canoeing, boating, and fishing are also available. Archery, bicycling, and lawn games are offered, and a full program of evening activities includes beach parties and floodlit pool and game areas. For an extra fee boat rides on Lake George are offered. In winter advanced skiers go to Gore Mountain, but Hidden Valley has its own slope with a chairlift for beginners, a ski school, rentals, snowmaking, and snack shop. Ice skating and snowmobiling are also offered. ☐ **Mod** ☐ **R, W, F, S**

Holl's Inn, *Inlet, NY 13360, (315) 357-2941* ☐ *Open July to Labor Day* ☐ *Off SR 28* ☐ Operated by the Holl family since 1935, Holl's is a friendly, 150-acre enclave on Fourth Lake. The main hotel, a dark-wood structure with white trim, overlooks the lake and offers rooms with one to four beds and private bath. The 25 rooms in a motel-style annex have picture windows with lake views; most have two double beds. This resort operates in AP, and in the main hotel are the lakeside dining room, Tyrolean cocktail lounge, and lounges. You can swim and fish on the 2000-foot lakefront, where canoes and rowboats are provided free; other boats are available at nearby marinas. Also offered are a tennis court, badminton, and babysitters. Golf, tennis, horseback riding, and other activities are available in Inlet or Old Forge. ☐ **Inexp to Mod** ☐ **W, F**

Interlaken Lodge, *15 Interlaken Ave., Lake Placid, NY 12946, (518) 523-3180* ☐ *Open all year* ☐ In the center of town, this is a Victorian hotel with a European atmosphere. Rooms are spacious, and some have four-poster beds and other antiques. The restaurant features Swiss food and wines, and there is a cocktail lounge. Rates include continental breakfast. The lodge is convenient to Lake Placid's watersports and ski areas. ☐ **Mod**

Lake Placid Resort Hotel, *Mirror Lake Drive, Lake Placid, NY 12946, (518) 523-3361* ☐ *Open all year* ☐ This large complex offers some 400 guest rooms and suites on Mirror Lake. The restaurant has colonial decor, does its own baking, and often has dancing. This resort has a pool as well as a bathing beach on the lake, canoe and sailboat rental, 45 holes of golf including two 18-hole courses, 14 tennis courts, outdoor paddle tennis, indoor squash courts, skeet and trapshooting, horseback riding, and other activities. Rental equipment and instruction are available. In winter, 12 miles of cross-country trails are maintained and there is ice skating. □ **Mod to Lux** □ **W, S, T, G**

Mirror Lake Inn, *35 Mirror Lake Dr., Lake Placid, NY 12946, (518) 523-2544* □ *Open all year* □ *½ mi W off SR 86* □ This attractive complex, on eight-acre hillside overlooking Mirror Lake, offers a variety of units in lodges and cottages. Many units have terraces and lake views. The pleasant restaurant, overlooking the lake, has a salad bar and specializes in beef and seafood. Also here are an outdoor cafe and cocktail lounges.

On the lake the inn has a sandy swimming beach and a dock for fishing and boat rental. Also on the premises are a pool, wading pool, tennis courts, outdoor game areas, and a health spa. The inn is convenient to Lake Placid's famous downhill and cross-country ski areas and bobsled run. □ **Mod to Lux** □ **W, F, S**

Mohawk Inn, *Fourth Lake, NY 13420, (315) 357-2401* □ *Open late May to mid-Oct, and late Dec to mid-Mar* □ *Off SR 28* □ On Fourth Lake in the Fulton Chain of Lakes, this rustic inn has been in operation since 1896. The three-story inn offers 28 rooms and on the wooded grounds are 15 cottages with kitchens, fireplaces, and one, two, or three bedrooms. There is a restaurant and bar. On the lake is a private sand beach and a fishing dock. Mohawk provides tennis, putting green, and indoor and outdoor recreation areas; golf and riding are available nearby. This is a good place for families, with supervised play and many planned activities in summer. Canoeing and lake cruises are available, and the Old Forge ski area is ten miles south. □ **Inexp to Mod** □ **W, F, S**

Placid Manor, *Box 870, Lake Placid, NY 12946, (518) 523-2573* □ *Open late May to Labor Day* □ *3 mi NW of Lake Placid off SR 86* □ Set on a private stretch of beach with a beautiful view of Lake Placid and Whiteface Mountain, this cluster of motel-style rooms and cottages offers more than 40 units, a coffee shop, cocktail lounge, cardrooms, and lounge with TV. The lake has a swimming beach, good fishing, and boats and waterskiing are available. □ **Mod** □ **W, F**

Roaring Brook Ranch & Tennis Resort, *Lake George Village, NY 12845, (518) 668-5767* □ *Open all year* □ *2½ mi S on SR 9N* □ This AP ranch offers rates for nonriders and riders, and accommodation in 95 air-conditioned rooms in one- and two-story modern lodges. On the 500 acres are, in addition to the stables, five tennis courts with pro services, three swimming pools (one indoor), saunas, indoor and outdoor game areas, and a restaurant bar. Golf can be arranged nearby. In summer there are also cookouts and children's programs. In winter, activity focuses on Gore Mountain, 27 miles away. □ **Inexp to Mod** □ **R, T**

Rocky Point Inn, *P.O. Box 447, Inlet, NY 13360, (315) 357-3751* ☐ *Open mid-May to mid-Oct* ☐ *Off SR 28* ☐ Hugging the shores of Fourth Lake, Rocky Point's 175 forested acres culminate in a gently curving little peninsula rimmed with a natural sand beach. The inn is a large, four-story country-style house with 21 doubles and four singles upstairs; at ground level are the dining room (dress code evenings), game room, cocktail lounge, and other lounges. To either side of the inn are 28 cottages of dark shingles and white trim with paneled walls, fireplace, a sitting area, and two double beds, plus room for extra cots. All units have private bath. The AP rates include tennis and other facilities as well as scheduled activities for adults and teens and a supervised program for children. Behind the inn are five Har-Tru tennis courts and a pro shop operated by a teaching pro. There is swimming from the 300-yard beach, fishing, a 27-hole putting green, lawn games, and a game room. The inn has a waterskiing center with drivers and instructors, plus rentals of canoes and other crafts. Golf courses are nearby. ☐ **Mod** ☐ **T, W, F**

Sagamore, *Green Island, Bolton Landing, NY 12814, (518) 644-3131* ☐ *Open mid-May to Labor Day* ☐ *½ mi E of SR 9N* ☐ On the shore of Lake George, with a beautiful view of wooded islands and mountains, this resort offers 192 air-conditioned rooms, a few of which share a bath. On the grounds are indoor and outdoor olympic-size pools, a health club, tennis courts, beauty and barber shops, a restaurant, and a bar with dancing and entertainment. At the private beach you can rent boats. The 18-hole golf course climbs above the lake, offering lovely views. Horseback riding is available nearby. ☐ **Mod** ☐ **W, G, F**

Wawbeek Inn, *Tupper Lake, NY 12986, (518) 359-3800* ☐ *Open late May to Labor Day* ☐ *9 mi N of Typper Lake on SR 30* ☐ On Upper Saranac Lake, this small resort offers EP accommodations in an older two-story inn and newer annex as well as in 10 cottages; a dozen units have kitchenettes. The inn has a restaurant, bar, and game room. On the lake guests will find a private sand beach, fishing, and boat rental. The grounds of the inn also have tennis, a playground, and lawn games. Riding stables and an 18-hole golf course are nearby. ☐ **Mod** ☐ **W, F**

Whiteface Chalet, *Wilmington, NY 12997, (518) 946-2207* ☐ *Open all year* ☐ *OFF SR 86, 1½ mi from Wilmington* ☐ This alpine-style chalet with wooden balconies offers 20 spacious rooms and suites (some units will sleep a family of six) on MAP as well as a two-bedroom chalet with kitchen. The lodge has a parlor, a dining room with mountain view, a playroom, bar, and lounges. On the 10-acre wooded grounds are a large outdoor pool, sauna and whirlpool, a playground, lawn game areas, a tennis center with five all-weather courts, teaching pro, and rentals. In winter the highlight is the proximity to Whiteface Mountain Ski Center with 7 chairlifts, 28 trails, and a 3216-foot vertical drop. There is snowmaking, a ski school, rentals, and a large base lodge. Cross-country trails, dogsledding, ice skating, tobogganing, snowmobiling, and special events such as ski jumping and bobsledding are also available either here or at nearby Paleface Ski Center. In summer the nearby attractions include High Falls Gorge on the Ausable River, toll road and chairlift to Whiteface summit, and hiking. □ **Mod to Lux** □ **S, T**

BEAR MOUNTAIN: Bear Mountain Inn, *Bear Mountain State Park, Bear Mountain, NY 10911, (914) 786-2731* □ *Open all year* □ *Off SR 9W* □ Within the boundaries of Bear Mountain State Park, over 50,000 acres of mountains, woods, fields, and streams rising above the majestic Hudson River, this inn has an unusually vast "backyard" and is close to many historic sights, including West Point and Fort Clinton. Large numbers of tourists flock to Bear Mountain in summer, the season when it is possible to cruise to the inn on an excursion boat from New York City. The park is less crowded in spring and for fall foliage, but in winter many visitors come for ice skating, sledding, cross-country skiing, ski jumping, and carnivals. Within walking distance of the inn is a public swimming pool and a zoo, as well as hiking trails. Lakes in the Harriman section, which have swimming, fishing, and boating, and the Silver Mine Ski Area are within driving distance. The inn consists of a log and stone hunting-style lodge with massive public rooms and five modern lodges. All buildings have large fireplaces in the lounges. About half the rooms are rustic and half are modern. The beamed restaurant has butcher-block tables and cane-seat chairs. Weekends feature a smorgasbord and dancing. □ **Inexp**

THE CATSKILLS: Over the years the name of this mountain range, within a hundred miles of New York City, came to designate a group of resorts such as Grossinger's and the Concord–large-scale operations offering full service, European-spa atmosphere, and everything from golf to tennis to horseback riding. But these hotels were especially noted for their nightclubs, and because they offered kosher as well as American menu, entertainers called it the "Borscht Circuit." Many well-known performers got their start here, and up-and-coming as well as big-name entertainers are still featured. The mountains themselves are noted for scenic lakes, rolling hillsides, and fall color. Catskill Park, a 700,000-acre mountain preserve, offers many recreation sites, trout fishing, hiking trails, and the Belleayre Ski Center, as well as a game farm and other family attractions. Ice Cave Mountain (four miles south of Ellenville on SR 52) has unique geological formations–ice caves, a balanced rock, a lost river, and a mountaintop lookout.

Concord Hotel, *Kiamesha Lake, NY 12751, (914) 794-4000* □ *Open all year* □ *On SR 42, 2 mi N of Monticello* □ The Concord is among the largest resorts in the world. Everything about the complex is outsized–the three mammoth, contemporary hotel wings housing 1200 doubles and 170 suites are the center of 300 wooded acres and immediately surrounded by 40 tennis courts (16 indoor), 45 holes of golf, shops, gardens, lakes, even a ski area. More a town than a resort, the Concord offers a complete one-stop vacation, winter as well as summer. The ski area, not terribly awesome with its vertical drop of only 150 feet, is nevertheless serviced by four T-bars and two rope tows, snowmaking equipment, and a base lodge. An 820-foot toboggan run, indoor and outdoor ice skating, a pond for ice-fishing, and separate snowmobile and cross-country trails round out the cold-weather facilities.

Summer, of course, provides even more things to do. The three golf courses (two 18-hole, one 9-hole) include the stupendously tough Joe Finger championship course, one of the most challenging in the world. Probably the trickiest hole is the third, 562 yards with water hazards from tee to green. Besides dining facilities and pro shop, the course's clubhouse has 40 guest rooms for avid golfers. The second 18-hole golf course, the International, has its own clubhouse. All three golf courses border or overlook Kiamesha Lake. Although indoor courts are open in winter, tennis comes into full swing in summer, when several instructional programs are offered. During warm weather, horses are available for trail rides, and the outdoor, 280-foot-long swimming pool is open. Guests will also find basketball courts, softball diamonds, skeet shooting range, and badminton, handball, and volleyball courts.

All year round, the health clubs keep their whirlpools whirling and their steam rooms steaming, and every night the three nightclubs and ten dining areas present guests with great quantities of fine food and entertainment. The resort has social and sports directors and full services. ☐ **Mod to Lux** ☐ **T, G, R, S**

Golden Acres Farm & Ranch, *Gilboa, NY 12076, (607) 588-7329* ☐ *Open mid-May to Labor Day* ☐ *Off SR 30* ☐ Golden Acres, a family-run operation for three decades, is a 400-acre farm that is especially appealing to families or single parents. The nursery cares for infants, a day camp is for kids up to twelve, and there is a teen program. Children will find goats, duck, ponies, playgrounds, and outdoor games areas. EP and meal plans are available. Accommodation is in rooms and suites in a complex consisting of the modernized farmhouse buildings or in a separate cluster of motel units; kitchenette apartments are also available. In the knotty-pine dining room, the kosher kitchen specializes in homemade soups and breads. The 20 saddle horses are available by the hour, for lessons, or for trail rides. Tennis on the two lighted macadam courts and rowboats and pedal boats for the pond, which is stocked with bass, are free. Also on the grounds are a heated outdoor pool and a recreation hall. In the evenings square dances, cocktail parties, parlor games, and other activities are held. ☐ **Inexp to Mod** ☐ **R**

Granit Hotel and Country Club, *Kerhonkson, NY 12446, (914) 626-3141, (212) 563-1881* ☐ *Open all year* ☐ *Off SR 55, W of New Paltz* ☐ At Granit city folk can escape to the country without suffering culture shock, because the 350 rooms and suites are in a high-rise tower. Most rooms have balconies and all have room service and air conditioning. Operating on AP, the Granit serves meals at round tables in a spacious dining area reminiscent of a luxury liner dining room. Also featured are two cocktail lounges and a nightclub. On the 400 valley acres, with mountains rising in the distance, are indoor and outdoor pools, sauna, outdoor game areas, free tennis on eight courts, paddle tennis, free rowboating and fishing on the stocked (bass) lake, and nature trails. At the 18-hole golf course designed by L. Block and complemented by a putting green and driving range, instruction is available at the pro shop. Horseback riding is an extra fee. In winter, activities range from downhill and cross-country skiing, tobogganing, and indoor and outdoor ice skating. ☐ **Mod** ☐ **T, G, W, S**

Grossinger Hotel, *Grossinger, NY 12734, (914) 292-5000* ☐ *Open all year* ☐ *Off SR 17 at exit 100* ☐ This gigantic, well-known resort has 600 air-conditioned rooms in a variety of hotels and lodges. The 800-acre estate includes 27 holes of golf redesigned by Joe Finger (the 18-hole Big G course and nine-hole Vista course); driving range and putting green; 12 all-weather outdoor tennis courts and several indoor courts; tennis and golf pro shops; large indoor and outdoor pools; wading pool; boating and fishing on the private lake; indoor miniature golf; Ping-Pong; volleyball; basketball; recreation room with electronic games; hiking and jogging trails; sport chalets with fireplace; small ski area with T-bar, rope tow, tobogganing, ice skating, and snowmobiling; shopping arcade; and health clubs with saunas, exercise room, and whirlpool. Most of these facilities are in the main complex.

The list of social activities and programs is equally impressive. Children four and older have a complete day camp and separate dining room in season. Babysitters are available for infants, and teenagers have their own program. The nightly all-star entertainment and the dancing in the Pink Elephant lounge and disco is legendary. Other highlights in the social calendar include dance lessons, weekly art auction, and make-up demonstrations. The resort operates on AP, which includes weekly cocktail party, special diets, and early and late continental breakfasts. ☐ **Mod** ☐ **G, T, W, H**

Kutsher's Country Club, *Monticello, NY 12701, (914) 794-6000, (212) 243-3112* ☐ *Open all year* ☐ *Exit 105B of SR 17 Quickway* ☐ Kutsher's has nearly everything a late-20th-century person on leave from everyday life could want. The AP accommodations include a variety of modern rooms and suites–some with balconies, patios, fairway views, or other features–in a tower and lodges scattered on the tree-shaded, 500-acre grounds by a small private lake. A chic dining room is supplemented by several cocktail lounges, some with dancing and professional nightclub acts.

The health club activities include whirlpools, exercise gym, massage, yoga, aerobic dancing, running, and other programs, plus a glass-enclosed indoor pool. The huge outdoor pool, with a large terrace for poolside service, has an adjacent wading pool. Tennis includes a dozen indoor, outdoor, and lighted courts, plus pro. The good 18-hole golf course is 7157 yards and comes with pro shop, instruction, driving range, putting green, and rental carts. The private lake has fishing and boating. Also available are horseback riding, playing fields, bicycling, and year-round ice skating. In winter you'll find gentle ski slopes, instruction, horse-drawn sleighs, tobogganing, ice fishing, and ski instruction. Kutsher's has a day camp and babysitting for small children and a clubhouse for teenagers. □ **Mod to Lux** □ **T, G, W, R**

Mohonk Mountain House, *Mohonk Lake, New Paltz, NY 12561, (914) 255-1000; (212) 233-2244* □ *Open all year* □ *6 mi W of New Paltz* □ This unique resort in the Shawangunk Mountains began in 1869 when the Smiley brothers, Quaker schoolteachers, bought a ten-room tavern on the shores of crystal-blue Mohonk Lake. Here they created a resort devoted to prayer, nature walks, and inspiring lectures. Drinking, dancing, and card playing were not allowed. For more than 30 years the Smileys expanded their hotel until it was a marvel of exuberant, eclectic architecture an eighth of a mile long. This wild, rambling building of wood, brick, and stone rises as high as eight stories in some sections, and is capped with a variety of towers, chimneys, and roof decoration that vary from Austrian to Italian in inspiration. Public rooms are also individualistic, and today dancing, smoking (except in the dining room, main parlor, and library) and card playing are permitted, and TV is available in the lounges. There is no bar, but cocktails and wine are served in the dining room, and bottled liquor is sold at the service desk. Few of the 300 guest rooms are alike, and the most coveted (and expensive) are the tower rooms. Some rooms share bath, more than 250 have balconies with lake or mountain views, and more than 125 have fireplaces. Also on the grounds are eight kitchenette cottages (with fireplaces) that sleep up to six people; these are open early May to late October.

Extra charges are levied for the pretty 9-hole golf course (carts available), tennis, fishing for trout in the stocked lake, canoes and boats, horseback riding, and carriage rides. Also on the grounds are a swimming beach, croquet and other lawn games, an 18-hole putting green, and a museum where demonstrations of old handcrafts are given. Hayrides, cookouts, clambakes, and other special events are offered. Winter activities include ice skating on the lake, snowshoeing, some 20 miles of maintained cross-country ski trails and another 20 miles of unmarked trails for the more adventuresome. Lessons, equipment rental, guided ski tours, and horse-drawn sleigh rides are also available. Mohonk also offers seasonal supervised play, babysitters, a beauty shop, and other services.

However, what makes Mohonk unique is the exceptional beauty of the site and the humanitarian heritage kept alive today. Under the Smiley twins, Mohonk became a conference center for people concerned about peace, the plight of Indians, conservation, and other social issues. Today Mohonk is a nature preserve and much of the former Smiley land is public park. The resort operates on AP, but many packages are available, and these frequently include special programs on stargazing, birding, holistic medicine, earth watch, nature seminars, and hiking and running holidays. The 2500 acres are laced with 60 miles of carriage roads and trails so extensive and contorted that you need a map; some trails are easy strolls through award-winning gardens dotted with benches and gazebos, but the terrain is steep and rocky enough to suit dedicated hikers. Experts conduct nature tours and garden walks. Caving expeditions are also available. Nearby are the Trapps, incredible cliffs offering the best rock climbing in the Northeast as well as spectacular hawk flights in the fall migration period. Just outside the resort, the public park areas offer ice caves, rock formations, waterfalls, and Lake Awosting for watersports. □ **Mod** □ **W, R, H**

Nevele, *Ellenville, NY 12428, (914) 647-6000* □ *Open all year* □ *1½ mi S of Ellenville on US 209* □ The main attraction at this 450-room resort is the full range of year-round sports activities. The pleasant 18-hole golf course (with cart and club rentals) plays around a lake roomy enough for the rental sailboats, rowboats, and paddleboats together with the few fishermen trying their luck. Besides several outdoor tennis courts, the resort boasts five indoor courts as well as racquetball and platform tennis courts. Bicycles and horses can be rented for rides around the 1000-acre property, which also contains within its boundaries ball fields, recreation rooms, miniature golf, and heated olympic-size pool. An indoor pool, sauna, steam room, and small gym are open all year. At the foot of the modest ski slope, near the chair lift, is the huge indoor ice skating rink. From the elegantly modern lines of the rink's roof trusses to the ten story high-rise hotel tower, the emphasis is on the contemporary. To keep things lively at night, three orchestras play in the ballrooms and lounges. Rooms are well furnished, up to date, and air conditioned. The Nevele runs on AP; kosher and American menus are available. Among the many services offered is a children's program of supervised activity. □ **Mod** □ **G, T, R**

The Pines, *South Fallsburg, NY 12779, (914) 563-3760* □ *Open all year* □ *Exit 107 of SR 17 Quickway* □ This complex of boxy modern buildings rising three and four stories above lawns and fairways offers 400 motel-style double rooms and four honeymoon suites on EP, meal plans, and packages. Units are air conditioned but a few rooms share a bath. The buildings are interconnected by covered sidewalks to the dining room, which is decorated in varying shades of red, the nightclub, and the lounges. Tennis facilities include four indoor courts (extra charge), eight outdoor courts, and a pro shop with rentals. There are indoor and outdoor pools plus a health club with massage and other services. Also here are a nine-hole golf course, an indoor ice-skating rink, and a disco roller rink. A toboggan run and ski slopes with artificial snow, chairlift, ski school and shop, and night lights keep you active in winter. For children there is a supervised day camp, nursery, playground, and separate dining room. □ **Mod to Lux** □ **T**

Rocking Horse Ranch, *Highland, NY 12528, (914) 691-2927* □ *Open all year* □ *Off SR 299* □ That this is a dude ranch is suggested by the attractive Western-modern architecture of its two-story wood-and-glass buildings and confirmed by the size of the stables. In the ranch's 119 double rooms and dorm units, rented on MAP and AP, color TV, air conditioning, and wall-to-wall carpets are standard. Family-style meals are served in the glass-walled dining room or on the patio, and there is a quiet cocktail lounge as well as the Round-up Room nightclub. Saddle horses, riding instruction, and trail rides are part of the story, but you can be severely allergic to hay and never at a loss for something to do. The ranch has a private lake for fishing, boating, and water-skiing. There is a glass-enclosed indoor pool, an outdoor pool with a slide, sauna, and poolside service. Other facilities include lighted, all-weather tennis courts with a pro, target shooting, archery, bicycle riding, mini-golf and other outdoor games, and a recreation room. A playground, sitters, and a children's program are available. In winter, cross-country skiing and snowmobiling. □ **Mod** □ **T, R, W**

Roundup Ranch, *Downsville, NY 13755, (607) 363-7300* □ *Open all year* □ *Off SR 206 3 mi N of Downsville* □ This family-oriented ranch offers 40 rooms in the main farmhouse (some rooms share bath), frame lodges, motel units, and log cabins scattered on a grassy slope in Catskill foothills. The various units sleep between two and six people. AP rates include all ranch activities. The dining room and a dining terrace overlook the outdoor pool. There is a bar and a lounge with entertainment. On the grounds are an indoor pool, saunas, separate game rooms for children and adults, a tennis court, and out-door game areas, but the main feature is year-round riding in the ranch's arena (where Saturday night rodeos are held through October). All-day picnic rides, cookouts, surrey rides, and riding instruction are also offered. In winter there is cross-country skiing and a slope (with rope tow) suitable for beginning skiers; instruction is free and equipment may be rented. There is also an indoor ice-skating rink and beside it a heated glass-enclosed lounge. Snow-mobiling and horse-drawn sleigh rides are also available. □ **Mod** □ **R**

Tennanah Lakeshore Lodge, *Roscoe, NY 12776, (607) 498-9900* □ *Open late May to mid-Oct* □ *Exit 94 on SR 17 Quickway* □ This elegant, old-fashioned lodge, on the shore of the largest lake in the Catskill Mountains, offers 180 rooms on MAP. Some rooms share a bath, some have lake views. The dining room, TV room, and card room are air conditioned. The resort offers lake swimming, an indoor pool, a health club with saunas and whirlpool, and lawn games. Rowboats are provided free, and the lake has pickerel and bass. □ **Inexp** □ **W, F**

Williams Lake Hotel, *Rosendale, NY 12472, (914) 658-3101* □ *Open all year* □ *Off SR 213* □ Surrounded by a private 600-acre preserve, this complex clusters around the shore of a small stocked lake. Almost 100 contemporary units are offered on AP in the farmhouse-style main building, a motel, and sleeping cottages. Meals are served family style in the lakeside dining room, and there is dancing in the cocktail lounge. Summer guests have access to a guarded swimming beach, a small fleet of canoes and Sailfish, three Har-Tru tennis courts, wide-ranging hiking paths, and facilities for lawn games. In winter there is polar-bear swimming, massage, sauna bath, a downhill ski slope, and snowmobiling, but the emphasis is on cross-country skiing. Rentals and instruction are available. ☐ **Mod** ☐ **W, S**

Winter Clove, *Round Top, NY 12473, (518) 622-3267* ☐ *Closed one week at Christmas* ☐ *Off SR 23* ☐ On the fringe of Catskill Mountain State Park, Winter Clove was originally a farmhouse that an ancestor of the current managing family transformed into an inn. This building has a wide porch running the facade, lined with rocking chairs and geraniums. Today capacity has been extended to more than 50 units with a motel annex. Some of the best of the farmhouse tradition has been preserved in the homemade breads, soups, and desserts that are part of the AP rates. Winter Clove has a tennis court, a putting green and nine holes of golf, indoor and outdoor swimming pools, hiking trails, six bowling lanes, areas for playing team sports and lawn games, and a recreation hall. Horse-drawn hayrides and bingo are among the activities. Horses can be hired nearby. In winter there is cross-country skiing. ☐ **Inexp**

COOPERSTOWN: Otesaga Hotel, *P.O. Box 311, Cooperstown, New York, 13326, (607) 547-2567* ☐ *Open early May–Oct* ☐ *On SR 80 (Lake St.)* ☐ A magnificent white-columned brick edifice reminiscent of the Dearborn Inn in Michigan or the Homestead in Virginia, the Otesaga is one of the northeast's great resorts. The public rooms are appropriately stately, done with chandeliers, girandoles, wing chairs, and the like. Besides the comfortable double and single rooms at the hotel, about thirty rooms are offered at the charming Cooper Inn, an elegant Federal-style building of yellow brick. MAP is offered at the hotel only, but the inn shares all of its other programs and facilities.

The focus of the resort is Otesega Lake. Verandas on the hotel overlook it, and the 18-hole golf course plays around it. A dock reaching far out into the lake is good for swimming, fishing, and sunbathing. On the grounds are tennis courts, game room, supervised play area, and heated pool. A babysitter list and programs for children over three are available.

The main dining room has hearty American fare and serves poolside buffets in season. The Hawkeye Bar and Glimmerglass Lounge are for cocktails, and the Templeton lounge has live entertainment and dancing most nights. ☐ **Mod** ☐ **G, W, F, T**

DEPOSIT: Scott's Oquago Lake House, *Deposit, NY 13754, (607) 467-3094* ☐ *Open late May to mid-Oct* ☐ *4 mi N of Deposit on Oquaga Lake* ☐ Six generations of the Scott family have been accepting guests on their 1000-acre lakeside retreat since 1870. The 110 double rooms (some share a bath) are in the old lodge on the lakefront and in a number of other lodges on the wooded grounds. The AP rates include family-style meals (salad and dessert bars at night), but there is no bar, and drinking is allowed only in your room. This is a homey resort where all facilities and activities are included in the rates and no tipping is allowed. The Scotts compare themselves to the Trapp family and often go on entertainment tours to publicize their resort. Entertainment is scheduled nightly; this may include dancing to a live band, square and folk dancing, stage reviews, singalongs, or talent shows. The Scotts also offer informal exercise classes, lake cruises on a small showboat, speedboat rides, van trips, and many other group activities, and guests often form softball teams or have shuffleboard tournaments. Supervised programs for children are offered. On the grounds are many outdoor game areas, a nine-hole golf course, four tennis courts (one indoor), bowling alleys, and hiking trails. The lake has swimming, fishing, waterskiing, and canoes, sailboats, and rowboats are supplied free. □ **Mod** □ **W**

THE FINGER LAKES: Finger Lakes State Parks, *R.D. #3, Trumansburg, NY 14886, (607) 387-7041* □ *May to mid-Oct* □ Spanning all or parts of 13 counties in western New York, the Finger Lakes Region is a beautiful area of shimmering lakes, magnificent glens, gorges, and rolling countryside. The State of New York maintains 17 parks in the region, seven of which rent rustic vacation cabins from late spring into autumn. All of the parks have swimming, most have fishing and naturalist programs, and three have marinas. All cabins have a porch, electricity, and refrigerators. Most have a stove and an indoor fireplace; a few have outdoor fireplaces and no stove. Cabins range in size from one room sleeping up to four people to those with three to five rooms sleeping five to seven people. In all, you supply bedding and household effects. Very few have inside toilets—communal comfort stations are the rule. □ **Inexp** □ **W, F**

Rainbow Cove Motel, *Himrod, NY 14842, (607) 243-7535* □ *Open May–Nov* □ *1 mi E of SR 14* □ On nine acres on Lake Seneca, this is a friendly family resort where the home-cooked food is passed on platters in the dining room. The 24 rooms in the two-tier motel have balconies and lake views. On the shaded lawns are outdoor game areas, private lake swimming beach, and fishing dock. The lake is noted for large lake trout, rainbows, and Atlantic salmon, and guides are available. Evenings there are campfires, sometimes the innkeeper entertains, and babysitters are available. □ **Inexp** □ **F, W**

Springside Inn, *Rte. 38 South, Owasco Lake, Auburn, NY 13021, (315) 252-7247* □ *Open all year* □ *1 mi S of Auburn via SR 38* □ A five-minute stroll from the shores of Owasco Lake, this three-story, eight-bedroom traditional inn is set on lawns and painted red with white trim. Flanking the original

Victorian structure are two recently added wings that house two dining rooms, lounge areas, and an open porch. Furnished in mid-Victorian style, a few of the air-conditioned bedrooms share a bath, and some are large enough to accommodate couples with children. A complimentary breakfast–fruit, homemade, buttered muffins, and coffee–helps start the day. The inn has a restaurant (closed Mondays) specializing in fish and beef; wine and cocktails are available as well as Sunday brunch. During July and August, the Springside Dinner Theater, a company of professional actors, performs nightly. Nearby are public swimming and boating facilities on Owasco Lake, downhill and cross-country ski areas, two riding stables, two ice skating rinks, tennis courts, and golf courses. ☐ **Inexp**

Taughannock Farms Inn, *RD 3, Trumansburg, NY 14886, (607) 387-7711* ☐ *Open Easter–Thanksgiving* ☐ *10 mi N of Ithaca on SR 89* ☐ This three-acre farm landscaped with gardens overlooking Cayuga Lake and Taughannock Falls State Park, offers only four rooms and one private bath, in a lovely three-story inn, with shuttered windows and a wide porch. The inn, built in 1873 and furnished with Victorian pieces, was once the home of a wealthy Philadelphian. Meals here are notable and the menu includes roast duck or turkey, prime rib, and lobster. Rolls are from a Norwegian recipe handed down by the family who built the house. There is fishing and swimming in the lake, and the state park marina offers boat rental and good birdwatching. Nearby Cornell University has test gardens, a bird observatory, and art galleries. ☐ **Mod**

LETCHWORTH STATE PARK: Probably the most dramatic state park in New York, Letchworth surrounds the spectacular 17-mile gorge of the Genesee River. The forested canyon walls rise 500 feet over cascades and waterfalls. The park is a 14,000-acre wildlife sanctuary, and woodchucks and other animals and birds can be seen along the scenic drive and on the 60 miles of hiking trails. Near Middle Falls is a museum of pioneer and Indian history. On a ridge behind the museum is the grave of Mary Jemison, a captive adopted by the Seneca Indians, and a Seneca memorial. At the north end of the park is an olympic-size mountaintop swimming pool with wonderful vistas and separate diving and children's pools. At the south end is another pool, near the Lower Falls, picnic grounds, and game areas. The park also offers several picnic areas and fishing, and rents cabins in wooded areas near scenic spots. Some have private bath and others share centrally located comfort stations. All cabins have stove, refrigerator, cots, mattresses, and chairs, but you must bring your own bedding, dishes, and cooking utensils. For park and cabin information, write Letchworth State Park, Castile, NY 14427. ☐ **Inexp** ☐ **H, F**

Genesee Falls Inn & Motel, *P.O. Portageville, NY 14536, (716) 493-2484* ☐ *Open all year* ☐ *On SR 436, just outside S entrance to park* ☐ This three-story inn, built in the 1850s, is set on pretty grounds on the banks of the Genesee River. It offers 19 rooms in the inn and five rooms in an air-conditioned motel annex. All units have private bath. The restaurant, in period decor, offers seafood, steak, a salad bar, and home baking, and there is a bar. □ **Inexp** □ **H, F**

Glen Iris Inn, *Castile, NY 14427, (716) 493-2622* □ *Open Easter to early Nov* □ *In park, 1 mi N of S entrance* □ Overlooking the grand, 107-foot Middle Falls of the gorge is Glen Iris Inn, named for the rainbows frequently seen rising above the cascades. Built in the mid-1800s, the inn was once the home of William Letchworth, who donated part of the gorge to the state. This stately building, with white columns supporting the balustraded roof two stories above the porch, still houses the Letchworth family art collection and furnishings. It offers 13 air-conditioned rooms and suites, all with private bath. The choice unit is the honeymoon suite, with a balcony over the falls. Nearby are seven motel units with kitchenettes; these sleep up to four or five people. The restaurant is charming and serves three meals daily, specializing in seafood, salad bar, and home-baked breads and pastries. Cocktails and wine are available. □ **Inexp to Mod** □ **H, F**

LONG ISLAND: Dering Harbor Inn, *13 Winthrop Rd., Shelter Island Heights, NY 11965, (516) 749-0900* □ *Open May-Nov* □ *SR 114 to Winthrop Road* □ Shelter Island is an asymmetric bit of land out in Gardiners Bay, which cleaves the tip of Long Island. Linked to the mainland by ferries, the island caters to homeowners and appeals to those who want to slow down rather than live it up. Perched on a hill overlooking Dering Harbor, the inn offers 30 rooms and suites in several modern cedar-shingle buildings shaded by tall, stately trees. All units have private porches, air conditioning, color TV, and wood paneling; most have kitchenettes and dining areas, and some have fireplaces. The largest units have two bedrooms and two baths, but even the one-bedroom units will sleep three or four people. Rooms are rented EP, and those on the waterfront are priced highest. The restaurant–steak and seafood–is unpretentious, yet solid, and dress code is enforced at dinner. On the premises is a large freshwater pool and two all-weather tennis courts. Guests are permitted to play at a nearby 18-hole golf club. Rental sail and motorboats, and charter fishing, are also available locally. □ **Lux**

Gurney's Inn, *Montauk, NY 11954, (516) 668-2345* □ *Open all year* □ *On the Old Montauk Highway* □ This luxury inn is a charming complex of singled Cape Cod beach cottages, living lodges, and communal buildings set on landscaped grounds along a 100-foot private beach on the Atlantic Ocean. The 128 units are varied–many have patios, terraces, refrigerators, sunken living rooms, fireplaces, or other extras. Despite color TV, air conditioning, and modern furnishings, an attempt has been made to give all lodgings a casual, woody feel. Four dining rooms offer sea views, buffets, and specialties such as roast duckling, lobster, and local seafood. Beach, pool, and room service are available. Early risers may arrange for breakfast as early as 4 am. The cozy bar has entertainment. There are lounges, beauty and barber shops, sports and social directors, and other services. Within a few miles are marinas with rental boats, charter boats, party boats, and guide service as well as stables offering beach and trail riding and good bird watching areas. Gurney's also has an ultramodern health and beauty spa with saunas, herbal wraps, whirlpools, steam rooms, massage, gymnasium, and a number of therapies and cosmetology services. The spa has its own glass-enclosed lounge and swimming pool, and on the roof is a terrace with lawn games, jogging track, and sun decks. This resort complex is geared to adults, and does not provide cots or other special services for children under twelve. The spa facilities are not included in the MAP rates. ☐ **Lux**

Montauk Yacht Club & Inn, *Star Island, Montauk, NY, 11954, (514) 668-3100; (212) 895-6446* ☐ *Open all year* ☐ *2 mi N of SR 27* ☐ A knob of dunes almost entirely surrounded by the peaceful waters of Lake Montauk, Star Island is the setting of the Montauk Yacht Club & Inn, an elegant port of call. The Club provides MAP and EP, plus a sizable marina for guests who want to dock their boats and even live on board. In the center of the complex stands an operating 60-foot high lighthouse, and following the curve of the shoreline are low-lying, naturally weathered cedar-shingle buildings, discretely modern in style. Some units face a crescent-shaped beach and others the marina. All have floor-to-ceiling glass doors leading to private terraces or patios. The chic brown-hued interiors feature wood paneling and oversize beds. In addition to sophisticated dining and bar with entertainment indoors, there are informal deck areas for snacks and drinks. The attractive indoor pool has whirlpool, saunas, exercise room, and other health club facilities, and there is also an outdoor pool. The marina provides sailing, waterskiing, and fishing services. Golf, tennis, and horseback riding are available nearby. This adult resort has no facilities for children. ☐ **Lux** ☐ **W, F**

Westhampton Racquet Club, *Dune Road, Westhampton Beach, NY 11978, (516) 288-3096* ☐ *early May to early Oct* ☐ *Off SR 27A* ☐ For skilled and would-be practitioners of the fine art of tennis, this 4-acre waterfront resort offers 15 Har-Tru outdoor courts and a clubhouse under the direction of a USPTA national senior champion. The club boasts a sizable seasonal membership in addition to its modest complement of guests who occupy the ocean-view double rooms and four suites in the modern gray-toned lodge. Some units have balconies or kitchenettes, and all have room service. On the grounds are an outdoor swimming pool and a marina. ☐ **Lux** ☐ **T, W**

SALAMANCA: Allegany State Park, *RD 1, Salamanca, NY 14779, (716) 354-2535* ☐ *Open all year* ☐ *7 mi W of Salamanca on SR 17* ☐ Allegany State Park, 65,000 primitive, peaceful acres, is bordered on the north by the Allegany River and set among well-forested rolling hills with ponds and streams that provide a habitat for many varieties of animals and birds. The park provides nearly 400 primitive cabins without indoor plumbing. The basic single-room cabin has a porch, four cots with mattresses, a picnic table and two benches, cupboard, and a wood stove. Some have electric lights and kitchenettes, and gas heaters. You must bring your own bedding and cooking utensils. The restaurant is open all day on weekends and midday during the week. The park has swimming, boat rental, trout and bass fishing, bicycle rental, two tennis courts, hiking trails, and interpretive nature programs. In winter there is a downhill area with a rope tow, cross-country trails (equipment rental), ice skating on the lake, and sledding. ☐ **Inexp** ☐ **W, F, S**

SARATOGA SPRINGS: Gideon Putnam Hotel, *Saratoga Springs, NY 12866, (516) 584-3000* ☐ *Open all year* ☐ *Exit 13N off I-87* ☐ In its Victorian heyday, high society came to take the waters at the mineral springs, to enjoy thoroughbred racing, and to attend countless parties and balls. Times have changed, and of the several grand hotels that once lined Broadway, only the Gideon Putnam remains to keep alive the spirit of gracious living. Set on shaded lawns, the red-brick Georgian-style hotel offers the formal, albeit relaxed, atmosphere of an exclusive club. Public rooms are large and expertly decorated for maximum comfort. The four dining rooms, with chandeliers and fine woodwork, also exude unhurried ease. The cuisine is largely continental, and luncheon is a lavish buffet. The bar often has dancing and entertainment. The 125 attractive rooms and suites are tastefully furnished to create a homelike ambience, and all units have color TV, private bath, and air conditioning.

The Gideon Putnam is in 2200-acre Saratoga Spa State Park, which has lakes, springs, and geysers, and which still provides mineral baths, massage, and other health club facilities. Swimming pools include the 75-meter Peerless Pool. For golfers there is a nine-hole course and an 18-hole PGA championship course, and for tennis buffs there are four all-weather courts. The well-known Performing Arts Center schedules concerts, dramas, ballet, and other performances all summer. Saratoga Lake has swimming, boating, fishing, waterskiing, and amusement parks. The Casino, on Broadway, is now a museum of Saratoga Springs history. The seasonal harness and thoroughbred racing is as famous as it was in the old days, and the track has a museum of racing. In winter there are 20 miles of cross-country trails, ice skating, and a speed-skating rink. ☐ **Lux** ☐ **G, T, W, S, F**

THOUSAND ISLANDS: This chain of 1800 islands speckling a 50-mile segment of the St. Lawrence River varies from islands several miles long to some that are merely rocks jutting above the water. Clayton and Alexandria Bay are the headquarters, offering a variety of cruises through the islands. Some of these stop to see the 100-room Boldt Castle, begun by the dishwasher who became owner of the Waldorf-Astoria Hotel but left unfinished at his death. Also avaialble are guided fishing trips. Anglers have an excellent chance of hooking pike, muskellunge, perch, and bass, and ending the day by cooking the catch as a "shore dinner" is traditional. More than half a dozen state parks on the mainland and on the islands offer swimming, fishing docks, and marinas with boat rental. Clayton also has historical and nautical museums.

Pine Tree Point, *1000 Islands, Alexandria Bay, NY 13607, (315) 482-9911* □ *Open mid-May to mid-Oct* □ *4 mi NE of I-81 off Sr 12* □ On a 40-acre forested peninsula on the St. Lawrence, this resort offers 164 double rooms and suites in contemporary motel-style lodges, plus a three-room cottage that sleep up to six. Most units have balconies and many have water views. The dining room and terraces overlook the river, and the cocktail lounge has entertainment. An attractive pool is shaded by pines and there is also river swimming in a quiet cove. Guests may use the marina and other facilities at Pine Tree Point's companion resort, Capt. Thompson's Motor Lodge, on James Street. Capt. Thompson's lacks the pretty wooded setting of Pine Tree Point, but the 67 paneled rooms all have balconies overlooking the seaway. □ **Mod** □ **W, F**

Riveredge Resort, *1000 Islands, Alexandria Bay, NY 13607, (315) 482-9917* □ *Open May to late Oct* □ *17 Holland St.* □ On landscaped grounds on the seaway, this resort offers 50 motel-style units with knotty-pine paneling; many rooms have private patios or balconies. The Steak Pit restaurant has indoor and outdoor dining and does its own baking. The resort has a dock, swimming pool, and babysitter list. □ **Mod** □ **W, F**

Thousand Islands Inn, *335 Riverside Dr., Clayton, NY 13624, (315) 686-3030* □ *Open late May to early Oct* □ Opened in 1897 when the Thousand Islands were a popular playground for the rich and famous, this inn across the street from Clayton's public docks preserves early-American decor in its public areas. Of the 18 single and double rooms, 13 have private baths and 6 have water view. Specialties in the dining room, which does not use convenience foods, includes Carolina quail, pork chops flambe, and seafood, and there is a cocktail lounge. Guided fishing charters and package fishing trips with shore dinners are available. Babysitters and other services are offered. □ **Inexp** □ **F**

North Carolina

ASHEVILLE AND THE BLUE RIDGE MOUNTAINS: Since shortly after the Civil War the Asheville area has been a vacation spot popular for its clean, cool air and lush green mountain scenery. George Washington Vanderbilt constructed the famous Biltmore Estate here at the turn of the century. The grounds and gardens of the 11,000-acre estate and the chateau with 250 rooms of priceless antiques and works of art constitute one of the area's major tourist attractions. Novelist Thomas Wolfe was born and raised in Asheville, and the city is called "Altamont" in his autobiographical novel *Look Homeward, Angel.* His family residence, known as "Dixieland" in the novel, is open to the public. Other attractions in the city include the Biltmore Homespun Shops, which preserve the old handicraft methods of dying, spinning, and weaving wool, the Asheville Art Museum, and Colburn Mineral Museum.

The Smokies are among the chief glories of the Appalachian Range. A half-hour's drive west from Asheville is Great Smoky Mountains National Park. The city is in fact surrounded by beautiful wilderness areas including Pisgah, Cherokee, and Nantahala national forests. These areas offer excellent opportunities for hiking, climbing, fishing, hunting, and mineral collecting. Most of North Carolina's ski slopes are in this area, as well as many excellent golf courses. The Blue Ridge Parkway, a scenic, tortuous mountain road, ends (or begins) in the Great Smokies park and follows mountain crests along the Blue Ridge range, taking in some of the most spectacular scenery in the East along its 469 miles.

Beech Mountain Resort, *P.O. Box 261, Beech Mountain Parkway, Banner Elk, NC 28604, (704) 378-4246* □ *Open all year* □ *W of Banner Elk on SR 194* □ This modern residential development offers accommodations in condominium villas, town houses, and chalets. These vary from studios to one to five bedrooms, and many have decks, saunas, or other extras. Many are within walking distances of the slopes. The village at the base of the slopes offers restaurants and lounges, shops, and an ice skating rink. Beech Mountain has a gondola, five chairlifts and other tows, 22 trails, night skiing, snowmaking, ski shop, rentals and an extensive racing program. A Kinder Klub provides lessons and activities for children aged five to nine. Day care is available for younger children, and evenings there are babysitters. □ **Mod to Lux** □ **S**

Cataloochee Ranch, *Maggie Valley, NC 28751, (704) 926-1401* □ *Open May to late Oct* □ *3 ml NE of US 19, watch for sign* □ This mile-high, 1500-acre working ranch at the edge of Great Smoky Mountains National Park was established in 1932 as a wilderness retreat. Although electricity, heat, and plumbing have been added since that time, the ranch still caters to those who prefer to meet nature halfway, without amenities such as deep-pile carpeting or TV. For this reason, the ranch does not accept children under six. The main ranch house, formerly a barn, consists of two-foot-thick stone walls and heavy, hand-hewn logs. The wide, random-width floor boards were all planed and cured on the ranch from huge oak and ash trees felled here. In the evening guests congregate around the big stone fireplace. The main house has ten bedrooms, all with either private or connecting baths. Five nearby hand-hewn log cabins with exposed beams have two bedrooms each, baths, open stone fireplaces, wood floors with handmade hooked rugs, and some antique mountain furnishings. The AP rates include three ranch-style meals with fresh vegetables and ranch-grown beef. The ranch provides setups but you must bring your own liquor. Horseback riding is extra, and the fee includes a guide. With advance notice the ranch also outfits park trips of two days to a week or longer into the national park. A tennis court, fishing, and swimming pond round out the recreational facilities. Whitewater rafting and golf are available. □ **Inexp to Mod** □ **R, F**

Eseeola Lodge, *Linville, NC 28646, (704) 733-4311,* □ *Open June to mid-Sept* □ *On US 221, SR 105 and 181, 2 mi W of Blue Ridge Parkway* □ Eseeola Lodge, which dates back to 1936, is a distinctive two-story structure with chestnut-bark siding set near the foot of Grandfather Mountain. Each room at the AP lodge has its own porch supported by rough poplar pillars. The rooms are paneled with chestnut and most of the furnishings were handcrafted on the premises. Adjacent to the lodge are seven guest cottages with living rooms equipped with TV. The Swiss chef prepares continental cuisine, and there is a cocktail lounge and lounges for TV and cards.

The lodge boasts an 18-hole golf course with bent greens, a putting green, clubhouse restaurant, and pro instruction. Lessons are also available at the tennis courts and riding stable. Also on the grounds are a heated pool, hiking trails, trout fishing, beauty and gift shops, and seasonal programs for children. Caverns, waterfalls, and other attractions are nearby. □ **Mod to Lux** □ **G, T, R, F**

Great Smokies Hilton,*One Hilton Dr., Asheville, NC 28806, (704) 254-3211* □ *Open all year* □ *Take Hilton Inn Drive exit off US 19 & 23 Bypass* □ On 120 landscaped acres in the foothills of the Blue Ridge Mountains, the modern, three-story main lodge has three guest wings which echo the contours of the hills. Guest rooms are large, and many have balconies overlooking the golf course. Huge executive suites are also available. The restaurant, Aunt Minnie's Wisdom, serves American and continental foods and looks out over the golf course. The wood-paneled Library Lounge serves cocktails and frequently offers live entertainment. The 18-hole, 6300-yard golf course surrounds the resort. The tennis complex has four indoor and four outdoor courts, ball machines, videotape, and a pro. There are two pools and a sauna. ☐ **Mod** ☐ **G, T**

Grove Park Inn and Country Club, *290 Macon Ave., Asheville, NC 28804, (704) 252-2711* ☐ *Open April–Nov* ☐ *Take US 26 to Charlotte St. exit, left at light to Macon Ave., then follow signs* ☐ Set on 140 acres of rolling country with beautiful mountain views, this magnificent mansion began to draw presidents, celebrities, and European nobility soon after its completion in 1913. The unusual and imposing main building is of roughhewn granite from the Smokies and is surrounded on three sides by groves of grand old trees. Covered terraces along the facade overlook lawns, gardens, and a large free-form swimming pool of red tile that matches the mansion's red-tile roof. Huge granite fireplaces and the original oak furniture warm the lobby and public rooms. The 252 air-conditioned guest quarters are spacious and expertly decorated, and parlor suites are available. Six dining rooms run the gamut from formal elegance and continental cuisine in the traditional Plantation Room to informal terrace dining, a breakfast room, and poolside service. The private club provides beer, wine, and setups as well as dining and entertainment.

Separated from the inn by stands of evergreens is the French-inspired Country Club, open all year, which originally started in 1893 as a hunt club. Here there is another outdoor swimming pool, a putting green, and the 18-hole, par-71 golf course. Tennis buffs will find eight courts, two indoors and four lighted. Also on the grounds are Biltmore Homespun Shops, established by Mrs. George Vanderbilt to preserve Old World methods of wool manufacture; demonstrations are given all year. Grove Park Inn is especially noted for gracious, deft service. ☐ **Mod to Lux** ☐ **G, T, H**

High Hampton Inn and Country Club, *Cashiers, NC 28717, (704) 743-2411* ☐ *Open mid-Apr to early Mar* ☐ *2 miles SE of Cashiers on SR 107* ☐ This former estate of 2000 acres is set in a stunning valley of the Blue Ridge Mountains, 3600 feet above sea level. The lakeside three-story inn looks most appropriate to the rugged mountain setting, with its chestnut-bark exterior and rough-log pillars. Inside, the inn is as uncompromisingly rustic and simple as outside, with stone fireplaces, wide-plank wood floors with braided scatter rugs, exposed beams, and tall ladder-back chairs. The owners seem to feel that a resort in the mountains of North Carolina shouldn't look exactly the same as one on the beach in Florida. As a result, High Hampton is one-of-a-kind. If TV and thick carpeting are essential to your notion of the good life, this is not the place for you. The 120 accommodations in the inn and the surrounding cottages are comfortable and simple. Rooms are finished with rough-pine boards and handcrafted furniture. All have private baths.

The most notable attraction is the setting of lush forest, and wildflowers, spectacular cliff formations, waterfalls, and clear lakes. The 18-hole golf course occupies the gently rolling valley beneath the mountains. The course was designed by George W. Cobb and is noted for the par 3, 137-yard eighth hole with its island green, as well as for its natural beauty. Tennis buffs may use eight courts as well as a tennis shop. Guests can swim, sail, canoe, or fish on the three lakes. Other sports include trap and skeet shooting, archery, croquet, hiking, horseback riding and trail rides, and birdwatching, all on the premises. In winter, skiing is at Sapphire Valley, seven miles away. Seasonal recreational programs are offered for children from early morning through evening, and teenagers have their own lounge. High Hampton also schedules many adult activities such as cookouts and cocktail parties. In the Rock Mountain tavern setups are provided, but you must bring your own liquor. Rates are AP, with most sports extra. ☐ **Inexp to Mod** ☐ **G, T, W, R, H**

Hound Ears Club, *P.O. Box 188, Blowing Rock, NC 28605, (704) 963-4321* ☐ Nestled against wooded slopes in a high mountain valley, Hound Ears calls itself a "non-resort" because it takes pride in cultivating an intimate clublike atmosphere. Exceptionally attractive accommodations in the modern hillside lodges consist of 24 double rooms, suites with bedroom and sitting room, and suites with two bedrooms and setting rooms. All have picture windows and private balconies providing sweeping views across the golf course to the mountains. Also available are private chalets in the woods. The lodge offers meal and package plans and has two glass-walled dining rooms and a lounge.

The beautiful 8-hole, par-72 golf course was designed by Cobb for 6165 yards, and has fairways bordered with stands of old trees, putting green, and clubhouse services. Also available are four Grasstex tennis courts with teaching pro, a heated swimming pool, trout fishing, and hiking. Riding stables are a few miles away. Winter offers ski slopes suitable for beginners and intermediates, with snowmaking, ski school, rentals, and a lodge with restaurant. ☐ **Mod** ☐ **G, T, F, S**

Lake Toxaway Country Club, *Lake Toxaway, NC 28747, (704) 966-4488 (summer); (704) 884-4916 (winter)* ☐ *Open late April to late Oct* ☐ *19 mi SW of Brevard via US 64 and SR 281* ☐ Set on a thickly wooded peninsula jutting into Lake Toxaway, this resort consists of a grand turn-of-the-century club building offering double rooms and suites. Also available are housekeeping cottages and apartments with one to four bedrooms. Many units have lake views. The clubhouse has a dining room. There is also an indoor and outdoor grill overlooking the fairways at the pro shop. Golf is a feature here, with an 18-hole, par-72 course, and there are also six tennis courts. Lake Toxaway has fine swimming, waterskiing, and fishing, and rentals and other services are available at the marina. Also on the grounds are an outdoor pool and nature trails. Weekly cookouts, and other activities and services are provided ☐ **Inexp** ☐ **W, G, T, F**

Maggie Valley Country Club and Motor Lodges, *P.O. Box 126 Maggie Valley, NC 28751, (704) 926-1616* ☐ *Open all year* ☐ *Off US 19* ☐ This modern complex, set on 145 acres of rolling hills in the shadow of the Great Smokies, has an 18-hole, 6500-yard golf course with bent-grass greens, and a heated swimming pool. Fishing and horseback riding are available nearby. Motel units are in several lodges stepped up a wooded hillside giving each room a private balcony and view. The more luxurious Fairway Villas are one- or two-bedroom units with cooking facilities and maid service overlooking the golf course. There is a dining room and either EP or MAP is available. In winter, Maggie Valley offers ski package plans for nearby (six miles) Cataloochee Ski Slopes. ☐ **Inexp** ☐ **G**

Pisgah View Ranch, *Rte. 1, Candler, NC 28715, (704) 667-9100* ☐ *Open May–Nov* ☐ *From Candler, take SR 151 8 miles S* ☐ Set in a high valley beneath the mile-high mountain from which it takes its name, this ranch offers frame cottages and log cabins with pine paneling and private bath. Some units have fireplaces and private decks. The AP rates include all facilities except horseback riding. Family-style meals are served in the century-old farmhouse, with fried chicken, country ham, and homemade breads among the specialties. The ranch sits on 1700 acres that can be explored on horseback. You can swim in the large outdoor pool or take a boat out on the pond, which is stocked with rainbow trout. At night there is entertainment, often square dancing. ☐ **Inexp** ☐ **R, F**

Sapphire Valley Resort and Fairfield Inn, *Star Route 70, Box 80, Sapphire, NC 28774, (704) 743-3441* ☐ *Open all year* ☐ *From Cashiers 3 mi E on SR 64* ☐ The combination of rugged mountains and broad valley floor make this 5700-acre complex a beauty. The Fairfield Inn, built in 1896, sits on a slope facing a mountain cliff and overlooking Lake Fairfield. This gracious alpine structure with a broad veranda and patio has 62 rooms and suites. Also available are rustic cottages with one or two bedrooms, living room, kitchenette, and sun porch. More luxurious are contemporary one-, two-, or three-bedroom villas with kitchens and terraces; many have a fireplace. Rates are EP but MAP is available.

One of the two 18-hole golf courses was designed by George Cobb so that each tee presents the golfer with a scenic vista. Lake Fairfield is a beautiful setting for swimming, fishing, and boating, and the inn has two heated pools on the premises. The tennis complex has 12 courts (10 clay and two hard; some lighted) a shop, and full instruction program. Seasonal tennis and golf clinics are available. Trail rides into the mountains begin at Sapphire Stables and wander on miles of trails leading up from the valley. Serious hikers and climbers will find challenges in the scenic mountains. Also here are gem hunting in the resort's mine, formerly worked by Tiffany, birdwatching, lawn games, stream fishing, and children's programs. From mid-December to mid-March, three ski runs are open, with ski shop, ski school, sleigh rides, and toboggan runs. The dining room features mountain cooking and continental cuisine and serves three meals a day. The private club provides setups. ☐ **Mod** ☐ **G, T, W, H, R, S**

Snowbird Mountain Lodge, *Joyce Kilmer Forest Rd. Robbinsville NC 28771, (704) 479-3433* ☐ *Open mid-May–Oct* ☐ *10 miles NW of Robbinsville on Joyce Kilmer Forest Rd.* ☐ This 130-acre secluded mountain hideaway in Nantahala National Forest commands a spectacular view of the surrounding Snowbird Mountains and of Lake Santeetlah in the valley. The view alone is worth the trip. The lodge is an attractive and appropriately rustic structure made of chestnut logs and stone. Hardwood paneling, polished hardwood floors, and handcrafted furniture make for a naturally cozy interior. Both the lounge, with library, and the dining room have large stone fireplaces. The 16 AP guest rooms are finished in a variety of native woods, and all have handmade furniture and private bath. The recreational focus here is on the appreciation of nature. Hundreds of trails from the lodge and in Joyce Kilmer Memorial Forest are available for hiking and climbing. Nearby stream and lake fishing is excellent. A mountain stream pool one mile from the lodge serves as swimming hole. There's also shuffleboard, table tennis, archery, croquet, and horseshoes. Note: this resort does not accept children under twelve. ☐ **Inexp** ☐ **H, F**

Sugar Mountain Resort, *P.O. Box 369, Banner Elk, NC 28604, (704) 898-4521* ☐ *Open all year* ☐ *On SR 184 between Linville and Banner Elk* ☐ In a beautiful high mountain valley, this resort offers 32 motel-style rooms with picture windows and one- to four-bedroom chalets and condominium apartments with fireplaces and kitchens. Winter activity centers on Sugar Mountain, with a vertical drop of 1200 feet, three chairlifts and surfacelifts, and 12 trails. Services include ski school, shop, rentals, snowmaking, patrol, a racing program and Nastar races, a nursery, a cafeteria, and a restaurant. In summer there is golf on an 18-hole course, and tennis on nine courts (four clay, two all-weather, three indoors), a large outdoor swimming pool, and alpine slide, and a recreation program. ☐ **Mod** ☐ **S, T, G**

Waynesville Country Club Inn, *Waynesville NC 28786, (704) 456-3551* ☐ *Open all year* ☐ *1½ miles SW on US 19, 23* ☐ The attractive, gray-stone and clapboard three-story inn sits in a Blue Ridge valley. Since the late twenties the inn has been drawing golfers, and 18-hole golf, against a beautiful mountain setting, remains at the center of activities here. The inn also has two-all-weather tennis courts and a heated swimming pool, with skiing, hiking, and horseback riding nearby. Rooms in the older main lodge and Woodcrest buildings have twin beds, color TV and private bath. The newer, two-story Fairway and Brookside buildings feature private balconies overlooking the golf course, air conditioning, and two double beds. The separate cottages overlooking the ninth green are similarly furnished, and have private porches. The dining room serves breakfast and dinner (dress code), while the less formal, cozy Taproom is open for lunch and snacks. ☐ **Inexp to Mod** ☐ **G**

Wolf Laurel Resort, *Rte. 3, Mars Hill, NC 28754, (704) 689-4111* ☐ *Open all year* ☐ *From Asheville take US 23 27 miles N to Wolf Laurel turnoff* ☐ This vacation community embraces 5000 acres of wooded hills. Rental accommodations are available in the rustic-modern, four-story Wolf Laurel Inn, and in privately owned one-to five-bedroom log cabins, cottages, and A-frames, the latter coming completely furnished and with cooking facilities. The Mountain Top restaurant next to the inn serves three meals a day all year. Wolf Laurel's 18-hole golf course takes golfers up to an altitude of 4785 feet. There are two all-weather tennis courts, a large, W-shaped swimming pool, and riding stables on the premises. Winter recreation centers on the Wolf Laurel Ski Area, with double chairlift, Poma lift, and rope tow serving a natural ski bowl. Rentals, ski school, and a restaurant are in the rustic ski lodge on top of the slopes. □ **Inexp to Mod** □ **G, R, S**

PINEHURST/SOUTHERN PINES: This is golf country. Golf came to the Sandhills in 1900 when Donald Ross, who was to become famous as a designer of some 600 championship courses around the country, came from Scotland to Pinehurst to design a nine-hole course. Ross subsequently designed four of the 18-hole courses at Pinehurst and his Number Two course, called "the St. Andrews of America," is regarded as one of the finest in the world. Overlooking the course is the World Golf Hall of Fame (open to the public), and the Colgate Hall of Fame Golf Classic is played here in August. Today there are 27 courses within a 15-mile radius of Pinehurst, some designed by Robert Trent Jones, Ellis Maples, George and Tom Fazio, and other notables.

Pinehurst is a privately owned corporation with a charming Village Green and many estates boasting Georgian residences. The parks and roads were laid out by Frederick Law Olmsted. Southern hospitality allows you to visit the grounds–Pinehurst Hotel, the Village Green, craft shops, the golf museum, Claredon Gardens, and the pottery–but the fabulous facilities of Pinehurst Country Club are open only to members or to guests of Pinehurst. Adjacent Southern Pines is the nonexclusive sister village, with motels and restaurants, as well as shops and historic homes.

The Sandhills also have well over a hundred tennis courts (Pinehurst again has the best facility) and hundreds of lakes and ponds for fishing and swimming. Lakes at Pinehurst, Whispering Pines, Lake Surf, Seven Lakes, and Foxfire are large enough for sailing. This is also horse country, with many breeding farms and training centers for trotters, pacers, and jumpers. Fox hunting to hounds is at a number of private clubs. The Moore County Hound Hunters Trials are held in early March, and you may reserve a parking space along the course. The Stoneybrook Steeplechase Races are in April at Southern Pines.

Charlton Motel, *Box 245, Southern Pines, NC 28387, (919) 692-2232* □ *Open all year* □ *1 mi S on US 1* □ Set back from the highway and shaded by trees, this is a good choice for those who want to enjoy the Sandhills on a budget. The 44 rooms are air-conditioned, paneled, and cozy, and most open onto a large porch overlooking pretty grounds with a swimming pool, two little lakes (one stocked), and a playground. Golf can be arranged. ☐ **Inexp**

Foxfire Golf and Country Club, *P.O. Box 711, Pinehurst, NC 28374, (919) 295-5555* ☐ *Open all year* ☐ *From Pinehurst take US 211 NW then follow signs* ☐ Opened in 1968, Foxfire offers rows of condominium town houses offering two- and three-bedroom units, multi-unit villa apartments with private decks, and individual, glass-walled houses among trees along the fairways. The 18-hole course, plus a nine, were designed by Gene Hamm, and a full instructional program is conducted. A driving range and practice greens are near the clubhouse, which has a dining room and bar and an outdoor lounge for informal meals. Also nearby are tennis courts and a swimming pool. Swimming, fishing, and sailing are available on 20 acre Lake MacKenzie. ☐ **Mod** ☐ **G, W, F, T**

Mid Pines Club, *Southern Pines, NC 28387, (919) 692-2114* ☐ *Open all year* ☐ *From Pinehurst circle take SR 2 E 4 mi* ☐ This sprawling yet dignified brick and clapboard three-story Georgian-style inn overlooks fairways and is surrounded by groves of trees and flowering shrubs. It offers elegant public rooms, including a quiet dining room and the informal Sun Room that has an unusual ceiling of painted tile. The bar provides wine, beer, and setups. Accommodations are AP in either the 63 air-conditioned inn rooms, in the modern golf lodge with beamed ceilings and picture windows, or in cottages secluded among the trees (these sleep up to 12 people). The 6501-yard course was laid out by Donald Ross. Low handicappers will find a challenge in the narrow fairways of the lower holes. Also on the grounds are four tennis courts, an outdoor swimming pool, and a health club with saunas. ☐ **Mod** ☐ **G, T,**

Pinehurst Hotel and Country Club, *Box 4000, Pinehurst, NC 28374, (919) 295-3131* ☐ *Open all year* ☐ *1½ mi W of US 15 and 501* ☐ A New England soda fountain manufacturer named James Tufts, who found that the balmy climate of the Carolina Sandhills improved his delicate health, established Pinehurst in 1895 as a retreat for those suffering from climate-related afflictions. Within a couple of years dairy workers on the grounds were complaining of golfers who harassed the cows. And since the early part of the century the primary tonic Pinehurst has offered is golf. With six 18-hole courses, Pinehurst is one of the finest golf resorts in the world. But Pinehurst offers far more than golf. It is without doubt one of the nation's grandest resorts. As you approach the hotel along beautifully landscaped Midland Road, skirting the Georgian village center, you already sense you are being lulled by the graciousness of a bygone era. There is no ostentation. Although the white, four-story Pinehurst Hotel houses some 300 rooms and suites, the overhanging roofs, green awnings, and immense grounds (10,000 acres) make it look surprisingly small. Skillful design and decoration, and the use of muted colors, keep the scale of public rooms intimate. And of course, in the finest tradition of Southern hospitality, your welcome is warm and the service unflaggingly deft. Rooms in the hotel have traditional furnishings, and the best suite, on the top floor, has four bedrooms and a parlor. More contemporary accommodation is in more than 300 Golf Villas which vary in design from glass-walled treehouses to wooden lodges with sloping shingled roofs to concrete-and-glass structures. Secluded among pine groves lining the fairways, these villas offer between one and three bedrooms, living room/dining room, equipped kitchen, and balconies or decks.

Public rooms include the sedate Carolina Room overlooking sunlit gardens, the informal grill, an English-style pub, and the Pinehurst Lounge. A social director schedules dancing and entertainment, and services include limousines to three airports and supervised programs for children.

Pinehurst Country Club opened in 1898, but the present clubhouse is a sprawling modern complex with glass-walled restaurant and lounges overlooking the fairways, driving range, and putting greens. Full services include rentals, caddies, and pros. Course One (6129 yards, par-70), the fabulous Course Two (7051 yards, par-72), Course Four (6905 yards, par-71), and Course Five (6946 yards, par-71) have bent greens, while Course Three (6013 yards, par-71) has Bermuda greens. Architects were Donald Ross, Ellis Maples, Peter V. Tufts, and Robert Trent Jones. Course Six, the newest, is by George and Tom Fazio.

In a shaded pine grove, The Tennis Club has 24 courts (18 clay, 6 Laykold, 4 lighted) and topnotch instruction with the latest equipment at the Australian Institute of Tennis. The Gun Club, where Annie Oakely was once an instructer, offers skeet and trap shooting, quail hunting, an archery range, and a full-service clubhouse. The riding stables offer instruction, ponies for children, carriage rides, trail rides on 200 miles of tree-shaded bridle paths, horseshows, even foxhunting. In addition to indoor and outdoor pools, there is a swimming beach on 200-acre Lake Pinehurst; here you can rent canoes and sailboats and fish for bass and bream. The health club has whirlpool, sauna, massage, Roman bath, and exercise room. And of course there are playgrounds, jogging trails, birdwatching, bicycle paths and rental and many tournaments and special events. □ **Lux** □ **G, T, R, W, F, H**

Pine Needles Lodges and Country Club, *Southern Pines, NC 28387, (919) 692-7111* □ *Open all year* □ *¼ mi off US 1 on SR 2* □ With only 67 rooms, Pine Needles is one of the smaller and less formal golf resorts in the Sandhills. Accommodations (AP) are in two- and three-story lodges set in rows among pines. The units are finished in paneling and have floor-to-ceiling windows, and many have balconies. Sizes range from double rooms to suites with four bedrooms. The dining rooms, lounges, and bar feature wood paneling and brick. In the evenings there is music in the main lounge. Pine Needles is designed primarily for the golf enthusiast. The 18-hole course, by Donald Ross, has a putting green, a driving range, and pro shop. A game room, heated swimming pool, and two tennis courts round out the recreational facilities. □ **Mod** □ **G**

SANFORD: Carolina Trace, *P.O. Box 2250 Sanford, NC 27330, (919) 499-5103* □ *Open all year* □ *Just off SR 87 south of Sanford* □ This modern, 1500-acre resort focuses around a golf course and lake in the Sandhill country. Two types of vacation rental are available here. Glassy contemporay patio villas, clustered in wooded groves, have one to three bedrooms, full kitchen, fireplace, separate dining room, and private patio or deck. Two-story contemporary condominiums with private deck are also available. All units have color TV and daily maid and linen service. The 27-hole golf course was designed by Robert Trent Jones. The glass clubhouse, overlooking the fairways and lake, has a complete pro shop and a restaurant. A heated, olympic-size outdoor swimming pool, and eight lighted, all-weather tennis courts are nearby. Trace Lake has seven miles of lakeshore for sailing, fishing, and swimming. □ **Inexp to Mod** □ **G, T, W**

WINSTON-SALEM: Tanglewood Park, *Clemmons, NC 27012, (919) 766-6421* □ *Open all year* □ *12 miles SW of Winston-Salem on US 158, Tanglewood Exit* □ The former estate of William N. Reynolds (of the tobacco company family), Tanglewood Park is 1100 acres of rolling countryside and landscaped grounds containing lodging and recreational facilities. The stately, colonial Manor House of the estate accommodates a limited number of guests in 11 large rooms with dignified colonial-style furnishings, fireplaces, original paintings, air conditioning, TV, and telephone. Adjacent are the Motor Lodge's 18 more contemporary units with private porch or patio overlooking a wooded area of the park. Several cottages on the shores of Mallard Lake sleep up to five people and are available on a weekly basis. The restaurant at the Manor House serves three meals daily, including buffets.

The park is noted for its two 18-hole golf courses. The west course was designed by Robert Trent Jones and was the site of the 1974 PGA National Championship. Also on the grounds are an 18-hole pitch-and-putt, driving range, and miniature golf. The tennis center has ten courts and a program of instructions. Mallard Lake is the scene for canoeing and fishing. For swimming there's a large outdoor pool. Trail riding and riding lessons are offered at the park's stables, and every April the annual Tanglewood Steeple Chase is held here. The historic Moravian village of Old Salem, Reynolda House's fine arts collections, and other attractions are in nearby Winston-Salem. □ **Inexp to Mod** □ **G, T, R, W**

Ohio

AVON LAKE: Aqua Marine Lodge, *216 Miller Rd., Avon Lake, OH 44012* ☐ *(216) 933-2000* ☐ *Open all year* ☐ *9 mi E of Lorain off US 6* ☐ This deluxe two-story motel has 250 double rooms and luxury suites, many with oversize beds, balconies, and rustic knotty-oak paneling, all set on 60 acres of lawns and trees. The restaurants include a steakhouse, and dancing and entertainment are available in the cocktail lounges. The indoor pool is olympic size, and the health club has whirlpool, sauna, steam bath, massage, and exercise room. Outdoors is another olympic-size pool and a wading pool, lawn game areas, an 18-hole golf course, a par-3 course, and a putting green. Other facilities include indoor and outdoor tennis with pro instruction, pool table, bowling, recreation room, and playground. Babysitters are available. ☐ **Mod to Lux** ☐ **G, T**

DELLROY: Atwood Lake Lodge, *Dellroy, OH 44620, (216) 735-2211* ☐ *Open all year* ☐ *On SR 542, between SR 39 from Carrollton and SR 212* ☐ Set back from an arm of man-made Atwood Lake, this lodge offers 102 motel-like guest rooms with balcony or terrace in a two-story building spreading among lawns, as well as 17 lakeside cottages with four bedrooms and equipped kitchens. The attractive public areas have a lot of glass, varnished wood, and stone. Separated from a lounge by a transparent wall is the heated indoor swimming pool with its own wall of windows. Besides the lobby and lounge areas, there are whirlpool, sauna, a game room, gift shop, snack bar and deli, and a dining room where steak is the specialty.

On the 125 acres are an 18-hole golf course with cart rental, a 9-hole course, a putting green, driving range, five lighted tennis courts (with instruction available), an outdoor pool, and a playground. On the lake there's swimming from a sand beach, fishing from an 80-foot floating dock, several types of small pleasure boats, and lake cruises. The lake has three large marinas. In winter, Atwood opens its beginning and intermediate ski slopes and several miles of cross-country trails. Rental equipment is available. Sledding and ice fishing are offered. Within 20 miles are the interesting reconstructions at Zoar Village and Schoenbrunn state memorials, the Warther Museum, and Ft. Laurens ☐ **Mod** ☐ **W, T, G, F, S**

MASON: Kings Island Inn, *5691 Kings Island Dr., Mason OH 45050, (513) 241-5600; Ohio Toll-free Number: (800) 582-3051; elsewhere (800) 543-4301* □ *Open all year* □ *25 mi N of Cincinatti on I-71* □ Indeed a one-of-a-kind phenomenon, Kings Island boasts it is the Midwest's largest family entertainment center. On the 1600 acres are six different theme areas, a wild-animal safari, the Jack Nicklaus Golf Center, many rides, restaurants, shops, and special events–all open May through October. A free bus shuttles the two miles between the amusement park and the inn, a Swiss-style chalet complex surrounded by landscaped grounds. The 300 double rooms and suites are large and some have private balconies. You can eat in the park's numerous restaurants or in the Swiss-style restaurant and bar, the Alphenhorn, which livens up every evening with entertainment. The inn has two tennis courts, miniature golf, two swimming pools, a playground, and a game room, as well as planned children's activities. Extra-special is the nearby Jack Nicklaus Golf Center with two 18-hole courses designed by Nicklaus himself: the 6500-yard, PGA tour Grizzly and the 3736-yard Bruin, lighted for night play. The Golfer Center is complete with pro shop, rentals, teaching pro, and restaurant. □ **Mod** □ **G**

OHIO STATE PARKS: *Ohio State Park, Division of Parks and Recreation, Fountain Square, Columbus, OH 43224* □ Six of Ohio's state parks offer top-notch lodges and housekeeping cabins, plus recreational facilities reserved for guests as well as all the park facilities. The cabins have four rooms with four twin beds and a hideaway bed so that they sleep six; all have heat, a screened porch, cooking equipment, and linens. What makes these lodgings such a family bargain is that lodge rooms and cabins are rented by the unit–thus one to four people may occupy the lodge rooms and up to six people may sleep in the cottages, for the same nightly EP rate. All of the lodges have restaurants serving three meals a day, and all of the parks have seasonal naturalist programs.

Hueston Woods State Park Lodge, *Rte. 1, College Corner, OH 45003, (513) 523-6381* □ *Open all year* □ *5 mi NW of Oxford* □ The first state-park resort facility sits in a densely wooded 3596-acre park containing a 200-acre preserve of virgin beech and maple trees. On a bluff above Acton Lake, a huge, glass-fronted A-frame contains the lodge lobby, with a 100-foot high sandstone fireplace, decorated with Miami Indian artifacts. To one side are the three-story lodging units with 94 rooms in Indian motif and with private patios or balconies. Farther down the lake shore are 25 cottages in the woods and 34 simpler one-room housekeeping cabins that sleep up to four people. The lodge has indoor and motor pools, a wading pool, a putting green, lawn games, lighted tennis courts, a recreation room, a playground, and picnic grounds on the opposite shore of the lake are the 18-hole golf course, a pioneer farm museum, a nature center, a marina with boat rental, and a swimming beach with scuba diving. Hiking trails surround the lake. Stablès outside the park rent horses for the park's bridle trails. Fishing is good for largemouth bass, walleye, and crappie. In winter there is sledding, skating, ice fishing, ice boating, and cross-country skiing. ☐ **Inexp to Mod** ☐ **W, F, G, T**

Punderson State Park Manor House, *Rte. 87, Newbury, OH 44065, (216) 564-2201* ☐ *Open all year* ☐ *Off SR 87 between Cleveland and Middlefield* ☐ Overlooking a glacial lake, this 890-acre park offers an unusual Tudor Mansion that was once privately owned. The public rooms radiate Western Reserve warmth and elegance, and there are only 35 guest rooms. In the woods nearby are 26 cottages. The manor house offers an outdoor pool, a playground, a recreation room, shuffleboard courts, and hiking trails. Strung out among the lakeshore are a swimming beach, a marina for boat rentals and fishing (bluegill, bass, trout, and catfish), and tennis courts. Inland is an 18-hole golf course with putting green and pro shop. Within the park are two other lakes and a nature center. In winter there is ice fishing, snowmobiling, ice skating, sledding, and a sports chalet. ☐ **Inexp to Mod** ☐ **W, F, G**

Salt Fork State Park Lodge, *P.O. Box 550, Cambridge, OH 43725, (614) 439-2751* ☐ *Open all year* ☐ *8 mi NE of Cambridge off US 32* ☐ With a cathedral-ceiling lobby and sloping roofs over the three- and four-story lodging wings, this lodge offer 148 air-conditioned rooms, many with refrigerators, balconies, and oversize beds. In the woods nearby are 54 cottages. The inn offers indoor and outdoor pools, saunas, game room, tennis courts, bicycle rental, putting green, and outdoor game areas. The 20,000-acre park has an 18-hole golf course, a lake swimming beach, fishing and a marina with boat rental. ☐ **Inexp to Mod** ☐ **W, F, T, G**

Shawnee State Park Lodge, *P.O. Box 98, Friendship, OH 45630, (614) 858-6621* ☐ *Open all year* ☐ *On SR 125 NW of Portsmouth* ☐ Located in one of America's greenest areas near the Ohio River, this 1165-acre park offers a modern three-story lodge of wood, stone, and glass housing 50 motel-style rooms, as well as 25 cottages in the woods nearby. The lodge has indoor and outdoor pools, tennis, putting green, playground, outdoor game areas, picnic areas, and a recreation room. The park also offers nine-hole golf, fishing, two swimming beaches, rental boats, and bridle trails. Adjoining the park is 60,000-acre Shawnee State Forest, with six lakes, extending to the banks of the Ohio, where there is a marina. The forest has seasonal hunting. ☐ **Inexp to Mod** ☐ **W, F**

PAINESVILLE: Quail Hollow Inn, *Painesville, OH 44077, (216) 946-7990* ☐ *Open all year* ☐ *Off I-90 and SR 44* ☐ On 200 acres of attractively landscaped lawns and woods, this resort offers 168 comfortable rooms in four-story modern lodges; all units are carpeted and have color TV and floor-to-ceiling windows; some have balconies. The Quail Wagon restaurant features quail and seafood. An outdoor grill offers patio dining. The coffee shop at the clubhouse opens at 6:30 am. The cocktail lounge, in hunting motif, has dancing and entertainment.

The 18-hole, 700-yard golf course was designed by Bruce Derlin and features rolling fairways and water on five holes. Also here are a pro shop, cart rental, driving range, putting green, all-weather lighted tennis courts, a large outdoor pool, and a playground. The circular indoor pool has a whirlpool and saunas. In winter, miles of cross-country trails are groomed and there is ice skating too. Services include babysitters. ☐ **Mod** ☐ **G, S,**

WARREN: Avalon Inn, *9519 E. Market St., Warren, OH 44484, (216) 856-1900* ☐ *Open all year* ☐ *5 mi E of Warren* ☐ This modern motor inn offers 90 air-conditioned rooms and suites, many with refrigerators, two dining rooms, and a cocktail lounge with dancing and entertainment. The large pool is indoors and has saunas. Indoor and outdoor tennis, with instruction, is available at an extra charge. There are also two 18-hole golf courses with driving range, putting green, and services. ☐ **Mod** ☐ **G, T**

Oklahoma

GRAND LAKE O' THE CHEROKEES: Shangri-La, *Rte. 3, Afton, OK 74331, (918) 257-4204* ☐ *Open all year* ☐ *SR 125 on Monkey Island* ☐ In the center of 60-mile-long Grand Lake O' the Cherokees, out on the tip of Monkey Island, is Shangri-La. Although not quite the mysterious paradise where no one can grow old, this Shangri-La does as much as possible to make certain its guests enjoy their all-too-short vacation lives. This is a good-size resort—large enough to have its own airfield, an 18-hole, 7000-yard golf course with pro and clubhouse services, a recreation center with bowling alley and health spa facilities, indoor and outdoor tennis (with pro shop), swimming, full-service marina with guides, and bird-hunting trips for which guides and dogs are supplied. It also features several different "environments" that make for variety in dining and socializing—the Tahitian Terrace for tropical drinks and light food amid tropical plants, and occasional luaus: the Hogan, an octagonal pavilion built around a central fireplace for barbecues; and the Angus Room for prime ribs. For dining and dancing, the Golden Eagle presents a more formal, more continental atmosphere.

Built of stone and capped with a short mansard shingle roof, the zig-zagging main lodge has 273 guest rooms, including some one-, two-, and three-bedroom suites. Other accommodations include the nearby two- and three-bedroom condominiums that look out on the golf course. All rooms come with color TV, private bath, plush carpet, and tasteful furnishings. The suites are luxury apartments with private patio, wetbar, living and dining rooms, complete kitchen, and a bath for each bedroom. Lakeview accommodations have a patio or balcony. ☐ **Mod to Lux** ☐ **T, F, G,**

OKLAHOMA STATE PARKS: Oklahoma State Lodges, *Oklahoma Department of Tourism and Recreation, 500 Will Rogers Bldg., Oklahoma City, OK 73105; from Oklahoma, call (405) 521-2464; from Oklahoma City or out of state, call toll-free (800) 522-8565* ☐ *Open all year—but some restaurants or facilities may close in Dec or Jan for renovation* ☐ A land that not long ago was known as the Dust Bowl now attracts vacationers with its abundance of large lakes. Created by damming important rivers that run through Oklahoma, these lakes are natural sites for parks, and Oklahoma has developed seven state resort parks on or near their shores. Each Oklahoma state resort is different. Besides their unusually reasonable rates, all resorts have air-conditioned, comfortable, contemporary lodges and housekeeping cottages operated on European Plan. Sports and recreational opportunities are either directly connected to the lodge or available within the park. Most activities are free; when fees are charged, as greens fee for golf, they are low. These resorts are geared to families–cribs are free, children under twelve in your room are free, and most of the lodges will supply cots at reasonable cost. The restaurants usually have children's plates and will pack up box lunches. All of the resorts have recreation rooms, and social directors schedule activities especially for children. The management can usually arrange babysitters. To make reservations, contact the address above or the individual parks directly, at the addresses given below. Most of the resorts provide a free airport bus.

Arrowhead Lodge, *Box 57, Canadian, OK 74425, (918) 339-2711* □ *18 mi N of McAlester on US 69* □ On the south shore of 102,000-acre Lake Eufaula, Arrowhead Lodge is constructed of native stone with huge expanses of glass and offers 104 double rooms, many with private terraces. Bridal suites and sitting rooms are also available. The 104 cottages here sleep two or four people and have kitchenettes (no equipment) and fireplaces. There is a nice restaurant, and the private club has a unique underwater view of the huge swimming pool. Recreational facilities include a nine-hole golf course, putting green, tennis, riding, and outdoor game areas. A supervised play program for children operates from spring through summer.

The park has a marina with boat rental and other services. Fishing is good for black bass, white bass, and crappie, and there is a covered fishing dock. Archery, badminton, bicycling, and playgrounds are available, and there are hiking trails in the park and at Army Corps of Engineers recreation sites. Also on the lake are several historic sites belonging to the Creek Nation. (Fountainhead Lodge, see below, is also on this lake.) □ **Inexp** □ **W, F, G, T, R**

Fountainhead Lodge, *Checotah, OK 75740, (918) 689-2501* □ *7 mi S of I-40 on SR 150* □ On the tip of a wooded peninsula on the north shore of Lake Eufaula, Fountainhead is slightly better endowed than its sister resort, Arrowhead. The lodge entrance has gardens and fountains, and inside you'll be greeted by a red carpet. The five-story concrete lodge is constructed in a sweeping arc, and looks like something you'd expect to find in Miami Beach. But the design gives all of the 180 large rooms and suites private balconies and exceptional views of the lake or park. The poshest unit is a suite with a private pool and a sundeck overlooking the lake. Many units have refrigerators. The 22 cottages here are nice and come with kitchenette and one or two bedrooms, sleeping two or four adults. The restaurant is rather elegant, with background music, and specializes in steak. A seafood buffet is featured on Friday night.

There is also a bottle club, and many weekends you'll find dancing and entertainment.

The large pool is oval shaped and has a sunning apron and wading pool. Also here are an 18-hole golf course with putting green, lighted tennis courts, archery, bicyle rental, a coin laundry, a kennel, and a 3000-foot airstrip. For other park facilities, see Arrowhead Lodge, above ☐ **Inexp** ☐ **W, F, G, T, R**

Lake Murray Lodge, *Ardmore, OK 73401, (405) 223-6600* ☐ *7 mi S of Ardmore on I-35 and 2 mi E of SR 775* ☐ This Y-shaped one-story lodge of native timber is on a lovely site previously designated for the vacation retreat of Oklahoma governors. It has a sleek ranch look, and the 50 rooms and suites have color TV and room service. The 88 cottages are nicely spaced. The smaller cottages, sleeping two adults, have kitchenettes, but larger ones accommodating up to six people are for sleeping only. Dining here is by candlelight on specialties such as catfish or steak, and the bread is homemade. The private club has dancing. There is an especially large outdoor pool, and this resort even has its own chapel.

The 13,000-acre park has a marina with waterskiing, boat rental, and other services. Fishermen will find two fishing piers, a heated dock, and two ice docks. The crystal-clear lake has a swimming beach with bathhouses. Other facilities include tennis, golf, riding, a grocery, a 2500-foot lighted airstrip, and a museum of minerals. ☐ **Inexp** ☐ **W, F, R**

Lake Texoma Lodge, *Box 248, Kingston, OK 73439, (405) 564-2311* ☐ *On US 30 15 mi W of Durant* ☐ Oldest and most popular of Oklahoma's resorts, Texoma was built in grand style. On spacious lawns, on a rise with a commanding view of the lake, the white concrete lodge has a circular center with the 100 rooms and suites strung out to one side. All units have color TV and room service, some have lake views, and two suites have wetbars. A 20-room annex is less desirable. The cottages here sleep up to eight people and come with color TV and kitchenettes (no equipment); many have fireplaces. You can even rent a houseboat. The attractive glass-walled restaurant serves catfish and roast beef to background music, and there is a private club for drinks. In summer, a snack bar also operates.

This resort has two swimming pools, a supervised play program, barber and beauty shops, a coin laundry, and a paved 3000-foot airstrip. The surrounding park covers almost 2000 acres, and most of what is not owned by the state is maintained by the Army Corps of Engineers, which has lined the wooded shores with fine swimming beaches and other facilities. The park marina has rentals, waterskiing, fishing guides, and other services, and there is a fishing dock. The nine-hole golf course has a driving range and a putting range. The swimming beach has a bathhouse. Tennis, shuffleboard and other games, trapshooting, playgrounds, riding, and a grocery are also available. ☐ **Inexp** ☐ **W, F, G, T, R**

Quartz Mountain Lodge, *Lone Wolf, OK 73655, (405) 563-2424* ☐ *20 mi N of Altus on SR 44* ☐ This lodge is set on the shore of Lake Altus but differs from Oklahoma's northeasterly lake resorts due to the terrain. Instead of wooded shores, Lake Altus has exceptionally craggy, rugged hills with mesquite, prickly pear, and other vegetation characteristic of the Southwest. The blue of the 6700-acre lake contrast sharply with the stark shores. The lodge here is built of stone that harmonizes with the texture of the surrounding rock outcroppings, and the woods used are a color similar to the hills. The 46 rooms and suites, in a two-story wing beside the main lodge and lawn, have color TV and room service. The 16 cottages sleep up to six people and have kitchenettes (no equipment). The restaurant looks across lawns to the lake and has lulling background music. Specialties include Southern fried chicken and a Sunday night buffet, and there is a private club.

In addition to a key-shape outdoor pool, there is an indoor pool. The 11,000-acre park has boat rental, waterskiing, an ice dock, an enclosed dock, and a swimming beach with bathhouse. Fishing is excellent for bass, crappie, and catfish. Also here is a nine-hole golf course, and riding horses can be rented a little over a mile away. ☐ **Inexp** ☐ **W, F, G, R**

Roman Nose Lodge, *Box 61 Watonga, OK 74467, (405) 623-7281* ☐ *7 mi NW of Watonga on SR 8 and SR 8A* ☐On the heavily wooded bluff above Lake Boetcher, Roman Nose is the most intimate of the state park resorts, with only 20 rooms and 10 cottages (each sleeping four). These cottages are set along bluffs and have lovely views. The scenic canyons were once the home of the Cheyenne, and fittingly, American Indian motifs are carried throughout the decor of the lodge. The restaurant serves traditional country cooking and does its own baking; there is background music but no private club. Outside is a large, freeform swimming pool. This small park–only 500 acres–has two jewel-like, spring-fed lakes edged with green; higher still are colorful bluffs. The lakes are stocked with bass, trout, bluegill, and catfish. Also on the grounds are a nine-hole golf course, riding, playgrounds, and paddleboat rental. ☐ **Inexp** ☐ **F, G, H**

Western Hills Lodge, *Box 509, Wagoner, OK 74467, (918) 772-2545* ☐ *18 mi W of Tahlequah on SR 51* ☐ On the shore of 19,000-acre Fort Gibson Lake, this ultramodern lodge emphasizes Oklahoma's Western heritage. Brick and timber construction give the lodge an easygoing ranch style atmosphere. Most of the 100 doubles and studios face either the pool or the lake and have private patios or balconies; a bridal suite and 11 cabana units are also available. All units have room service. The 62 cottages have one to three bedrooms and sleep up to six people; roughly a third have kitchenettes (no equipment). The restaurant has Western decor and will arrange special barbecues, but steak is always in the menu. There is a private club with weekend dancing and entertainment.

The lodge's two-story buildings ring a large pool and wading pool surrounded by lawns. Along the lakefront is an 18-hole golf course with pro. Activities include hayrides, stagecoach rides, lighted tennis, bicycle riding, and horseback riding. The lodge is set in 2875-acre Sequoyah State Park, which has a marina with boat rental, fishing docks, waterskiing, and other services. Lake Ft. Gibson is stocked with bass, catfish, and crappie. Other facilities include a swimming beach, playgrounds, a grocery, and a 3,200-foot airstrip □ **Inexp □ W, F, G, R, T**

Oregon

BEND: Deschutes National Forest, southwest of Bend, has spectacular scenery that has often been the setting for Westerns. A marked, 100-mile auto loop along Century Drive and Cascades Lakes Highway passes magnificent vistas—Tumalo and other waterfalls, Bachelor Butte, the grand peaks of Three Sisters Wilderness, many alpine lakes with good fishing and boat rentals (seven lakes are reserved for fly-fishing, and Elk Lake is noted for sailing), the osprey and waterfowl refuge at Crane Prairie, the Newberry volcanic crater, and a large lava area with a visitor center, lava fields, tunnels, caves, and a lava-cast forest. Along this route are many national forest recreation areas and state parks offering watersports, hiking, and rockhounding. Mt. Bachelor Ski Area (22 miles SW of Bend on Cascades Lakes Highway) has seven chairlifts serving 18 runs, ski school, rentals and repairs, racing programs, and special seminars, clinics, and programs. A nordic center offers 20-miles of marked trails, some into Three Sisters Wilderness, rentals, lessons, tours, picnics, and races. Three lodges offer cocktails and lunch. A day care center accepts children aged from six weeks to six years. In addition to resorts described below, Mt. Bachelor Village (Century Drive, Bend, OR 97701, (503) 389-5900) which has condominium rentals, and a number of motels near the slopes offer packages plans; For information, write Mt. Bachelor, P.O. Box 1031, Bend, OR 97701.

Cultus Lake Resort, *Box 262, Bend, OR 97701, Winter: (503) 382-2494; Summer: (503) 977-3903* ☐ *Open mid-May to mid-Oct* ☐ *From Bend, take the Cascade Lakes Highway SE 50 mi* ☐In Deschutes National Forest on a hillside overlooking 1200-acre Cultus Lake, this cottage colony is a good choice for the fisherman or family. Many of the duplex cottages have fireplace, kitchen, knotty-pine paneling, and a view of the lake. In the lodge are a restaurant, a bar, a coffee shop, a grocery, and a tackle shop. Canoes, kayaks, fishing boats, and motors are available for those who want to explore the lake or go after its rainbow trout or mackinaw. ☐ **Inex** ☐ **F, W**

The Inn of the Seventh Mountain, *P.O. Box 1207, Bend, OR 97701, (503) 382-8711* ☐ *Open all year* ☐ *7 mi W of Bend on Century Dr.* ☐ Set in spectacular Cascades country, sprawling away from a bluff above the Deschutes River, this is an attractive condominium resort that operates on EP and package plans. The complex consists of clusters of three-story ultramodern lodges, with the accent on wood (on exteriors, decks, many paneled interior walls, and furnishings). About 300 units, most with good mountain views, are offered. In the main lodge are double rooms and studios that sleep up to four people. The other lodges have suites with one to three bedrooms, a living room, an equipped kitchen, a fireplace, color TV, and a private balcony. The complex has a restaurant, a bar with nightly entertainment, a snack bar, a grocery store, two outdoor swimming pools, a wading pool, whirlpools, saunas, and a game room.

A recreational staff oversees special children's activities and plans adult recreation on and off the premises. On the resort's 60 acres are riding stables offering trail rides along the river, to lava beds, and to scenic areas, as well as cookouts, special rides, and pack trips. Tennis buffs will find Plexi-pave courts, rentals, and instruction. Golf can be arranged at the Bend Golf and Country Club. The resort has a private stocked lake, fishing in the Deschutes River, bike rental, an ice and roller skating arena, and hiking trails. The staff organizes a number of trips that often include lunch–rafting on the Deschutes River, all-day bike trips, and guided hikes to the summit of Mt. Bachelor, to a secluded lake, or to a giant snowfield. ☐ **Mod** ☐ **R, T, F, S**

Rock Springs Guest Ranch, *64201 Tyler Rd., Bend, OR 97701, (503) 382-1957* ☐ *Open all year* ☐ *From jct US 97, take US 20 NW to the Emporium, turn W and follow to Tyler Rd. on left* ☐ Lodgings at this fine, pleasantly landscaped ranch run from the Owl's Nest studio–with sofa bed, queen-size bed, kitchen, bath, and deck–up to two-bedroom cottages and suites. Each of these contemporary cottages has kitchen, fireplace, and private deck. Three meals are served each day at the handsome, modern lodge, and in good weather a cocktail hour is held on the wide veranda. From mid-June to mid-September, the ranch runs on AP that includes horseback riding, fly-fishing in the ranch ponds, and use of the two lighted tennis courts, horseshoe and volleyball areas, and small barbell-shaped pool. The rest of the year the cottages are available at an EP rate without the extras. Even though the ranch provides part-time babysitters for preschoolers, it has an unusual offer families will appreciate: any full-time babysitter brought by guests will be provided with accommodations and meals free of charge. Mt. Bachelor Ski Area is a half-hour's drive west, and Lava River Caves State Park, with a one-mile-long lava tunnel, is 12 miles south on US 97 ☐ **Mod** ☐ **R, F**

Sunriver Lodge, *Sunriver, OR 97701, (503) 593-1246* ☐ *Open all year* ☐ *Take US 97 15 mi S of Bend, then 2 mi W on private road* ☐ Surrounded by mountains and pine forest, this resort sprawls over 5500 acres of alpine meadowland by the side of the Deschutes River. A huge wood and stone lodge in the center of the complex houses the dining room, lounge (dancing and entertainment), recreation room, and sauna facilities. More than 200 one- to two-bedroom accommodations are spread out around the lodge in multi-unit modern wooden lodges of one or two stories. In addition, 125 privately owned one- to four-bedroom houses rent by the week or month. The larger units have kitchens and living and dining rooms, but even the least elaborate units feature such niceties as stone fireplaces, exposed-beam ceilings, and private decks.

The marina supplies boat and equipment rental for fishing or sport on the Deschutes River. You can also swim in the olympic-size pool, bicycle on the 17 miles of paved bike paths, or hike or take guided horseback tours on the nature trails. Riding lessons are available. A resident golf pro gives instruction on the 18-hole course, and there is a putting green and driving range. Tennis lessons are available on 14 Plexipave courts. And there are numerous indoor and outdoor game areas. In winter, cross-country ski equipment rental and lessons are offered and there are sleigh rides. Miles of nordic trails extend into the adjoining forest. Alpine skiing is at Mt. Bachelor, to which the lodge operates daily bus service. Extensive facilities make Sunriver a self-contained village, complete with its own 4500-foot airstrip and a shopping mall. Besides the main restaurant in the lodge, smaller informal eating places are located throughout the grounds. A monthly newspaper announces each day's entertainment and activity schedule, from nature walks and rockhounding to movies and dancing in the Owl's Nest Lounge. There is a teen club, and a supervised play program operates in summer. ☐ **Lux** ☐ **G, T, R, S, F, W**

COTTAGE GROVE: The Village Green, *Box 277, Cottage Grove, OR 97424, (503) 942-2491* ☐ *Open all year* ☐ *Off I-5 at exit 174* ☐ This resort consists of luxuriously appointed one-story motor lodges connected by walkways and set on spacious, pleasantly landscaped grounds. The attractively furnished rooms and suites are air-conditioned and have views of the golf course or pool. Some have private patios or refrigerators. A few of the one-, two-, or three-bedroom suites also have fireplaces. The Iron Maiden restaurant serves excellent continental cuisine, either indoors before the fireplace, or outdoors on the terrace if the weather permits. Sports facilities here include a nine-hole pitch-and-putt course, playground, several tennis courts, and swimming and wading pools. A nine-hole regulation course is in town and an 18-hole course is within ten miles. Right behind the lodge is the depot for a steam train that leaves on regularly scheduled excursions. The resort will provide booklets for drives to scenic points in the area, including five covered bridges, the gold- and silver-mining areas around Bohemia City ghost town, and Umpqua National Forest. ☐ **Mod to Lux**

CRESCENT LAKE: Crescent Lake Lodge and Resort, *P.O. Box 73, Crescent Lake, OR 97425, (503) 433-2505* ☐ *Open all year* ☐ *From US 97 take SR 58 15 mi W, then 2½ mi SW on Lake Rd. to lodge* ☐ This is a homey colony of well-maintained rustic cabins clustered along the sandy shore of Crescent Lake, amid the pines and firs of Deschutes National Forest. Guests have their choice of one-, two-, or three-bedroom cabins with good views; seven of the larger ones have stone fireplaces, and all but one have a kitchen. The lodge supplies bedding, linens, and dishes. There is good fishing in the lake for rainbow, mackinaw, and kokanee. If you've got your own boat with you, the lodge can provide moorage; if you don't have a boat, they'll rent you one. In the winter you can rent a snowmobile and tool around on the lodge's groomed trails; or bring your cross-country skis and explore the forest. The central lodge building houses a grocery store, comfortable lounge, and a restaurant of which the managers are especially proud: all the food is cooked and baked there, and the resort even has an old-fashioned soda fountain for sundaes in the pines. ☐ **Inexp** ☐ **F, W**

GEARHART: Gearhart-by-the-Sea, *P.O. Box C, Gearhart, OR 97138, (503) 738-8331* ☐ *Open all year* ☐ *From US 26/101, take Golf Course Rd. or Pacific Way E to resort on Marion Ave.* ☐ All 99 units in this beautiful seaside complex are one- to three-bedroom suites with fireplace, kitchen, private lanai, and color TV. They are individually decorated, most with contemporary furnishings, and all but the single-bedroom units have two baths. Besides the spectacular coastline in front of the high-rise main building, the resort's top attraction is the adjacent 18-hole golf course. The first nine was laid out in the early 1890s, making it the oldest course in Oregon. A pro shop and the Sandtrap restaurant and lounge adjoin the course. Gearhart also offers a small indoor pool and whirlpool. A playground is on the grounds, and the resort will supply a list of sitters. A short walk away are two unlighted tennis courts (no charge) and a riding stable. ☐ **Mod** ☐ **G, W**

GLENEDEN BEACH: Salishan Lodge, *Gleneden Beach, OR 97388, (503) 764-2371* ☐ *Open all year* ☐ *20 mi N of Newport on US 101* ☐ Oregon millionaire John D. Gray wanted happy memories of the spot on the Oregon coast where he and his family had vacationed. So he decided to buy the place. In 1964, after he'd acquired a beautiful chunk of bayside forest at the mouth of the Siletz River, he instructed his architects to design a resort that would "look as though the lodge complex has been dropped into the woods." What they came up with–Salishan Lodge–has been called one of the the most beautiful modern hotels in the world. The architects have emphasized simplicity and clarity of line. Most of the 150 guest rooms and suites are housed in two strings of two-story wooden lodges which drift gracefully out from either side of a massive central building. Their rough-sawn cedar siding and cedar-shake roofs echo on a smaller scale the design of the master building. Covered walkways link the buildings, and the whole is enclosed by the bay on one side and the rolling, manicured lawns of the golf course on the other.

Inside and out, the buildings allude to the ecology of their setting and the traditions of the Northwest. Wood is everywhere: nearly two million board feet of cedar, Douglas fir, and hemlock were used in construction. The imposing doors of the main lodge–200 pounds each–are solid oak. Cedar shakes for the roofs were made in a small mill nearby and native stones were assembled to fashion the huge fireplace in the lodge's cedar-paneled lounge. Plantings throughout the grounds are almost exclusively of indigenous trees and shrubs: salal, huckleberry, kinnikinnick, and pine, spruce and hemlock. The understated landscaping suggests the influence of Japanese garden design. Although the furnishings of the rooms are modern, guests are in no danger of forgetting that they are at the meeting point of two kinds of wilderness: forest and ocean. The buildings are arranged so that the terrace of each room gives an excellent view of either the bay and ocean beyond, or of the forested hills. All rooms have works of art by Oregon artists from the collection assembled expressly for Salishan by the Grays. (Part of the lodge itself was, in fact, designed to house this growing collection.) Cedar paneling, brick fireplaces, and exposed-beam ceilings lend a rustic note to the clean spaciousness of the lodgings. Even in the most elaborate suites–in the Chieftan and Tennis Houses–the architects have preserved Salishan's characteristic balance of luxury and simplicity.

The excellent par-72 golf course has been carefully laid out to fit into the larger design of the resort; the first nine holes lead golfers on a tour of lodge and lake; the back nine head out to the dunes which separate the ocean from Siletz Bay. The fully equipped tennis center features three indoor Plexipave courts, glare-free lighting, Viewer's Lounge, and pro shop with pro. The 20-by-60-foot heated swimming pool is enclosed in its own cedar-paneled pavilion, which also contains sauna (complete with cold plunge), hydrotherapy pools, and men's and women's gyms. Salishan has a private beach–three miles of it. Guests can spend the day whale-watching or rockhounding (for agate and jasper, coral, pink granite, and quartz). A guide book is available for Salishan's 750-acre private forest preserve with nature trails. Fishermen need travel only the five miles to Depoe Bay for excellent deep-sea fishing; Salishan will arrange charters for you. River fishing is in the nearby Siletz and Salmon Rivers, and Drift and Schooner Creeks, inhabited by salmon, cutthroat trout, steelhead, blueback, and cod.

Exhausted by all this relaxation, guests may not have the energy to lift the Gourmet Room's nine-page menu, and this would be a shame. It includes a filet mignon stuffed with a Pacific oyster, duckling a l'orange, and grenadine de boeuf aux morilles. The wine list is enormous. All of this is served in the quiet splendor of the three-level Gourmet Room. Less elaborate meals can be found in the Sun Room coffee shop, or in the Cedar Tree (open in the summer only), for steaks and seafood cooked over mesquite charcoal. Then its up to the Attic Lounge for a nightcap amid the pines. ☐ **Mod to Lux** ☐ **G, T, F**

GOLD BEACH: Gold Beach is doubly blessed–not only is it on the famed Oregon coast, but it lies at the mouth of the magnificent Rogue River, which has earned classification in the Wild and Scenic River system. The town was named for placer mining on the Beach, and gold panning and agate hunting are still popular on the river. Today, however, the emphasis is on fishing—clamming and surf casting on the beach, salmon and steelhead trout in the river—and on scenic river trips. Sightseeing trips up the river were inaugurated in 1895, when tourists boarded the mail boats. Today the trips are by jetboat, and two excursions are offered by several operators north and south of the Rogue River Bridge. The 64-mile round-trip tour ventures into the Siskiyou National Forest wilderness and takes about 6½ hours, including a 2-hour lunch stop at Agness. The 104-mile round-trip tour includes whitewater rapids in the upper Rogue and takes seven or eight hours (with a lunch stop). Overnight trips are also available. Unlike whitewater rafting, whitewater jet trips are safe for children. (See also Grants Pass.) Cape Sebastion and other scenic state parks on the coast are especially pretty when azaleas and wild flowers are in bloom.

Ireland's Rustic Cottages, *Box 774, Gold Beach, OR 97444, (503) 247-7718* ☐ *Open all year* ☐ *L mi S on US 101 to 1120 S Ellensburg* ☐ This is a beautiful spot, right on the beach, offering 20 secluded cottages on attractive grounds. Some of these quaint cottages have decks, ocean views, or fireplaces, and a few have a kitchenette. There is no restaurant, but several seafood restaurants are within walking distance. ☐ **Inexp**

Jot's Resort, *Box J, Gold Beach, OR 97444, (503) 247-6676* ☐ *Open all year* ☐ *At N end of Rogue River Bridge on US 101* ☐ This two-story motel along the riverfront has 75 units, about a third with kitchenettes, some of which have balconies overlooking the river, the bridge, and the marina. The restaurant specializes in local seafood, and there is a bar. The marina has a sports shop, boat rental, charter fishing, and guide service. Reservations should be made in advance for the whitewater river trips that include lunch or overnight in wilderness lodges. **Mod** ☐ **F, W**

GRANTS PASS: Best Western Riverside, *Box 1128, Grants Pass, OR 97526, (503) 476-6873* ☐ *Open all year* ☐ *971 SE 6 St.* ☐ Overlooking the upper end of the Rogue River (see Gold Beach), this comfortable motel offers 107 rooms and suites, many with balconies overlooking the river, and two swimming pools. The restaurant and dining terrace also overlook the river. The resort offers several sightseeing jetboat trips between mid-May and late September. These include two-hour tours with a barbecue stopover, and whitewater tours. Longer river trips, for fishing or sightseeing, last from one to five days; some go the entire length of the river to Gold Beach, while others offer wilderness lodging. Along the river are many public recreation areas for swimming, fishing, and boat rentals. Much of the river flows through Siskiyou National Forest, which has old mining camps, unusual vegetation, waterfalls, and magnificent scenery. Trail rides and pack trips are available locally. Grants Pass has an 18-hole golf course that is open to the public. Oregon Caves National Monument is 48 miles south. ☐ **Mod** ☐ **W, F**

IDLEYLD: Lemolo Lake Resort, *P.O. Box 147, Idleyld Park, OR 97447, Call Roseburg mobile operator, then ask for (503) JS 2- 7060* ☐ *Open all year* ☐ *5 mi E of SR 138 on Lemolo Lake Rd, 75 mi E of Roseburg* ☐ This modest cottage colony, on the 8.3-mile shoreline of Lake Lemolo in the high country of Umpqua National Forest, is 15 miles north of the entrance to Crater Lake National Park. It is most likely to appeal to those seeking a simple, secluded retreat in the pines. Accommodations are in eight modern A-frame cabins: four of them, right on the lake, have equipped kitchenettes, separate bedrooms and living rooms, and patios. A central lodge, in matching chalet style, houses a restaurant–steak and shrimp are specialties, and wine and beer are served–and a grocery and fishing-tackle store. Fishing (for German brown trout) and boating are the main summer activities here. Canoes, motorboats, sailboats, and fishing equipment can be rented, and waterskiing is permitted on the lake. A stable five miles away offers trail rides. In winter the resort is the site of the Lemolo Lake (cross-country) Ski School, and snowmobile rentals are offered for use on the more than 400 miles of nearby trails. ☐ **Inexp** ☐ **F, W, S**

MOUNT HOOD NATIONAL FOREST: *Red Lion at Bowman's, Welches, OR 97067, (503) 622-3101 or 224-7158* ☐ *Open all year* ☐ *1 mi S of US 26 on Welches Rd.* ☐ In a spectacular valley west of Mt. Hood, the modern buildings of this resort community nestle against pine-covered slopes and give you private patios or balconies looking out over fairways and meadows with wildflowers, streams, and ponds. Two low-slung motels offer 30 doubles and suites, some with a refrigerator. Clusters of two-story condominium lodges and cottages have more than 60 apartments with kitchens and one to five bedrooms. These are roomy, with beamed ceilings or other warm touches, and all but the one-bedroom units have large stone fireplaces. The restaurant has specialties such as chicken breast florentine, and the bar has dancing and entertainment. Outdoor salmon barbecues and other special events are scheduled too.

The golf course offers 27 holes, with putting green, driving range, and pro shop with rentals and services. Pro service is also available for the six lighted, Plexi-pave tennis courts. Also on the grounds are an indoor swimming pool, a wading pool, saunas, whirlpool, exercise room, recreation room, and bicycle rental. Stables providing trail rides and pack trips are nearby. Steelhead trout are plentiful in the Salmon and Sandy rivers. In winter there is a free bus shuttle to the three full-service ski areas–Timberline, Mt. Hood Meadows, and Multorpor-Ski bowl. These provide chairlifts for day and evening skiing, ski school, rentals, and bars and cafeterias. Palmer Snow Field in the Timberline area makes skiing possible into summer. The Meadows complex has cross-country trails. Bowman's provides babysitters and sports instruction programs for children. The Mt. Hood Loop Highway circles the mountain and passes many public recreation sites for swimming, fishing, and hiking along the Columbia River. From many of these you can hike to remote high-altitude lakes. Cascades Locks Marine Park, east of Bonneville, has full marina services including sightseeing cruises. ☐ **Mod to Lux** ☐ **G, T, H, S, F**

NEWPORT: Embarcadero Resort Hotel and Marina, *1000 SE Bay Boulevard, Newport, OR 97365, (503) 265-8521* ☐ *Open all year* ☐ *1 mi S of US 20 on SE Moore Dr., 1 mi E of US 101* ☐ Dungeness crab and salmon are the local celebrities in Newport, and the streets of the Old Town restored fishing village offer ample opportunities for the vacationing gourmet to stroll and feast. But for the do-it-yourself seafood lover, or the dedicated fisherman with a taste for luxury, Embarcadero, at the end of Old Town, offers an ideal combination. The huge complex of modern brown-shingle buildings, encircling its own marina on Yaquina Bay, includes 140 one- and two-bedroom suites with kitchens, fireplaces, and sun decks. The additional 70 "patio suites" are one-room units without kitchens. Appointments are thoroughly and elegantly modern, but leather and wood combine with exposed-beam, knotty-pine ceilings and expansive views of the bay to emphasize the Embarcadero's nautical theme. The resort operates on EP. The restaurant, with a panoramic view of the bay is open from 7 am to 10 pm and specializes in local crab and salmon. The cocktail lounge has dancing nightly.

The moorage facilities–including gas dock and storage lockers–can accommodate boats of almost any size, and make this a convenient starting point for short cruises to some of the best salmon fishing waters in the Northwest. You can rent boats and fishing equipment by the day, week, or month; take charter trips; or fish and crab off the piers outside of the recreation center. Within the recreation center is the glass-enclosed swimming pool, directly adjacent to the bay, with sauna and whirlpool. A children's play area, fish cleaning station, crab cooker, and barbecue pit are provided.

Newport has undersea gardens, two public golf courses, hiking in Siuslaw National Forest, shopping and sightseeing in Old Town, and just walking and building sand castles along the miles of nearby ocean beach. ☐ **Mod to Lux** ☐ **F, W**

TOLOVANA PARK: Tolovana Inn, *P.O. Box 165, Tolovana Park, OR 97145, (503) 436-2211* ☐ *Open all year* ☐ *On US 101 Beach Loop, 2 mi S of Cannon Beach* ☐ This cluster of three-story shingle-style condominiums sits right on seven-mile-long Cannon Beach and offers 150 modern units. Most desirable are the studios and one- and two-bedroom apartments with kitchens, fireplaces, and private balconies directly over the beach. Identical but cheaper units are set back from the water, at an angle to the beach, offering views of the ocean and Haystack Rock, one of the largest monoliths on the Oregon coast. Cheapest of all are conventional double rooms in the rear; these do not have ocean views, but look across the parking lot toward the coastal mountains to the east. The oceanside restaurant specializes in local seafood, and is open for breakfast, lunch, and dinner in the summer, and lunch and dinner in the winter. The bar has entertainment. On the grounds are a heated indoor swimming pool, sauna, whirlpool, and billiard room. The beach has swimming, surfing, surf fishing, and wonderful beachcombing and tide-pool exploring. Four miles north is Ecola State Park, where a hike along cliffs offers spectacular views of the coastline as well as bird and sea lion rocks and clamming. ☐ **Inexp to Mod** ☐ **F**

UMATILLA: Best Western Columbia Inn, *705 Willamette, Umatilla, OR 97882, (502) 922-4871* ☐ *Open all year* ☐ This motel sits on the edge of Wallula lake, formed when McNary Dam was built across the Columbia River, overlooking the river and the dam. The motel offers 70 air-conditioned rooms, about half with kitchenette in a two-story design that gives the units either private balconies or patios. On the grounds are a swimming pool, a golf course, lawn games areas, a restaurant, and a bar, but most activity centers on the Columbia River. The fish ladders and other dam areas are open to the public daily, and guided tours are offered in summer. The lake has fishing, swimming, and full-service marinas where waterskiing and boat rentals are available. ☐ **Inexp** ☐ **W, F**

WARM SPRINGS: Kah-Nee-Ta Vacation Resort, *Warm Springs, OR 97761, (503) 553-1112* ☐ *Open all year* ☐ *16 mi NW of Madras on US 26* ☐ This is a topnotch resort in the hilly desert of sunny Oregon.

How's that again? Sunny Oregon? One of the pleasantest surprises in the Northwest is that once you've passed the rainy coast and then the lush Cascades you find yourself in land that could just as easily be the Southwest. The dramatic transition is reason enough to make a point of approaching Kah-Nee-Ta from the west. Owned and operated by the Confederated Tribes of Warm Springs Reservation, this superb resort is laid out by the side of the Warm Springs River. The V-shape main lodge, in a two- and three-story stepped design set on a bluff, offers 144 air-conditioned rooms and suites with private balconies giving valley views. Decor in guest quarters and public rooms is based on Indian motifs. The candlelit Juniper restaurant offers continental cuisine and wines. The more colorful Appaloosa Room features Indian favorites such as fresh huckleberries and smoked salmon for breakfast, terrace lunches, and dinner specialties that might include charcoal-broiled steak, game hens cooked in clay, or buffalo, venison, or bear specialties. Outdoor salmon bakes are also scheduled. The lodge also has a bar, a lounge with entertainment, a bakery, a gift shop, a swimming pool, and full services from babysitting to airport bus.

Farther down the river is the village, where two dozen air-conditioned cottages offer bedrooms, living rooms, and kitchens. You can also camp out in authenic (unfurnished) Indian teepees. This complex centers on hot mineral springs that feed a large swimming pool, children's pools, Roman-style baths massage rooms, and saunas. The River Room restaurant has a balcony over the river. Also in the village are a snack bar, mini-golf, and an Indian crafts shop. Near the main lodge, an 18-hole golf course spreads along the river bank. It has a clubhouse with restaurant, bar, and pro shop. The lodge also has tennis (instruction available), fishing along the river as well as in nearby Lake Simtustus, and riding stables that offer lessons and rides. You can rent a bike, hike, or jog along trails, some of which run along the river. Costumed Indian festivals, rodeos, and other special events are held spring through fall. ☐ **Mod to Lux** ☐ **G, R, F**

Pennsylvania

BEDFORD: Bedford Springs Hotel, *Bedford Rd., Bedford, PA 15522, (814) 623-6121* ☐ *Open May–Oct* ☐ *2 mi S on US 220* ☐ On 2800 green, hilly acres, this resort has been a spa since magnesia springs were discovered here in the 1790s and was President Buchanan's summer White House. The three-story hotel provides 250 air-conditioned rooms and suites on MAP. Some rooms have TV, oversize beds, or refrigerators. The restaurant will provide a box lunch, and there is a snack bar. The cocktail lounge has dancing and entertainment. The 18-hole golf course, bordered by stands of ancient trees, has pro services, a pitch-and-putt course, and putting green. Pro service is also available for the tennis courts. Also on the grounds are a stocked lake, indoor and outdoor game areas, a playground and a trapshooting field. You can also rent bicycles or take a surrey ride. The springs feed the outdoor swimming pool, and there is an indoor pool as well. Supervised children's programs are available seasonally. ☐ **Mod** ☐ **G, T, F**

BOILING SPRINGS: Allenberry, *Boiling Springs, PA 17007, (717) 258-3211* ☐ *Open Feb–Dec* ☐ *6 mi SE of Carlisle on SR 174* ☐ Originally part of a land grant by William Penn, Allenberry is 55 peaceful acres of meadows and woods on winding Yellow Breeches Creek. The 32 contemporary accommodations are in the two-story motel-type Meadow Lodge. More traditional lodgings are in the 15-room, colonial-decor Stone Lodge, built in 1812. Scattered about under trees are a half-dozen housekeeping cottages with one to four bedrooms. All units have private baths, and lodge units are air-conditioned. The resort operates on EP and offers two restaurants. Stone Lodge Carriage Room with beamed ceiling and a fieldstone hearth is the elegant buffet room in charming Fairfield Hall, a building dating from 1785 that also houses one of the cocktail lounges. From spring until autumn, this pleasant resort offers an outdoor olympic-size pool, four tennis courts, lawn games, trout fishing in the creek, a playground, and a recreation room. Allenberry Playhouse, a professional summer-stock theater on the grounds, offers plays and musicals from the end of April into November. ☐ **Mod**

CHAMPION: Seven Springs Mountain Resort, *P.O., Champion, PA 15622, (412) 352-7777* ☐ *Open all year* ☐ *10 mi SE of PA Turnpike Exit 9 (Donegal)*

☐ Set in the beautiful Laurel Hills, not too far from where Frank Lloyd Wright built his famous house, Falling Water, this large complex spreads near the base of Seven Springs Mountain. In a sprawling layout that gives most units a private balcony, patio, or deck, the resort offers 392 rooms and suites (with wet bars) and a modern ten-story lodge. Clustered nearby are two dozen chalet suites with one to three bedrooms and kitchens. Less than a mile away are 80 apartment units, some of them duplexes, with one to four bedrooms and kitchens. Maid service is provided in the lodge and is available at extra cost for the chalets and apartments. The restaurant, which has a lovely mountain view and a fireplace, specializes in seafood smorgasbord. The cocktail lounge has dancing and entertainment. There is also a 24-hour coffee shop.

This complex provides indoor and outdoor swimming pools, a wading pool, whirlpool, saunas, a recreation room, bowling, and indoor and outdoor game areas. Tennis is also offered indoors and outdoors. The golf course has pro services, a driving range, and a putting green. There is a lake with fishing and a swimming beach. The chairlift operates in summer and for fall foliage, and there is an alpine slide. The Seven Springs ski area has snowmaking and opens in December. It provides daily day and evening skiing until March. The vertical drop is 940 feet, and the longest run 1½ miles. The runs are served by two triple-chairs and surface tows. Full services include ski school, rentals, and two cafeterias. ☐ **Mod to Lux** ☐ **G, T, W, S, F**

HERSHEY: Hotel Hershey and Hershey Motor Lodge, *Hershey, PA 17033, (717) 533-2171 and (717) 533-3492* ☐ *Open all year* ☐ *Near jct US 422 and US 322* ☐ During his many trips to Cuba, Milton B. Hershey, the chocolate magnate, acquired a fondness for that sun-drenched island and its graceful architecture. In the early 1930s, Hershey decided to add a grand hotel to his domain and, not surprisingly, built one whose Spanish-colonial architecture, interior design, and landscaping reflected the Old World atmosphere of Old Havana. Located on a beautifully landscaped hilltop amid 90 acres of rolling countryside, Hotel Hershey accommodates guests on AP or MAP in 250 double rooms and six plush suites. The spacious public areas include the beautiful Fountain Lobby, several lounges, a game room, and luncheonette. Meals emphasizing French cuisine are served in the Garden Terrace and in the Circular Dining Room, which is surrounded by 13 stained-glass windows. Free from support pillars, which was quite an architectural feat in the 1930s, the Circular Dining Room was designed according to Mr. Hershey's wish that "no guest would ever have to be seated in a corner."

In addition to a directed recreation program for adults, the hotel has a full range of sports facilities, the gem of which is the Country Club with five golf courses totaling 72 holes. Guests may also play tennis, swim indoors and outdoors, go horseback riding, and explore the miles of hiking, cycling, and cross-country skiing trails that thread through the hotel's vast acreage.

Near the hotel is 460-room Hershey Motor Lodge, a long, white, two-story building overlooking a lake filled with swans and ringed with weeping willows. The resort offers an untaxing but wide array of sports and health facilities including indoor/outdoor pool, saunas, whirlpool, game room, tennis and paddle tennis courts, and nine-hole pitch-and-putt course.

Within a mile of Hersheypark, an amusement park, Hershey's Chocolate World, and Hershey Museum of American Life, the hotel and motor lodge make an especially good choice for families. **Mod to Lux ☐ T, R, G,**

LANCASTER: Host Farm, *2300 Lincoln Hwy. E, Lancaster, PA 17602, (717) 299-5500* ☐ *Open all year* ☐ *4 mi E of Lancaster on US 30* ☐ In Pennsylvania Dutch Country, this MAP farm offers more than 500 motel-style, air-conditioned rooms, a restaurant, a bar with dancing, entertainment, and dinner theater. On the grounds are a 27-hole golf course with a driving range and putting green, indoor and outdoor tennis with instruction available, two indoor and two outdoor swimming pools, a wading pool, a health club with saunas, a recreation room, and bicycle trails. There are also two ice rinks (one indoors), a ski area with a chairlift and a rope tow, and tobogganing. Host Farm is associated with Host Town, an in-town motel at 30 Keller Ave., where you have the option of staying on EP or MAP and using the farm facilities. ☐ **Mod to Lux ☐ G, T**

Willow Valley Farms, *2397 Willow St. Pike, Lancaster, PA 17602 (717) 464-2711* ☐ *or (800) 323-1776* ☐ *Open all year* ☐ *4 mi S. of Lancaster on US 222* ☐ Owned by a Mennonite family native to Lancaster Country, this farm provides 170 rooms in modern motor lodges that have many balconies overlooking the pond, the indoor pool, or the nine-hole golf course and putting green. Also here are an outdoor pool, sauna, tennis, recreation room, and a playground. Rowboats are available for the pond, which in winter is used for ice skating. Bus tours are conducted of Pennsylvania Dutch country. The restaurant is noted for its Lancaster County recipes, with many ingredients raised on the farm. The frequent smorgasbord is popular. The farm also runs a bake shop. ☐ **Inexp to Mod**

POCONO MOUNTAINS: This scenic mountain area with some 200 lakes and many waterfalls, offers year-round recreation. In summer there is swimming and fishing, boating, golf, tennis, horseback riding, and hiking, as well as canoe trips on the Delaware River. In fall there is splendid colored foliage and some hunting. In winter a dozen ski areas open. At Stroudsburg, you'll find a wild animal farm and Quiet Valley, a living historical farm, and throughout the mountains are family attractions, art galleries, antique shops, and restaurants. The hundreds of accommodations vary from sedate old inns to honeymoon resorts famous for such accoutrements as heart-shaped beds to ordinary motels to family-oriented resorts in all price ranges. For a complete list, write Pocono Mountains Vacation Bureau, 1104 Main St., Stroudsburg, PA 18360.

Skytop Club, *Skytop, PA 18357, (717) 595-7401* ☐ *Open all year* ☐ *15 mi N of Stroudsburg off SR 390* ☐ This imposing ivy-covered five-story lodge, which looks like an English country castle, would make a sovereign feel at home. Set on 550 acres on a rise above a forest-rimmed lake, it offers more than 200 rooms with mountain views and air conditioning (some connecting rooms share a bath), in the gray-stone main building, the nearby Garden House, or in studio cottages. The last have a combination living room/bedroom with twin beds and a private porch and entrance, and are available in suites of one to four rooms. All units are available on AP or package plans. Public rooms are especially nice, and include a formal and elegant dining room (dress code evenings), cocktail lounge, taproom, a paneled lounge, and other lounges and porches, as well as many areas for cards, pool, billiards, and other indoor games.

The grounds have lovely lawns and gardens, and include areas for archery, quoits, lawn bowling, croquet, and putting green. Except for horseback riding, (which can be arranged eight miles away), Skytop provides most sports, and instruction in many. The 18-hole golf course has water and other hazards, and a little unpretentious clubhouse with cart rentals. For tennis there are six Har-Tru and one all-weather court. The upper part of the lake has a boathouse with canoes and rowboats, and fishing licenses are available. There is also stream fishing. On the lower lake is a swimming beach with diving towers and a shallow pool for children. You can also swim year-round in the tiled pool, which has a glass wall leading to a terrace. The sports pavilion has movable walls and provides shuffleboard and other games in summer; November through March it becomes an ice skating rink. Around the lakeshore are marked nature trails. Also available are bicycling, motorized hayrides, summer and winter cookouts, and many special events such as films and concerts. Seasonal supervised play for children, and evening babysitters, are also provided. Winter brings snowmaking on the modest downhill slopes, cross-country trails, a ski school, a rental shop and sledding. ☐ **Mod to Lux** ☐ **W, F, G, T**

Split Rock Lodge, *Lake Harmony, PA 18624, (717) 722-9111* ☐ *Open all year* ☐ *Off SR 940* ☐ On the forested shore of 135-acre Lake Harmony, this attractive mountain resort was once a private enclave for coal company executives, and the rustic main lodge overlooking the lake, which houses the restaurant and cocktail lounge, dates from that time. The 50 double guest rooms are in two three-story modern lodges that also face the lake. Set back in the woods are 16 rustic cottages, dating from the era of the main lodge, with fieldstone fireplaces and knotty-pine interiors. The lake has a swimming area, fishing, and a dock where you can rent canoes or other crafts or go waterskiing. A stable provides guided trail rides. Also on the grounds is a glass-enclosed swimming pool, eight Har-Tru and all-weather tennis courts with instruction available, a skeet and trapshooting range, archery, outdoor game areas, and bicycling. Golf is available nearby. Split Rock has its own downhill ski area with snowmaking and a lift, plus cross-country trails, and rentals, as well as ice skating, sledding, and snowmobiling. Its ski packages also include skiing at the nearby Big Boulder and Jack Frost complexes. Seasonal surpervised play is available for children. The lodge operates on MAP and package plans. ☐ **Mod** ☐ **W, T, F, R**

Tamiment Resort and Country Club, *Tamiment, PA 18371, (717) 588-6652* ☐ *Open all year* ☐ *Take US 209 to Bushkill; follow signs 5 mi N to resort* ☐ This lavishly equipped resort occupies a lovely 2200-acre site by the side of a 90-acre lake in the Poconos. The main lodge and golf course cut a clearing in the woods at the edge of the lake, but all around is dense forest. Recreational facilities here are first class. The golf course, designed by Robert Trent Jones, is 6700 yards of rolling fairways by the lake and woods. There are ten all-weather lighted tennis courts and six handball courts. You can take your pick of pools: an enormous heated outdoor L-shaped affair by the side of the lake, or an olympic-size indoor pool surrounded by a tier of balconies. Sauna and steam rooms are also available. You can swim in the lake, too; or rent a sailboat, rowboat, or paddleboat. Other facilities include basketball, volleyball, shuffleboard courts, a miniature golf course, and game rooms with table tennis, pool, and electronic games. The winter list of activities is equally impressive, with downhill and cross-country skiing, ice skating, tobogganing, snowmobiling, and snow hayrides. Many planned activities include full-time day camp for children in the summer.

The variety of accommodations includes rustic cottages, in the woods, the most economical, that feature air conditioning, heat, and private bathrooms. The Tamiment House and Clubhouse rooms are more luxurious, with wall-to-wall carpeting, refrigerators, color TV, and coffee makers. Some suites are available. The MAP rates include free tennis and golf (during the week) and two meals a day in Tamiment's excellent dining rooms. There's big-name entertainment in the Las Vegas-style showroom, and dancing, cocktails, and more entertainment well into the night in the hotel's cocktail lounges. ☐ **Mod** ☐ **G,T,W**

SHAWNEE-ON-DELAWARE: Shawnee Inn, *Shawnee-on-Delaware, PA 18356, (717) 421-1500* ☐ *Open all year* ☐ *Off I-80 and SR 209* ☐ The tree-bordered grounds of the pleasant, quiet Shawnee Inn lie on the western bank of the Delaware River, just north of the Delaware Water Gap. The inn, a white, four-story, red-roofed building constructed in 1905, offers 86 doubles with private bath, color TV, and climate control. The inn complex contains an airy dining room (dress code), a coffee shop, a cocktail lounge, game rooms, a billiard room, and a gift shop. Also on the premises is Shawnee Playhouse, a summer repertory theater in the building from which Fred Waring and his Pennsylvanians used to make their radio broadcasts. At the edge of the Delaware are two large outdoor pools, and another large pool indoors. Near the inn and pools are six all-weather courts, lighted at night, a driving range, and a putting green. Golf here consists of 27 holes, playable in nines, and the beginning of one course and the end of another surround two sides of the inn. Most of the holes, however, are across a foot bridge on a tree-ringed island in the Delaware River. The inn arranges canoeing on the Delaware, rents mopeds, and schedules children's activities and entertainment. Shawnee Inn provides transportation to Shawnee Mountain, which has beginner to advanced ski trails, four double chairs, base and summit lodges, and a ski school. □ **Mod** □ **G, S**

STATE COLLEGE: Toftrees Country Club and Lodge, *State College, PA 16801* □ *(814) 237-5311* □ *Open all year* □ *Off US 322* □ Set on a rise in rolling woodland, between the Pennsylvania State University campus and state gamelands, this lodge is a cluster of rambling modern buildings with angled roofs and rustic decor. It offers 69 double rooms, 54 units with a double bed and a studio bed, a bridal suite with a circular bed and sunken bath, and a waterbed room. Some rooms have balconies. In addition to the airy main restaurant, Le Papillon, which has a cocktail lounge featuring live music, there is informal patio dining and poolside services. The EP rates include use of the swimming and wading pools. Surrounding the lodge complex is an 18-hole, 6983-yard championship golf course with rental, carts, a driving range, and a putting green. Also available are four all-weather tennis courts (two lighted), two illuminated courts for platform tennis, saunas, and bicycle rental. The pro shop can arrange lessons and rentals for several sports. □ **Mod** □ **G, T**

UNIONTOWN: Mt. Summit Inn, *Box T, Uniontown, PA 15401, (412) 438-[illegible]4* □ *Open mid-Apr to mid-Nov* □ *6 mi E of Uniontown on US 40* □ Sitting on the top of Mt. Summit, with forested hill panoramas, this three-story, sprawling country inn offers 123 rooms and suites on MAP. This is a good family resort, with pleasant but untaxing sports facilities–indoor and outdoor pools, a wading pool, a whirlpool, tennis courts, a nine-hole golf course, a putting green, a playground, and outdoor game areas. This pretty Laurel Hills area has caverns, old furnace sites, historic sites, and other attractions. Ohiopyle State Park has canoeing and hiking in beautiful Youghiogheny Gorge, which has several waterfalls and scenic overlooks. □ **Mod**

South Carolina

CHARLESTON: Kiawah Island, *P.O. Box 12910, Charleston, SC 29412, (803) 559-5571* □ *Open all year* □ *From US 17, 6 mi E of Charleston, take Main Road 17 mi S to island* □ It sounds like one of the Hawaiian chain, but it's a barrier island just south of Charleston. The island is the province of only one resort development, whose stated goal is to create a resort community with the least ecological disturbance. One of the most essential steps in carrying this out was to plan all necessities and recreation areas within walking distance of each other. With facilities thus concentrated, the remainder of the island's 10,000 acres could be left as a sanctuary for birds and wildlife. Some notable animal inhabitants of the island are white-tailed deer, bobcats, alligators, and the giant loggerhead turtle which comes to nest each summer. Jeep tours through the interior and along the shoreline are scheduled all year long.

This still left plenty of room for sports. The first golf course, designed by Gary Player, has water hazards on 13 holes. By contrast, the second course, designed by Jack Nicklaus, uses the natural terrain–dunes, moguls, and berms–as hazards. The 11 composition and two lighted Grasstex tennis courts are supervised by pros who offer tennis camps from late-March to mid-December. The island has 10 miles of beaches, perfect for swimming and, on more isolated stretches, horseback riding. Hobie Cat and Sunfish rentals and instruction are also available on the beach. If you prefer freshwater swimming, a large pool overlooks the sea. And you may rent fishing gear to go after channel bass, drum, bluefish, trout, or crabs. Supervised play for young children, and a teen center, are available.

Lodgings come in three basic types–inn rooms and privately owned villas and homes. Inn rooms and suites, contained in four separate cypress-and-cedar-clad lodges, have double beds and private balconies. The one- to four-bedroom villas come with fully equipped kitchens; some have fireplaces, sun decks, and screened porches. Homes are set either on the river or are a short walk from the beach. Though usually less elaborate than the villas, and without daily maid service, all homes have complete kitchens. Of several restaurants on Kiawah, the main one is the Jasmine Porch. An airy and spacious room decorated with hanging planters and ceiling fans, it overlooks the pool and ocean beyond. The menu features Low Country seafood specialties such as she-crab soup, lobster, and broiled fish. Sunday brings a brunch with fresh-fruit crepes, eggs Benedict, and a variety of cold salads and meat and fish dishes. The intimate Charleston Gallery has continental and gourmet cuisine, and there are informal restaurants and snack bars in the shopping mall, the clubhouse, and at poolside.

One of Kiawah's great attractions is it's proximity to Charleston–a historic and cultured city of many beautiful old buildings. ☐ **Mod to Lux** ☐ **G, T, W, F**

Seabrook Island, *P.O. Box 32099, Charleston, SC 29407, (800) 845-5531* ☐ *Open all year* ☐ *23 mi SE of Charleston via SR 700* ☐ Occupying 2200 oceanside acres, this modern resort community offers over 200 rental units rated as standard, superior, and deluxe. These vary from one to three bedrooms but all are air-conditioned and have kitchens and balconies or patios. Most units have peaked or sloping shingled roofs and some have fireplaces, lofts, beamed ceilings, clerestories, or other features. Beach Club villas are clustered near the water; Golf Villas are under trees along the fairways; Spinnaker units are next to the tennis courts; Tarpon Pond cottages face lagoons and moors; and some super-deluxe individual homes offer beautifully landscaped private enclaves. All are rented EP, based on double occupancy, or on golf or tennis packages. The center of social life is the Beach Club, where the restaurant specializes in she-crab soup, local shrimp, and other Low Country cooking. The cocktail lounge is also here. Beside the club, separated from the beach by palm trees, are two large swimming pools and a wading pool. Spreading inland is the 6850-yard, par-72 golf course, curving around tidal marshes and bordered by stands of live oak and palmettos. Near the ocean are the seven Har-Tru tennis courts, with practice area and ball machines. The Treehouse Club offers lessons and a pro shop. Also available are an equestrian center with six horses for trail or beach rides, and rentals of bicycles, sailboats, and fishing and crabbing gear. The area has good birding, there are nature trails, and the ponds are stocked. ☐ **Lux** ☐ **G, T, W, F, R**

FRIPP ISLAND: Fripp Island Inn, *Fripp Island, SC 29920, (803) 838-2411* ☐ *Open all year* ☐ *22 miles SE of Beaufort on US 21* ☐ Not as large or developed as Hilton Head Island, Fripp Island is modeled after its neighbor to the south and is a similar year-round resort community in a beautiful, semitropical setting. Fripp was developed along a 3½-mile beach with respect for natural beauty, and the palms and live oaks are still here shading the new villas and tennis courts. The pirate Blackbeard sought refuge here in the 17th century, and Fripp Island is still a hideaway. You cross a bridge from Hunting Island State Park to the community of over 100 privately owned villas, some of which you can rent on a weekly or monthly basis, and the 500-room Fripp Island Inn. Single and attached town-house-style homes–set along the oceanfront, or a short walk from the beach along inland waterways and the fairways of the golf course–have an attractive, contemporary look, central air conditioning, all-electric kitchen, and will accommodate from 6 to 24 people. The two-story inn has a lean ranch look, and doesn't hog the skyline. Inside are attractively furnished, spacious rooms, all with private balconies, color TV, telephone, and room service.

Following the inland waterways and groves of live oak, sand dunes, and seashore, the 18-hole golf course designed by George Cobb is spectacular and hazardous. The 511-yard 18th hole is a monster flanked on one side by the Atlantic and on the other by sand dunes and sawgrass. The racquet club has eight soft and six hard-surface courts. The three pools on the island include one which is olympic size. The marina offers dockage, charter boats for deep-sea fishing, tackle shop, and other services. Crabbing, freshwater fishing, shelling, bicycling, a Parcourse (exercise trail), and supervised children's activities are also available. Social life centers on the beach club, with its Plantation Room restaurant, which serves seafood and continental cuisine, and the upstairs Sandbar Lounge with entertainment and spirits. ☐ **Mod to Lux** ☐ **G, T, W, F**

HILTON HEAD: Just a couple of decades ago Hilton Head Island was 45 square miles of unspoiled dunes, loblolly pines, palmettos, magnolias, live oaks, salt marsh, and 12 miles of brilliant white-sand beach. Fortunately, the Sea Pines Plantation Company, which first developed the island, proceeded with unusual respect for the natural surrounding. Deer, otter, alligator, and many varieties of birds still inhabit the island. Hilton Head–named for William Hilton, the British sea captain who discovered it in the 17th century–was at one time a rice and cotton plantation. Today, it is a recreational Eden. Four marinas cater to sailors, pleasure boaters, and fishermen, and small pleasure boats are for rent at many resorts. Offshore game fish include marlin, sailfish, king mackerel, wahoo, and barracuda, and there's excellent surf-fishing for channel bass, pompano, and croakers. The 11 golf courses include several of the best in the country. Jack Nicklaus helped Pete Dye design the Harbour Town course, which is widely regarded as one of the best in the world, and which hosts the annual Sea Pines Heritage Golf Classic in late March. The Palmetto Dunes is consistently rated as another of the country's finest courses. Among the uncountable tennis courts, the major complexes are at Shipyard, Palmetto Dunes, and Harbour Town in Sea Pines Plantation. Hilton Head also has riding, and hiking, with especially nice bridle and walking trails through four nature preserves. And of course, there are almost 100 restaurants, many bars, and numerous shops and boutiques. Many major facilities on the grounds of particular resorts are available no matter where you stay, and golf and tennis packages give you access to the best courses and racquet clubs.

Hilton Head, linked by causeway to the mainland via US 278, is 31 miles from Savannah. In addition to resorts listed below, the island has motels and rentals in villa complexes. For a complete list, write the Chamber of Commerce, P.O. Box 5647; for golf and tennis packages, write Box 5267—all Hilton Head, SC 29928.

Palmetto Dunes Resort, *Box 5606, Hilton Head Island, SC 29928, (803) 785-1161* ☐ *Open all year* ☐ This outstanding inland resort, within walking distance of the beach, offers an inn and villas clustered around golf courses and lagoons. At the center of the complex is an inn offering rooms with two double beds, suites with kitchenettes, and a restaurant and cocktail lounge. One-story or town-house villas are available nearby or along the fairways. These have from one to four bedrooms, equipped kitchens, and private balcony or patio. Each group of villas has its own swimming pool (there are 11 pools in all). The 18-hole Robert Trent Jones golf course has an excellent pro shop, a driving range, and a putting green. Nearby is the George Fazio Golf Course, rated one of the nation's best, with its own clubhouse. Lessons, clinics, and other services are provided. The tennis complex has 25 Har-Tru and Laykold courts (six with lights), a pro shop, instruction, clinics, and modern teaching aids such as videotape. Fish Creek Landing is a boat dock with restaurant and cocktail lounge where you may rent canoes, rowboats, pedal boats, and fishing equipment. Canoeing is especially pleasant along the resort's ten miles of interconnected lagoons and waterways. For the beach you can rent sailboats, sunfish, and catamarans. Also available are bike rentals, a supervised children's program, barbecues, outdoor buffets, dancing, and entertainment. Everything is within walking distance, and you can also walk to the adjacent Hyatt Hotel which shares the same strip of beach. ☐ **Mod to Lux** ☐ **T, G, F, W,**

Port Royal Plantation, *Hilton Head Island, SC 29928, (803) 785-3381* ☐ *Open all year* ☐ This resort with a private beach offers an inn with double rooms and villas with two bedrooms, living room, kitchenette, and screened porch. On the pine-dotted grounds you'll find an outdoor pool, four tennis courts (two with lights), bicycle rental, and a restaurant and bar. Golfers have access to two 18-hole championship courses. ☐ **Mod** ☐ **T, G**

Sea Pines Plantation, *Hilton Head Island, SC 29948, (800) 845-6131 or (803) 785-3333* ☐ *Open all year* ☐ Sea Pines set the standard for modern resorts. The result of creative and painstaking planning, for which it won numerous awards including one from the American Institute of Architects, Sea Pines Plantation consists of four resort villages and the Hilton Head Inn on a beautiful 5000-acre beachfront setting. The villas come with equipped kitchens and linens, and usually with maid service, although this is optional in some. Inn and villa units are rented EP or on family, holiday, or sports packages. Most sports involve extra fees unless they are included in a package plan.

The centerpiece of Sea Pines golf is the Harbour Town course, with its lighthouse known to television viewers as a backdrop for golf tournaments. The Sea Marsh and Ocean courses are also 18-holers. Cart rental, instruction, and clinics are offered at each pro shop. Sea Pines has 60 tennis courts at various locations, but the 30-court Sea Pines Racquet Club at Harbour Town is the best known because of many annual televised tournaments. Clinics and instruction are available for all levels of play, with videotape and other modern teaching aids, and a Tiny Tot program accepts children as young as five. The major marina is at Harbour Town (a smaller one is at South Beach) and offers full services including sailboat and motorboat rental, charter boats, a guided fishing program, head boats, fishing tournaments, and dockage. There is also a sailing academy. Fishing is so varied here–surf-casting, deep-sea fishing, crabbing, and inland in brackish and freshwater–that a guidebook is available. In addition to 4½ miles of beach, there are 14 swimming pools. Lawton Stables offers instruction, guided trail rides, hayrides, cookouts, pony rides, and equestrian events. Bicycles may be rented for the 13 miles of bike paths, and there are bridle and hiking trails through the 572-acre nature preserve where more than 200 varieties of birds have been recorded. A supervised youth recreation program is operated all summer and at some other holiday seasons for kids aged three through fourteen (large families get a bargain–the program only charges for your first three children).

Harbour Town, curving around a crescent shaped inlet with a waterside promenade, is marked by its 90-foot, candy-striped lighthouse. This is the liveliest area in Sea Pines, with five restaurants, lounges, boutiques, a swimming pool, and a playground. It is also the major sports center, with a full-service marina, the Harbour Town Golf links, and the Sea Pines Racquet Club all within walking distance. Available here are one- to four-bedroom villas, most with either water or fairway views. Interesting units here are the one- or two-bedroom octagonal sealofts which do not have panoramic views but are set on high pedestals among trees and come with a Japanese-style soaking tub. *Fairway Villas* are two- to four-bedroom units in cottages or town houses scattered along the three golf courses. They overlook fairways, marshes, lagoons, or woods. Depending on the location, you may need a bike or car to reach the beach, Harbour Town, or other facilities. *South Beach Villas* are isolated on a beach at the southern tip of the island and are not served by the shuttlebus, so you need a car if you don't want to bike or hike to other facilities. However, beach buffs may find all they need right here–a small-boat marina, 13 tennis courts, a landscaped swimming pool, a playground, a general store, a pavilion restaurant, a tavern, and fine beach. *Sea Pines Beach/Forest Homes* are not a village complex but single family dwellings scattered in several areas. They come as small as one bedroom and as large as six bedrooms and have oceanfront or forest views.

Hilton Head Inn is a beautifully landscaped complex of low lodges separated by lawns, palms, two swimming pools, and a wading pool from the 1000-foot beach. Within a flowered courtyard is a third swimming pool. Also here are a tennis complex with eight courts, a putting green, a playground, four restaurants and lounges (including a nightclub), and shops. The inn offers 202 rooms and suites, most with two queen-size or double beds and a refrigerator. These units overlook the ocean, the pool, or an oak grove. The inn also rents self-service villas with one to three bedrooms that are within 220 to 450 yards of the beach.

Sea Pines has a shuttle bus that runs mid-February through November from Hilton Head Inn to the Beach Club and Harbour Town. Mini-bus and taxi services are available all year, and limousines to Savannah Airport may be reserved. Sea Pines has 11 restaurants, ranging from a deli to continental cuisine in chic surroundings, many night spots, outdoor cookouts, and special events. ☐ **Mod to Lux** ☐ **G, T, F, W, R**

MYRTLE BEACH: The Grand Strand, a 50-mile stretch of unbroken white-sand beach with Myrtle Beach at its hub, is the state's most popular vacation area. It's a bustling family resort with more than 35,000 rooms for visitors, a state park, 27 golf courses, and nearly 100 tennis courts. While not as picturesque and unspoiled as Hilton Head Island and areas to the south, the Strand has more action, and accommodations for almost every budget. Family tourist attractions include seven amusement parks, the Myrtle Beach National Wax Museum, Castle Dracula, Ripley's Believe It Or Not Museum, and the Guiness Hall of World Records. Among the golf courses is that at the Dunes Golf and Beach Club designed by Robert Trent Jones, rated one of the best in the country. Arnold Palmer designed the 54-hole Myrtle Beach National. Most of the hotels are set side-by-side along a narrow strip between the beach and the busy main street, leaving room for only swimming pools on the premises. Golf and tennis are therefore either at inland public complexes or at private clubs at which the hotels have made special arrangements. On the beach, which is public, there is surf-fishing, and ten fishing piers are scattered along the coast. For a complete list of accommodations and facilities, write the Chamber of Commerce, Box 2115, 1301 N Kings Hwy, Myrtle Beach, SC 29577, (803) 448-5135.

The Breakers, *Oceanfront at 21st Ave. North, Box 485, Myrtle Beach, SC 29577, (800) 845-0688* ☐ *Open all year* ☐ This modern, high-rise hotel offers 144 units, about half with kitchenette, and membership at a number of golf clubs and tennis complexes. ☐ **Mod**

Driftwood Lodge, *Box 275, 16th Ave. at the oceanfront, Myrtle Beach, SC 29577, (803) 448-1544* ☐ *Open all year* ☐ The design here gives most rooms an ocean view and a balcony. More than half the units and family suites have kitchenettes. Driftwood offers golf and tennis privileges as well as fishing and sailing packages. ☐ **Mod**

Indigo Inn, *2209 Ocean Blvd., Myrtle Beach, SC 29577, (803) 448-1631* ☐ *Open early Jan to late Nov* ☐ The double rooms and efficiency apartments in this nine-story building all have private balconies and views. Public tennis courts are six blocks away, and golf is available. ☐ **Mod to Lux**

The Swampfox Motor Inn, *24th Ave. and South Ocean Blvd., Myrtle Beach, SC 29577, (803) 448-8373* ☐ *Open all year* ☐ This exceptionally long three- to five-story motel offers 179 units, including efficiencies, two-room apartments and deluxe efficiencies, many with private balconies and ocean views. One of the four pools is indoors, and golf and tennis are available. ☐ **Inexp to Mod**

Yachtsman Resort Inn, *P.O. Box 426, 1400 N Ocean Blvd., Myrtle Beach, SC 29577, (800) 845-0654* ☐ *Open all year* ☐ These twin 11-story towers offer 200 rooms, suites, and efficiencies, all with private oceanfront balconies. Golf and tennis are available. ☐ **Mod to Lux**

Tennessee

BUCHANAN: Paris Landing State Resort Park, *Rte. 1, Buchanan, TN 38222, (901) 642-4311* ☐ *Open all year* ☐ *17 mi NE of Paris off US 79* ☐ This beautiful wood-and-glass lodge has many rooms overlooking the large pool and Kentucky Lake, part of the Land Between the Lakes. Rooms are comfortably and tastefully furnished. The very good restaurant, offering lake view, serves catfish, country ham, and other regional specialties at reasonable prices. In winter, buffalo steak may be on the menu. There are swimming and wading pools, gift shop, lighted tennis, and an 18-hole golf course with rentals at the pro shop, a putting green and a driving range. The beach, covered dock, hiking trails, rental boats, fishing, archery, and other park facilities are also available to inn guests. Special events are scheduled, including instruction, crafts workshops, movies, and weekly entertainment at the outdoor theater. ☐ **Inexp** ☐ **G, F, W**

BURNS: Montgomery Bell State Resort Park, *Burns, TN 37029, (615) 797-3101* ☐ *Open all year* ☐ *33 mi W of Nashville on US 70* ☐ This contemporary EP inn of native stone offers 40 doubles overlooking scenic 35-acre Acorn Lake and modern two-bedroom cabins under trees across the cove. All rooms and suites have TV, telephone, air conditioning, private bath, and picture-window views. Cabins, sleeping six to eight, have all this plus fireplace and kitchenette. The inn has a restaurant serving regional specialties.

Swimming is in the small pool or off the sandy beach on the lake. Near the beach is a dock for rental of rowboats and dreamboats. Bream and bass inhabit Acorn Lake and the larger Lake Woodhaven, also in the park. The list of sports facilities is led by the 18-hole golf course with bent-grass greens for year-round play and a pro shop for rentals. Nineteen miles of hiking trails include one that passes the remains of iron ore pits and the furnace operated in the 19th century by Montgomery Bell, Tennessee's great ironmaster. Tennis, archery, and other indoor and outdoor sports are available. In summer there are guided tours, naturalist programs, hayrides, and other activities. ☐ **Inexp** ☐ **G, F, W**

CHAPEL HILL: Henry Horton State Resort Park, *Chapel Hill, TN 37034, (615) 364-2222* ☐ *Open all year* ☐ *On US 31A, S of Chapel Hill* ☐ Perhaps the biggest feature of this state resort park is the Buford Ellington 18-hole, par-72 golf course. With ample greens, defended by equally ample bunkers, the course is a challenge for the average golfer. The pro shop has dressing rooms, rental clubs, pull carts, and a line of golf equipment. A practice green is nearby.

The inn has 72 rooms, including four suites with kitchenettes. All rooms are carpeted, air-conditioned, and equipped with TVs and telephones. The five two-bedroom cabins, designed to sleep six people each, have fireplaces and fully equipped kitchens. A restaurant, specializing in hearty Southern cooking, an olympic-size pool with bathhouses, and tennis courts are nearby. The 1135-acre park is the former estate of Tennessee governor Henry H. Horton. Within the park are picnic grounds and fishing on Duck River, skeet and trapshooting range with rentals, and recreation center with equipment for badminton, basketball, archery, volleyball, Ping-Pong, shuffleboard, and croquet. Horse stables and canoe livery are just outside park. ☐ **Inexp** ☐ **G, F, W**

GATLINBURG: Cobbly Nob Resort, *RFD 3, Gatlinburg, TN 37738, (615) 436-9333* ☐ *Open all year* ☐ *11 mi NE of Gatlinburg on SR 73* ☐ Set in a beautiful valley high in the Smoky Mountains, this attractive MAP resort is composed of several ultramodern and rustic lodges with private balconies looking over woods, ponds, and streams. Rooms come in two sizes; each vista room has two queen-size beds; the studio room is more spacious and has woodburning fireplace, a queen-size bed, and a sofa-bed. Vista and studio rooms can be connected to form suites. There is a handsome clubhouse restaurant, a snack bar, and the private club has dancing and entertainment.

Sports facilities here include a landscaped freeform pool larger than olympic size. On a little rise near the pool are four clay tennis courts, supervised by a pro in season. Surrounding the resort is an 18-hole golf course designed by Gary Player. For fishermen, Webb Creek and nearby ponds have trout. Children will find a playground, recreation building, and a seasonal supervised play program. Horseback riding is available nearby. Year-round skiing, believe it or not, is available at Ober Gatlinburg ski area. One slope, covering approximately five acres, has a plastic surface open all summer long. The alpine slide and chairlifts also operate during the warm weather. In winter, three chairlifts and rope tows service snow-covered slopes, and there are full facilities including ski school, rentals, and ice skating. ☐ **Mod** ☐ **T, G, F**

NASHVILLE: Opryland Hotel, *2800 Opryland Dr., Nashville, TN 37214, (615) 889-1000* ☐ *Open all year* ☐ *From I-40 4 mi E of Nashville, take Briley Pkwy. 5 mi N* ☐ What began as a country-music radio show, broadcast from the cramped and unadorned Ryman auditorium in the heart of Nashville, evolved into a multi-million-dollar entertainment complex in the suburbs, combining amusement park, auditorium, and resort hotel all under the name, Opryland USA. Although the whole complex is set closely together, the razzle-dazzle and bright lights of the amusement park and auditorium don't extend to the hotel—which is grand in quite a different way than the Grand Ole Opry. In the tradition of plantation mansions, the huge building is fronted by a portico with tall white pillars. Inside, an enormous brass chandelier hangs above a Scarlet O'Hara staircase, and beyond that, in the middle of the building, is a glassed-in brick courtyard, the Galleria, surrounded by several clothing shops and restaurants. The restaurants offer everything from beans and barbecue to continental cuisine. All guest rooms are done in colonial style with modern additions such as message lights and wake-up alarm. Each room has wing chairs, double beds and quilts, dust ruffles, and electric blankets.

For the sports minded, Opryland Hotel has six lighted all-weather courts and a heated pool. With about 13 musical productions each night and 17 rides, the 217-acre Opryland park has plenty to see and do. Only five miles away is the Hermitage, Andrew Jackson's home. ☐ **Lux** ☐ **T**

PICKWICK DAM: Pickwick Landing State Resort Park, *Pickwick Dam, TN 38365, (901) 689-3135* ☐ *Open all year* ☐ *10 mi S of Savannah on SR 128* ☐ On the Tennessee River near the Alabama border, this modern, state-run inn houses guests in the lodge (75 doubles and 3 suites) or in housekeeping cabins. Lodge rooms have picture windows and recessed private balconies overlooking the kidney-shaped swimming pool and the lake; all have TV, telephone, private bath, and air conditioning. Cabins, which must be rented by the week beginning Monday, have two bedrooms, patio, TV, telephone, air conditioning, private bath, fireplaces, and kitchenettes. The restaurant provides regional cooking and a lake view. Facilities include a full-service marina, tennis courts lighted for night play, a championship 18-hole golf course with rentals at the pro shop, three beaches, archery, outdoor games, and nature trails. ☐ **Inexp** ☐ **G, F, W**

PIKEVILLE: Fall Creek Falls State Resort Park, *Rte. 3, Pikeville, TN 37367, (615) 881-3241* ☐ *Open Feb–Dec* ☐ *Take SR 30 18 mi W of Pikeville to turnoff for park* ☐ With 16,800 acres of land and 340 acres of lake, this is the second largest state park in Tennessee. Accommodations come in three types—inn rooms, cabins, and a dormitory-style lodge. There are 72 doubles and singles and one suite with kitchenette at the inn. The lodge will accommodate 104 people and is completely equipped for housekeeping and meal preparation and service. The standout lodgings here are the cabins; each of these modern units, with a porch extending out over the lake, and a fireplace, sleeps up to eight people. The dining room at the inn serves local specialties and reasonably priced family fare. The park is named for Fall Creek Falls, a 265-foot-high cascade that plunges into a shaded pool. Streams in the park are good for smallmouth, rock bass, and bream, and the lake has largemouth,catfish, and bream. Recreational facilities include an 18-hole golf course designed by Joe Lee, two tennis courts, swimming pool, playgrounds, canoe and fishing boat rentals, and, for those who really want to get away from it all, miles of hiking and backpacking trails cut through virgin forest. ☐ **Inexp** ☐ **G, F, H**

WATTS BAR DAM: Watts Bar Resort, *Box 600, Watts Bar Dam, TN 37395, (615) 365-9595* ☐ *Open Apr–Nov* ☐ *3 mi S of Spring City on US 27, take SR 68 8 mi E to resort* ☐ Nestled in the foothills of the Great Smoky Mountains, on the shore of Watts Bar Lake, are 52 cabins originally built for the Watts dam construction crews. These one- to three-bedroom EP cottages have since been carpeted, paneled, and air conditioned, and are widely spaced among lawns, gardens, and old shade trees. About half the units have equipped kitchenettes. All come with linens, and maid service is available. The smallest units sleep one or two people, but the largest cabins sleep six to eight. The restaurant has a garden and lake view, does its own baking, and will cook the fish you catch. The modestly priced menu includes local specialties as well as more exotic fare such as teriyaki. On the 200 wooded acres are a landscaped swimming pool, wading pool, two asphalt tennis courts, lawn game areas, gardens, and nature trails for birdwatching. At the marina you can rent canoes, sailboats, pontoon boats, or motorboats for waterskiing or fishing. You can also fish from the banks or the dock, and the 38,000-acre lake harbors bass, crappie, sauger, and catfish. The resort will find you babysitters. ☐ **Inexp to Mod** ☐ **F, W**

Texas

AUSTIN: Lakeway Inn and World of Tennis Resort, *101 Lakeway Dr., Austin, TX 78734, (512) 261-6200 (Inn), 261-6000 (Tennis)* ☐ *Open all year* ☐ *From Austin take SR 71 NW to SR 620, follow signs* ☐ Lakeway is two resorts–one of the world's best tennis resorts, the World of Tennis, set within an all-around sports and leisure community. The complex is set on 5500 acres of rolling Texas Hill Country 20 miles northwest of Austin, on a peninsula jutting into 65-square-mile Lake Travis. There are plenty of reasons to roll out of bed in the morning. Golf is one. There are 36 holes of it, plus driving range and putting green. Jack Nicklaus designed the newest course. Then there's Lake Travis marina, which rents all variety of craft and can take care of you whether you want to fish, waterski, or sail. Swimmers can choose among three pools. Trail rides and lessons are available at the equestrian center. The World of Tennis is a temple to the sport with numerous Laykold and Grasstex altars–33 courts, give or take a few. Two are indoors; more than half of the outdoor courts are lighted. The elegant clubhouse restaurant and cocktail lounge overlook the stadium courts, which are used for many professional and amateur tournaments. The clubhouse also contains a pro shop, saunas, and steam and whirlpool baths. Of course, a wide variety of instructional programs include clinics for all ages. And just so you don't forget where you are or what you came for, the swimming pool is shaped like a tennis racquet.

If you stay in one of the attached town houses on the World of Tennis grounds, you will probably overlook the courts. These weathered timber and stone units are luxuriously furnished with fireplace, equipped kitchen, and a private patio with a built-in grill and a refrigerator. Town houses sleep between two and twelve people. You can also stay at Lakeway Inn's attractive two-story apartments overlooking Lake Travis. All rooms have private balcony facing the lake and the nearby hills, and all have color TV and climate control. Units with fireplace and kitchenette are available. The inn also rents nearby octagonal cabins with two bedrooms and full kitchen. At the inn is the lakeside El Lago Dining Room. And the Yacht Club, also with a fine view of the lake, is a good place to revive yourself with a drink, dancing, or entertainment after a long day of sports. World of Tennis has its own restaurant and lounge. ☐ **Mod to Lux** ☐ **G, T, R, W**

BANDERA: The Wild West is the main industry in Bandera. Situated in the rolling Hill Country northwest of San Antonio, Bandera is a center for dude ranches. The major attraction in the town itself is the Frontier Times Museum which displays frontier artifacts and a collection of Western art. Wild-West-style nightclubs and dance halls keep the town swinging into the night. San Antonio, with many historic attractions including the restored Alamo, is less than an hour away by car. The LBJ National Historic Site is a short drive north.

Circle R. Resort Ranch, *Box 1376, Medina, TX 78055, (512) 589-2324* □ *Open Mar to mid-Sept* □ *16 m W of Bandera on SR 16* □ Unusual in having its own 2500-foot airstrip and its own ghost-town movie set, this ranch offers an attractive setting on the Medina River. The AP rates include accommodation in air-conditioned cottages and meals served family-style, as well as cookouts and organized activities such as trail rides and hayrides. Archery, trapshooting, a rifle range (hunting can be arranged), a large outdoor swimming pool, a recreation room, tennis courts, outdoor game areas, and a playground are provided. □ **Inexp to Mod** □ **R, F**

Dixie Dude Ranch, *Box 548, Bandera, TX 78003, (512) 796-4481* □ *Open all year* □ *From Bandera take SR 173 S to ranch road 1077. Ranch is 9 mi W on 1077* □ This is your basic dude ranch, where the AP rates include a ride a day and 750 scenic acres of Texas Hill Country to ride on. Guests stay in cottages with knotty-wood paneling, air conditioning, and private bathrooms with showers. Western meals like country-fried chicken are served family style or in a buffet, and there are frequent outdoor barbecues. The outdoor pool has underwater lighting, and there is a wading pool for children. Other activities include bingo, bumper pool, lawn games, and campfires. Guests can view fossilized dinosaur tracks close by. □ **Inexp** □ **R**

Mayan Dude Ranch, *Bandera, TX 78003, (512) 796-3312* □ *Open all year* □ *1½ mi W off SR 16* □ Sprawling on wooded bluffs above the Medina River, with fine views of the Hill Country, this is a working as well as dude ranch. The air-conditioned accommodations are in paneled motel-style rooms or in older two-bedroom cottages scattered under the trees. Some of the cottages are of native stone or have rough-cedar interiors, or fireplaces or wonderful porch views. In all units paneling, furniture bearing the ranch's brand, and other touches have been added to create a Western flavor. Friendly gestures such as coffee delivered to your cottage in the morning also help make this ranch special. The AP rates include two rides a day (except Sunday), use of the two tennis courts and other facilities, and all of the optional scheduled activities. The latter may include hayrides, cowboy-cooked breakfasts, a fish fry or a steakfry, or song fests. Children have their own activities, and at night they are often fed separately at poolside parties. The ranch is proud of its food and often schedules theme nights with Mexican, Italian, or other ethnic dinners. The dining room has beautiful picture-window views, and the cozy Peacock bar is named for the peacocks that wander the grounds.

You can swim in the river or in the olympic-size pool, and tubing and fishing are also popular on the river. The wranglers will teach you to ride on docile horses, and experienced riders have a choice of more spirited mounts. Hiking to hunt for fossils and arrowheads is also fun. A unique feature here is Mayan City, a Western town built as a movie set. Once a week the ranch has a Wild West show here, complete with can-can girls, and sometimes the Yellow Belly Bar is used for other evening activities. □ **Mod** □ **R, F**

Twin Elm Guest Ranch, *Bandera, TX 78003, (512) 796-3628* □ *Open Mar–Oct* □ *From Bandera take SR 16 W to SR 470. Ranch is 3 mi W on 470* □ A former cattle ranch on 400 acres overlooking the beautiful Medina River and Indian Creek, Twin Elm is a traditional AP guest ranch catering especially to families. Guests stay in rustic air-conditioned cottages. The dining room is ranch-style with benches, wide-plank tables, stone walls, and exposed beams. Everybody gets a horse from the stable and frequent hayrides. The Medina River is excellent for inner-tubing, fishing, and swimming, and there is an an outdoor pool by the ranch house. Mineral and fossil hunting is another popular activity. □ **Inexp** □ **R, F**

BROWNSVILLE: Rancho Viejo, *P.O. Box 3918, Brownsville, TX 78520, (512) 350-4000* □ *Open all year* □ *11 mi N on US 77* □ The name means "old ranch," and these 3000 acres were once part of a Spanish land grant. Later the land was planted in citrus, and today it is landscaped with palms, hibiscus, oleander, and bougainvillaea. The buildings are in the old hacienda style, with red-tile roofs, arches, little balconies, and flowered patios. Almost 300 rooms and suites with kitchens are offered in one- and two-story buildings. These vary from executive suites surrounding the swimming pool to villas with two or three bedrooms overlooking the fairways. Furnishings may be contemporary or Mexican, depending on the taste of the owners.

The sprawling clubhouse, beside a *resaca* that was once part of the Rio Grande, contains the Ranchero Room, which is open from breakfast through dinner as well as for snacks. It specializes in Mexican food. Also here are a cocktail lounge, pro shop, and clubhouse services. Two 18-hole, par-70 golf courses are laid out among citrus trees, palms, pines, and *resacas.* A driving range, a putting green, a halfway house, and clinics are available. Clinics and pro instruction are also provided at the eight screened tennis courts (four lighted). The large outdoor pool has a waterfall, swim-up bar, and poolside service. The Casa Grande Supper Club, which has roving Mexican troubadors and specializes in red snapper, is in the ranch's original hacienda: this landscaped enclave is isolated beside a *resaca* and is reached by a cruise boat from the clubhouse. On the three-mile trip, you have cocktails while you are serenaded by a Mexican band. The ranch can arrange charters for bay and deep-sea fishing. Two nearby attractions are the exceptional Gladys Porter Zoo in Brownsville and the Confederate Air Force Ghost Squadron Museum in Harlingen. □ **Mod to Lux** □ **G, T**

CONROE: April Sound, *P.O. Box 253, Conroe, TX 77301, (713) 588-1101* ☐ *Open all year* ☐ *From Conroe, take SR 105 W to resort* ☐ Only 50 minutes from Houston, April Sound has the feel of being completely in the country, partly because most of its 1000 acres overlook the vast expanse of Lake Conroe. The 22,000-acre lake, one the best bass fishing spots in Texas, is the focus of the resort. Many of the 27 holes of golf play alongside it, and one, the ninth hole of the White course, plays across a narrow part of its waters. Boats may be rented at the Moorings, the giant marina adjacent to the resort, for fishing and sailing. You may swim in the lake or in the three lakeside pools; one is a 25-meter pool and diving well. April Sound offers 17 tennis courts, nine lighted and four covered, along with pro shops and snack bar. Sauna and whirlpool are available to relax sore muscles after a game.

Guests are housed in privately owned one-, two-, or three-bedroom town houses or in lodge doubles or suites. These modern hotel-type rooms have contemporary furnishings, color TV, and sliding glass doors leading onto patios or verandas. The Fernery and Driftwood Club, on the lodge's upper level, offer fine dining and lounge area. ☐ **Mod to Lux** ☐ **T, G, W, F**

CORPUS CHRISTI: Sheraton Marina Inn, *300 N. Shoreline Blvd., Corpus Christi, TX 78403, (512) 883-5111* ☐ *Open all year* ☐ *6 blocks south of US 37 on Shoreline Blvd.* ☐ Every one of the 175 rooms in this modern, 11-story inn has a window wall and a private balcony, commanding a view of Corpus Christi Bay, about 200 feet from the back doors. The top-floor Spanish Main restaurant enjoys an especially good view of the Gulf and counts Chateaubriand and stuffed flounder among its specialties. Simpler meals can be ordered at the Mexican-style Cafe Ole. On a second-floor deck beside the lodging tower are tennis courts and an outdoor pool, but most guests prefer the Beach Club, on the bay about seven miles south of the inn. Facilities there include an olympic-size swimming pool, a sandy beach, a 600-foot fishing pier, and the clubhouse with a restaurant and bar. The full-service Corpus Christi marina is directly across shoreline Boulevard. The municipal golf course offers 18 holes, and 27 holes at the Corpus Christi Golf Center are also open to the public. ☐ **Mod to Lux** ☐ **W, F**

Island House Apartments, *P.O. Box 8729, Corpus Christi, TX, 78412, (512) 933-8160* ☐ *Open all year* ☐ *18 mi SE of Corpus Christi on Padre Island* ☐ On 1000 feet of the unsurpassed beach of Padre Island, this exceptionally nice three-story apartment motel is designed in a U-shape so that all units have a balcony with a view of the beach and gulf. The choicest units face directly on the water, but the motel is isolated on an underdeveloped strip of beach so that the two arms also have unobstructed views of dunes and water. Most units have two-bedroom, two-bathroom, kitchenette apartments, comfortably and attractively furnished, that run the width of the wing and open onto a second balcony overlooking the landscaped courtyard with swimming pool and parking area. One-bedroom and three-bedroom units are also available. The kitchens are equipped with cooking and eating utensils, and all units have color TV, telephones, linens, and maid service. Within a half a mile you can find restaurants and bars. A quarter of a mile away is the Padre Isles Country Club with tennis courts, an 18-hole golf course, and a restaurant, where you are granted guest privileges. A fishing pier is a half-mile away. ☐ **Mod to Lux** ☐ **W, F, T, G**

Padre Island Beach Hotel, *Box 8142, Padre Island, TX 78412, (512) 933-8041* ☐ *Open all year* ☐ *25 mi S of Corpus Christi on Windward Drive, Padre Island* ☐ On the magnificent beach of Padre Island, this 150-unit, six-story hotel puts you right next to the National Seashore Visitor Center (with many free programs) and facilities. Air-conditioned double rooms and suites are available; cribs are free and there is a small extra charge for cots. The hotel has a restaurant, a bar with entertainment, and a swimming pool. The hotel will arrange for golf and tennis at the nearby full-service Padre Isles Country Club. ☐ **Lux** ☐ **W, F, T, G**

GALVESTON: Marriott's Hotel Galvez, *Galveston, TX 77550, (713) 765-7721* ☐ *Open all year* ☐ *21st St. and Seawall Blvd.* ☐ Named for Count Bernado de Galvez, for whom the city was also named, this grand hotel was also known by the flamboyant nickname, "The Crown Jewel of the Gulf Coast." Its opening in 1911 marked the rebirth of Galveston after the devasting hurricane of 1900 that completely inundated the island and claimed over 5000 lives. The regal, Mediterranean-style hotel reached its heyday in the 1930s and 1940s and had its ups and down until the 1970s when it slumped and closed. In the summer of 1980 the completely refurbished hotel was reopened by the Marriott chain. All of the 228 rooms now boast color TV, telephone with message light, and AM/FM radio. Off the main lobby is a terraced dining area with adjoining bar. Close by, the main lounge has nightly dancing and entertainment. The dance area opens onto a wide semicircular terrace that overlooks the Gulf. At the hotel are an indoor-outdoor pool, sauna, and gift shop, and guests have privileges at the 18-hole Sun Meadow Golf Course.

The Galvez overlooks the fine beach and the ten-mile-long seawall that has protected the island from hurricanes since the turn of the century. Many fishing piers, some with restaurants at the end, jut into the ocean, and fishing charters are readily available along the waterfront. The town itself has an array of exceptionally fine Victorian commercial buildings and residences, the most notable of which is probably the Bishop's Palace, an ornate stone mansion designated one of the 100 outstanding buildings in the United States by the American Institute of Architecture. Families with children will appreciate the Marineworld and Seawolf Park, with its tours of a WW II submarine, tank, and destroyer escort. ☐ **Lux** ☐ **G, W, F**

JEFFERSON: Tejas Village, *Rte. 1, Jefferson, TX, 75657, (214) 777-2460* ☐ *Open Mar–Nov* ☐ *12 mi E of Jefferson on SR 726* ☐ In a beautiful setting on the pine-shaded shores of Lake o' the Pines, this 80-acre resort is a good buy for water-oriented families. Most of the buildings are of brick and redwood, with interior walls paneled in redwood, and all units are air-conditioned, have picture windows, and come with linens and maid service. The two-story motel has 51 large double rooms; 11 studios have a kitchenette and dining area; and there are more than a dozen one and two-bedroom housekeeping cottages. All units have either a deck or an outdoor sitting area, and the all-electric kitchens are equipped with cooking and eating utensils. Most units have lake views, but some are secluded in the pines and a few face the parking area. A lakeview restaurant specializing in seafood and a private club are available. Tejas Village has a swimming pool, a private beach on the lake, a fishing pier, and a grocery at the marina. At the fully equipped marina, the largest on the lake, you can rent boats or barges, or go waterskiing. Fishing is good for bass, crappie, and stripers.

The town of Jefferson, once a river steamboat port, has been charmingly preserved and is worth a visit; an annual festival and house tour is held in May. ☐ **Inexp to Mod** ☐ **W, F**

KERRVILLE: Inn of the Hills, *1001 Junction Hwy, Kerrville, TX 78028, (512) 896-2300* ☐ *Open all year* ☐ *1½ mi NW on SR 27* ☐ Although not, strictly speaking, a resort, this inn offers a good base for vacations of the western Hill Country. On landscaped grounds surrounding an olympic-size swimming pool, the two-story lodges are constructed of native materials and decorated in Spanish-colonial style. The 150 units range from double rooms to executive suites with one or two bedrooms to a prized six-bedroom house with a balcony overlooking a waterfall and pool. Many units have balconies, patios, refrigerators, wetbars, or other extra touches. The restaurant offers indoor or outdoor dining, does its own baking, and has frequent buffets. The bar has dancing and entertainment. Also available are lighted tennis courts, a putting green, lawn game areas, and a health club with steam bath and sauna. Golf is available at the 19-hole municipal course in Kerrville.

Some twenty miles north is the old German town of Fredericksburg and the LBJ National Historic Site. To the south is the colorful frontier town of Bandera. Four miles west is the huge Kerrville State Recreation area. Here large lakes have been formed by impoundments on the Guadalupe River, offering lake and river fishing, canoeing, swimming, and boating. ☐ **Mod to Lux**

MC ALLEN: Best Western La Posada, *113 N. Main St., McAllen, TX 78501, (512) 686-5411* ☐ *Open all year* ☐ *At Main Street and US 83 Business* ☐ Surrounded by citrus groves, McAllen is an oasis of palm-lined lakes, parks, and boulevards landscaped with bougainvillaea and flowering shrubs. And La Posada, in the center of town, is an oasis within an oasis. Opposite Archer Park, it offers more than 200 units in three-story Spanish-style motel buildings and in town houses across the street that are connected to the main complex with an elevated walkway. Many of the units here are suites with one to three bedrooms, and many have beamed ceilings, wetbars or kitchenettes, and balconies. The all-day restaurant and the elegant dining room for leisurely evening meals specializes in Mexican dishes, and the bar has Mexican music and entertainment. Apart from the swimming pool, there are no recreational facilities on the premises. McAllen, however, runs a unique Tourist Club that provides golf, tennis, swimming, and special programs from November through April. Fishing is excellent in nearby lakes and in the Rio Grande (eight miles south). On the river are Anzalduas Park, with a covered pier and boat dock, and beautiful Bentsen State Park, a nature reserve with fishing. Both of these parks are good for birdwatching, and the 2000-acre Santa Ana National Wildlife Refuge, also on the river, has recorded over 300 species of birds; it has a seven-mile auto tour, 14 miles of foot trails, and blinds at birdfeeders. Seasonal quail and dove hunting is extremely popular. ☐ **Mod to Lux**

PORT ARANSAS: Stretching north of Padre Island National Seashore, and linked by causeway with Padre Island and Corpus Christi, is Mustang Island, a rapidly growing vacation area famous for a beautiful white-sand beach and semitropical climate. If you tire of fishing in the bay, you can walk across to fish in the surf. Port Aransas is the jumping-off point for deep-sea fishing expeditions for marlin and other species. Shrimp trawlers still work out of the port. There are 18-hole golf courses nearby at Padre Island Country Club and the Corpus Christi Golf Center. Just north is Aransas National Wildlife Refuge (see Rockport).

Channel View, *P.O. Box 776, Port Aransas, TX 78373, (512) 749-6156* ☐ *Open Mar–Sept* ☐ *6 blocks E of SR 361 in Port Aransas* ☐ If Channel View were closer to the water it would need stilts. These modern, four-story apartment buildings look out over the Aransas Channel and, a few hundred yards to the east, the Gulf. All apartments have a water view and none are on the ground floor. One- to four-bedroom units with full kitchen are available by the day, week, or month. All are large and airy, furnished with elegant bentwood and painted rattan furniture. Penthouse suites have private sun decks. A marina is immediately adjacent to the property and Channel View's private pier is lighted for day and night fishing. A swimming pool is located in the tropical courtyard, and the Gulf beach is just a few minutes walk away. ☐ **Mod to Lux** ☐ **F, W**

Coral Cay, *Box 448, Port Aransas, TX 78373, (512) 749-5111* ☐ *Open all year* ☐ *1¼ miles S of Port Aransas* ☐ This island retreat is on the beach of the Gulf of Mexico. From any of the two- and three-story attached town house units you can practically roll out of bed onto the wide, white sandy beach. Up and down the coast are miles of good beach, popular for surfing, surf-fishing, and swimming. On the premises are two heated freshwater pools and four Laykold tennis courts. Charter boats are available nearby. The one- to three-bedroom apartment units come with living-dining rooms and electric, fully equipped kitchens. Rooms are airy and decorated with attractive, contemporary furnishings. Many have private balconies and ocean views. Maid and linen service are available, and there is a laundromat on the premises. ☐ **Mod** ☐ **T, W, F**

ROCKPORT: Rockport Racquet and Yacht Club Resort, *Rte. 1, Box 70, Rockport, TX 78382, (512) 729-0880* ☐ *Open all year* ☐ *5 mi N of Rockport on SR 35* ☐ Set on Aransas Bay among windswept live oaks–startlingly gnarled trees as asthetically attractive as cultured bonsai grown thousandfold–this resort offers good tennis facilities and fine modern accommodations. These 80 condominium units are available in one-, two-, and three-bedroom configurations with fully equipped electric kitchen. Most units have color TV, fireplace, and private patio or balcony (some of these overlook the bay). The courtyard rooms overlook the circular pool and the brick-paved sun terrace that surrounds it. Near the eight Laykold tennis courts (four lighted), the Net Result pro shop sells equipment, sports fashions, and gifts. Also on the 24 acres are a white-sand beach, another pool, saunas, 750-foot lighted fishing pier, and boat ramp. Redfish, flounder, and drum are often caught within the protected bay. The resort's restaurant specializes in seafood and steaks and the Saloon has drinks and dancing. ☐ **Lux** ☐ **F, T**

Sea-Gun, *Box 85, Rte. 1, Rockport, TX 78382, (512) 729-2341* ☐ *9 mi N of Rockport on SR 35* ☐ A haven for fishermen and wildlife enthusiasts, this unpretentious complex is set under trees on Aransas Bay and offers air-conditioned units in a motel and closely spaced cottages. About half the units have kitchenettes, but there is also a restaurant and bar. On the grounds are a swimming pool, lighted tennis courts, picnic grills and tables, a playground, and bicycle rental. However, the lure here is the marina with services that include a lighted fishing pier and fish cleaning facilities. Fishing trips with expert local guides are available all year. Equally popular are narrated boat trips on the *Whooping Crane* into Aransas National Wildlife Refuge. In winter, these are given in daylight and focus on birds, especially the rare whooping crane. In summer, the boat goes out at night in hopes of sighting nocturnal animals. Land tours by auto can also be made from the refuge entrance a few miles north of Sea-Gun. Yacht basins, fine sand beaches, and fishing piers are found all along SR 35. ☐ **Mod to Lux** ☐ **F, W**

SOUTH PADRE ISLAND: South Padre Island is a 35-mile barrier island of fine white sand, seashells, and grasses which stretches north from Port Isabel. For hundreds of years only pirates and sea turtles stopped here. Today the island attracts several million sunseekers a year. The southern tip of the island is an extensively developed vacation area which yields abruptly to Padre Island National Seashore, with beautiful empty beaches, dunes, and hardy vegetation. Hotels, motels, and restaurants–most of very recent vintage–are situated on either the Gulf or on Laguna Madre, which separates the island from the mainland. A free accommodations bureau at the end of the causeway will try to find you room if you have arrived without reservations. For a complete list, write South Padre Island, TX 78578, (512) 943-6433. South Padre is separated by a narrow channel from the northern end of the National Seashore (the northern section is reached from Corpus Christi).

The subtropic climate, the warm waters of the Gulf and lagoon, and the beautiful beaches make this a beach bum's paradise. Fishing here is great. Anglers can choose between surf, bay, and jetty fishing for redfish, speckled trout, drum, flounder, and tarpon. The marina at Port Isabel is the base for bay and deep-sea charters for marlin, blackfin tuna, barracuda, sailfish, and kingfish. The Texas International Fishing Tournament is held here in August. Sightseeing boat trips and waterskiing are also available, and there are four public fishing piers.

Bahi'a Mar Resort, *Park Road 100, Box 2280, South Padre Island, TX 78597, (512) 943-1343* ☐ *Open all year* ☐ *3½ mi N of causeway* ☐ This hotel-condominium complex is on the Gulf side. The modern 200-room hotel rises 12 stories, and rooms on the Gulf side all have private balconies and a window-wall view of the water. Angular, modernistic, attached town houses made of weathered-gray timber are clustered around a lagoon and the clover-shaped pool amid a landscaped semitropical garden. Singles, doubles, efficiencies, and suites are available in the hotel, while town houses come with one to three bedrooms and kitchenettes. The beach is the focus of Bahi'a Mar recreation, but there are also two tennis courts and a game room. The dining room overlooks the private lagoon and specializes in native seafood dishes. The Las Ventanas nightclub jumps with live entertainment nightly. ☐ **Mod to Lux**

Hilton Sea Island, *Box 2081, South Padre Island, TX 78578, (512) 943-2685* ☐ *Open all year* ☐ *1 mi N of causeway* ☐ On the Gulf side, this complex offers more than 230 units with private balconies or patios, more than half with kitchenette, plus two-bedroom apartments with kitchenettes. The seafood restaurant has dancing and entertainment. The grounds have tennis courts and three swimming pools. ☐ **Lux**

Las Brisas, *Box 2370, South Padre Island, TX 78597, (512) 943-5111* ☐ *Open all year* ☐ *3 mi N of causeway* ☐ On the bay side, these two- to three-story buildings offer 35 kitchenette apartments with one to three bedrooms. Many units have balconies and bay views. On the grounds are a swimming pool, tennis, lighted fishing piers, and docks. ☐ **Mod to Lux**

Sand Castle, *Box 2220, South Padre Island, TX 78597, (512) 943-1321* ☐ *Open all year* ☐ *½ mi N of causeway* ☐ These 65 modern, two-story apartment units on Laguna Madre offer a bedroom, living room/dining room, and kitchenette. Second-story units have balconies, and most ground-floor units have outdoor sitting areas. Some units overlook the bay, where there is a lighted fishing pier. ☐ **Inexp**

The Tiki, *Box 2398, South Padre Island, TX 78597, (512) 943-2694* ☐ *Open all year* ☐ *3½ mi N of causeway* ☐ In two- to three-story buildings made to look vaguely Polynesian, here are 144 kitchenette apartments with one to three bedrooms. On the landscaped Gulf-side grounds are two swimming pools and saunas. The restaurant and bar are also on a Polynesian theme, with shrimp tempura and teriyaki on the menu. ☐ **Mod to Lux**

WIMBERLEY: 7A Ranch Resort at Pioneer Town, *Rte. 1, Box 59, Wimberley, TX 78676, (512) 847-2517* ☐ *Open all year* ☐ *1 mi W of Wimberley on River Rd.* ☐ Pioneer town, a replica of an 1880's frontier town complete with summer melodramas, medicine show, a Gay Nineties revue, rides on a miniature steam railway, and a museum of Western art, is a good vacation spot in the Texas Hill Country for young families. The adjacent 7A Ranch Resort combines the unique entertainment features of Pioneer Town with recreational facilities that include an outdoor swimming pool, riding stable, fishing, tennis, playground, outdoor game areas, and a recreation room. The resort is set on 142 acres, with a mile of frontage on the Blanco River. One- and two-bedroom air-conditioned cottages along the river come with fully equipped kitchens. Singles and doubles are available in the quaint, Old Western-style Pioneer House and in the two lodges; some of these units also have kitchenettes. Picnic grills, crafts shop, and an arcade are also on the premises. Meals are available during summer at the Silver Spoon Cafe in Pioneer Town and all year at a nearby restaurant. The 7A Ranch makes a good base for exploring nearby Austin and San Antonio, and Hill Country attractions such as the LBJ National Historic Site. ☐ **Inexp** ☐ **W, R, F**

Woodcreek Resort, *1 Woodcreek Dr., Wimberley, TX 78676, (512) 847-2221* ☐ *Open all year* ☐ *2 mi NW of Wimberley on ranch road* ☐ In the Hill Country, this attractive 2500-acre resort offers about 200 rooms, suites, and kitchenette units in a two-story lodge decorated in early American style and in contemporary and Western-style town houses and creekside cabins. Most units have kitchenettes and either a porch deck, patio, or other outdoor sitting area. All units have climate control, and some have fireplaces. The restaurant, in Western motif, specializes in Friday night seafood, Sunday brunch, and buffets, and the bar has entertainment. In addition to river swimming, the resort has three outdoor pools. The golf courses total 36 holes and have pro shop services. The tennis complex has 14 lighted courts and pro instruction. Also provided are a health club, recreation room, outdoor game areas, nine horses for trail rides, pedal boats and canoes and bicycles. Fishing is available in stocked ponds and streams. ☐ **Mod to Lux** ☐ **F, W, R, T, G**

Utah

ALTA: Alta Ski Area: Set in Uinta National Forest are two of the finest ski resorts in the world: Alta and Snowbird (connected by shuttle bus). The base elevation is nearly two miles high, and the dry air brings Utah some of the best powder snow there is. (About 450 inches of snow falls here each year.) Located in Little Cottonwood Canyon (SR 210), half an hour's drive from downtown Salt Lake City and 45 minutes from Salt Lake International Airport, Alta is one of the country's most accessible ski areas. This full-service resort has six double chairlifts, three rope tows, 29 runs, and a vertical drop of 2000 feet. Two ski schools offer a full range of rentals and instructions in downhill, cross-country, racing and jumping. The variety of trails challenge skiers of all abilities, and experts can arrange helicopter trips to vast mountains of uncharted powder. You can ski directly to the lifts from the town's seven lodges. In summer, there's hiking, swimming, tennis, concerts, and art exhibits among some of the most beautiful landscapes in the country.

The lodges here offer a wide variety of accommodation from economy dorms to luxurious suites. In winter, MAP or package plans are offered, while in summer EP is available; some units may close in May and October. All inquiries should be addressed to whichever hotel, Alta, UT 84070, and the area code is 801.

Alta Lodge, *742-3500* □ This lodge began as an old-fashioned wooden hotel with a gabled, copper roof, and it offers four double rooms and dormitories. Two modern wings have an additional 43 rooms with floor-to-ceiling windows; some have private balconies and fireplaces. It has a dining room (with a beautiful view) that features fondue, chocolate souffle, and other continental dishes. For relaxing there's a fireplace-warmed lounge, and whirlpool and sauna. You may swim at the Rustler Lodge's pool. □ **Inexp to Lux**

Rustler Lodge, *742-2200* □ This balconied wood chalet has 55 rooms (25 with bedroom/sitting room combinations) two suites, and four dorms. Most units have balconies. The cozy, firelit dining room features continental cuisine, all-you-can-eat breakfasts, and Sunday brunch. The lodge has an outdoor heated pool, whirlpool, and sauna. □ **Mod to Lux**

Other lodges are: **Alta Peruvian** *(742-3000),* a shingle-roofed, wooden lodge with a dining room, bar and lounge, and a heated outdoor pool; **Blackjack**

(742-3200) and **Hellgate** *(742-2020),* condominiums with studios and one- to three-bedroom apartments with large picture windows, fireplaces, and kitchens; and **Goldminer's Daughter** *(742-2300)* and **Snow Pine** *(742-9943),* which also operate as public shelters, with public restrooms, restaurants, and lounges. ☐ **Inexp to Lux**

Snowbird Ski and Summer Resort, *Snowbird, UT 84070, (801) 742-2000* ☐ *Open all year* ☐ *On SR 210, 1 mi west of Alta* ☐ This resort development offers four modern, high-rise lodges, nestled at the base of 11,000-foot peaks, within walking distance of each other, Snowbird Center, and the ski slopes: 162-room Cliff Lodge, 181-room Iron Blosam Lodge, 160-room the Lodge at Snowbird, and 68-room Turramurra Lodge. Rooms in all these lodges have floor-to-ceiling windows and recessed private balconies. Units range from standard doubles with two queen-size beds to studios (with kitchen and fireplace) to suites (also with kitchen and fireplace) that sleep up to six or eight people. The lodges have the same rate schedule, and each has a swimming pool and sauna. The Lodge at Snowbird has a restaurant-lounge specializing in continental cuisine. Cliff Lodge has two private clubs, a Mexican restaurant, and the attractive Golden Cliff restaurant featuring seafood buffet on Friday evening and a hunt breakfast on Sundays. In the village you can also find a steak pit, cafeterias, fast foods, an ice cream shop, and other facilities, and lunch and snacks are available partway up the ski slopes.

Snowbird is a full-service ski area with five chairlifts and a tramway serving 32 runs. The vertical drop is 3100 feet. The ski school offers a full program of instruction for adults and children, twice-a-day free guide service for first-timers, guides for private or group tours off the trails, and helicopter skiing. Snowbird village has a covered mall with ski rentals and repairs, a grocery, a package store, a pharmacy, and other services. There are five heated outdoor swimming pools, a game room, and tennis courts. Child care for children over three is available at Cliff Lodge. A free bus shuttles around the village and to the slopes. There is also an ice skating rink with rentals. In summer the tram operates to the mountain summit for sightseeing, and an arts festival, run in conjunction with the University of Utah, offers outdoor concerts and art shows. ☐ **Lux**

DUTCH JOHN: Red Canyon Lodge, *Dutch John, UT 84023, (801) 889-3715* ☐ *Open May–Sept* ☐ *On Visitor Center Rd. off SR 44 W of jct SR 260* ☐ This lodge sits on the shores of Green's Lake in the center of the spectacular Flaming Gorge Recreation Center, which is a paradise for boaters, rafters, and geologists. The lake came to be when the Green River was dammed at Flaming Gorge. Guests who prefer still waters can take a rowboat or a canoe onto the huge lake. The daring can take a raft down the river below the dam, and the company renting the rafts will pick you up at the end of the ride. The "Drive Through the Ages," for those who want to understand the history of all the gorges, is an interpretive, 40-mile automobile tour between Flaming Gorge Visitor Center and Red Canyon Visitor Center.

Guests in Red Canyon Lodge stay in one of eight pine-shaded cabins on the shores of Green's Lake. The one- and two-room cabins are furnished and have small wood stoves for heating and cooking. Though some have attached bathrooms, most guests share the central bathhouse. Wood and linens are furnished, but you must bring your own dishes and cooking utensils. The handsome stone and shingle lodge building has a restaurant, a recreation room, fishing and hunting supplies, limited groceries, gifts, and gasoline. The lodge rents rowboats and canoes. ☐ **Inexp** ☐ **W, F**

FORT DUCHESNE: Bottle Hollow Resort, *P.O. Box 339, Fort Duchesne, UT 84026, (801) 722-2431* ☐ *Open all year* ☐ *2 mi W of Fort Duchesne, on US 40* ☐ The distinguishing characteristics of Bottle Hollow are its management–it is wholly owned and operated by the Ute Indian tribe–and its architecture–an unabashedly geometric, polygonal, and angular motor lodge. The resort is laid out around two hexagons. At the center of one is a large, hexagonal pool. At the center of the other is a large hexagonal plaza, where Ute dances and ceremonies are staged for the public. Some of the guest rooms border those two polygons, others are in lodges beside the main buildings. All the buildings are made from sand-colored concrete that blends nicely with the surrounding terrain. As you might expect, the architect's Euclidian passion for geometry was not abandoned in the design of the guest rooms, some of which have (yes, again) hexagonal floor plans and even hexagonal beds. In other respects they are typical modern resort rooms, all with color TV, some with kitchenettes and private patios

Bottle Hollow offers a gift shop selling goods manufactured by the Indians, a Ute tribal museum, a dining room that serves some authentic Indian dishes, a recreation room, a nursery, tennis courts, and a marina with boat and fishing rentals on Bottle Hollow Lake. The resort can also arrange float trips to its Florence Creek Lodge, accessible only by raft. ☐ **Inexp** ☐ **W, F**

KANAB: Parry Lodge, *Kanab, UT 84741, (801) 644-2601* ☐ *Open all year* ☐ *On US 89* ☐ Thirty-four of the lodge's 89 rooms carry the names of movie stars who stayed here while filming. The spectacularly colorful desert buttes and mesas of the area have been the backdrop for more than 50 movies and television series. Pictures of the celebrities, and sometimes the celebrities themselves, can be seen at the lodge. But most of the guests are tourists, here to see the countryside, not to be filmed in front of it. Within a two-hour drive are: Grand Canyon's North Rim, Coral Pink Sand Dunes, Lake Powell National Recreation Area, Glen Canyon Dam, Bryce Canyon, Zion Canyon, and many other parks and scenic spots.

The older part of the lodge, built in 1929, is a white, two-story building with black shutters framing the windows, a long porch in front, and other architectural details reminiscent of a New England inn. The inside, too, has the look of a tasteful country hideaway, particularly the dining room with its lace curtains and white tablecloths. Fifteen of the guest rooms are furnished with antiques; the remainder have contemporary furnishings. All rooms have telephone and air conditioning, and most have color TV. Nightly, except Sunday, the lodge stages a comedy melodrama in its theater, a converted barn. The lodge has a pool and gift shop and regularly schedules cookouts, covered wagon trips, guided Jeep tours, and mule and horseback rides through the Grand Canyon. □ **Mod**

LAKE POWELL: Almost all of 186-mile-long Lake Powell is in Utah, but Page, the major resort city on the lake, and the Glen Canyon Dam, which impounded the Colorado River and formed the lake, are in Arizona, giving that state a disproportionate share of the tourism generated by the lake. Surrounding the lake is Glen Canyon National Recreational Area, more than one million acres of ruddy sandstone spires and mesas. Vegetation is sparse, yet cottonwood, sagebrush, yucca, pinon, juniper, and mesquite grow in the desert soil.

Although the recreation area has some interesting sights–Rainbow Bridge, the world's largest natural stone arch, and Hole-in-the-Rock, the site of an early Mormon settlement–most time here is spent on the lake. Houseboats, powerboats, and complete marina service are available at the three principle recreation centers: Wahweap, Bullfrog, and Hite. The houseboats are very popular because they offer spacious private accommodations and freedom to explore the Indian ruins, petroglyphs, and geological curiosities within the canyon. Each center also has an airstrip for private planes (Hite's is limited to authorized planes only). For more information and reservations contact: Lake Powell Resorts & Marina, Box 29040, Phoenix, AZ 85038, (800) 528-6154.

Bullfrog Resort and Marina, *Hanksville, UT 84734, (801) 684-2233* □ *Open all year* □ *At end of SR 276 S* □ These modern mobile-home accommodations are available with and without housekeeping facilities. The housekeeping trailers have propane stove and oven, electric refrigerator, lines, and complete kitchen utensils, but no TV or telephone. Bullfrog has a restaurant and full marina, and offers boat tours to Indian ruins. □ **Inexp to Mod** □ **F, W**

Lake Powell Motel, *P.O. Box 1597, Page, AZ 86040, (602) 645-2477* □ *Open all year* □ *On US 89 at Wahweap jct* □ This 24-room motel is a good choice after Wahweap Lodge. Two miles away from the lodge, it offers similar rooms at a lower rate, a cafe, and a service station. □ **Inexp** □ **W, F**

Wahweap Lodge and Marina, *P.O. Box 1597, Page, AZ 86040, (602) 645-2433* □ *Open all year* □ *Off US 89, 5 mi NW of Glen Canyon Dam* □ The largest of the resorts on the lake, Wahweap Lodge offers 125 units, ranging from modern one- and two-bedroom chambers to more luxurious four-person suites. All rooms have air conditioning and TV. Although the layout and design of the resort has little appeal—its three, long, low-lying box structures flank a wide expanse of asphalt—it does have fine modern facilities. The glass-walled Rainbow Room has good food and exceptional views of the lake. The Driftwood Lounge has nightly entertainment and dancing in the summer, and a coffee shop provides quick breakfasts and lunches. Indian crafts and jewelery are available from the gift shop.

The lodge offers a wide range of boat or Jeep tours and package "adventures" from float trips to dinner cruises to trips to old movie set, mines, and Indian ruins. The huge marina has all types of rentals and dockage services. In addition, guests can play the nine-hole golf course at Page Country Club, and the tennis courts on the premises for free. ☐ **Inexp to Mod** ☐ **W, F**

MIDWAY: The Homestead, *Midway, UT 84049, (801) 645-1102* ☐ *Open all year* ☐ *5 mi W of US 40 in Midway, on Homestead Dr.* ☐ For years Simon Schneitter considered the hot springs on his farm a constant affliction, waterlogging his crops and turning his fields to mud. But when he realized that more and more of his neighbors were stopping by just to bathe in his "hot pot," he decided to stake his fortune on the springs and turn his farm into a resort. He built a pool out of wood, added a dining room to his homestead (now called the Virginia House), and called the place "Schneitter's Hot Pots."

The Homestead, as it is now called, looks like New England. The main lodge is white, Federal-style, with columns, flat roofs, and black shutters. Many of the original farm buildings have been tastefully converted into lodgings. All told, there are 44 guest rooms in seven buildings, ranging from the ivy-covered Virginia House to the newer Ranch House or Farm House. Rooms are nicely decorated but all are different—some have fireplaces, some have private balconies, some have air conditioning.

The mineral water from the hot springs is still a great attraction here. It is piped to two pools—one large outdoor pool kept at 80 degrees and one indoor pool kept at 99 degrees. The resort also has two all-weather tennis courts, horses for trail rides in summer, and sleigh rides in winter. The 27-hole Wasatch Park Golf Course is only one mile away. The Heber Creeper tour train, which passes through some spectacular scenery, is in downtown Heber City.

The main dining room, featuring country cooking, has a stone fireplace and a view of the stream and Heber Valley beyond. The Hot Pot is a more intimate room with steaks, prime rib, and seafood. ☐ **Mod to Lux** ☐ **R**

PARK CITY: This was always a boom town, first for silver, now for snow. The effects of the first boom haven't worn off yet. Many buildings from the silver mining era of the 1870s are scattered among the sleek modern homes and lodges, giving the town a bit of Old West Flavor. Even the mountain itself, covered with lifts and slopes, has its own legacy—600 miles of tunnels. This is

Utah's largest ski area, with a 3000-foot vertical drop, one four-passenger gondola, one triple chair, and nine double chairs. At the top of the lifts (10,000 feet) are the formidable Jupiter and Scott bowls with steep open faces for expert skiers. In between the top and the base are two restaurants, and several novice and intermediate runs, including a 1¼-mile lighted trail. Near the base is a ski area just for novices, restaurants, and nursery. Complete rentals, for both children and adults, as well as lessons for all ages and levels are also available here. The town has about 30 restaurants, 16 bars, and 30 lodges—food, drink, and lodging are not hard to find. A shuttle bus makes daily scheduled runs between the ski area, lodges, and town. For the ultimate in fresh powder skiing, Utah Powder Guides (649-9739) provides helicopter flights into the Wasatch and Uinta mountains. Cross-country rentals, lessons and tours are available through White Pine Touring Center (649-8701).

When the snow melts, the city's fine 18-hole golf course emerges at the base of the mountain. Also here are summer gondola rides, an Alpine Slide, and an art center. Rockport Lake and other water sports areas are within easy driving distance, and there is good hiking and fishing in the mountains.

In addition to accommodations listed below, Park City offers many chalet, apartment, condominium, and home rentals. A complete list may be obtained by writing Park City Ski Corp., P.O. Box 39, Park City, UT 84060.

Acorn Chalet Lodging, *Box 511, Park City, UT 84060, (801) 649-9313* ☐ *Open all year* ☐ *On Empire Ave.* ☐ These are called chalets, but most apartments are housed in one large modern building, an attractive structure with wood siding and large picture windows. The Blue Acorn, a private chalet 250 yards from the lifts that accommodates four to five people, and the Red Acorn, an older structure with two apartments, are also available. Some units have kitchenettes, gas fireplaces, TV, or ski lockers. ☐ **Mod to Lux** ☐ **S**

Blue Church Lodge & Annex, *Box 1720, Park City, UT 84060, (801) 649-8009* ☐ *Open all year* ☐ *On Park Ave.* ☐ These apartments were created by remodeling a turn-of-the-century church. Located in the heart of Park City's historic district, the elegant former church was thought worthy of being listed on the National Register of Historic Places. The 11 units, ranging from one to four bedrooms, have complete kitchen facilities and color TV. Some have two bathrooms, fireplaces, or in-unit laundries. A large whirlpool, common area, and Ping-Pong table are on premises for apres-ski lounging. ☐ **Lux** ☐ **S**

Prospector Square, *Box 1698, Park City, UT 84060, (801) 649-7100* ☐ *Open all year* ☐ *On Prospector Ave. off SR 248* ☐ Probably the biggest lodging complex in Park City, this modern facility contains a complete athletic club with four racquetball courts, four outdoor tennis courts, a huge indoor pool, exercise room, paddle tennis, gymnasium, nursery, saunas, steam room, large whirpool, and lounge. About 200 of the units are efficiency apartments—a single room with kitchenette, queen-size bed, convertible bed, and bath. Another 100 are standard rooms without kitchenette. More luxurious lodgings—one- and three-bedroom condominiums and deluxe houses sleeping up to eight—are also available. The Grub Steak restaurant features steaks, prime rib, seafood, and a huge salad bar, and a mini-mall in the complex has more restaurants, a deli, and assorted shops. □ **Lux** □ **S**

Skier's Lodge, *Box 778, Park City, UT 84060, (801) 649-8800* □ *Open all year* □ *On Norfolk Ave.* □ A modern building with double rooms and apartments, the Skier's Lodge is only 200 yards from the gondola lift. All rooms come with private baths, queen-size beds, color TV, room telephones, ski lockers, and covered parking. The apartments have woodburning fireplaces and full kitchens. □ **Lux** □ **S**

Vermont

BOLTON VALLEY: Bolton Valley Resort, *Bolton, VT 05477, (802) 434-2131* ☐ *Open late Nov to mid-Apr and mid-June to late Oct* ☐ *From US 2 in Bolton, take Bolton Valley Access Rd. to resort* ☐ The wintertime center of this year-round resort is the ski area. The slopes, with a 1150-foot vertical drop, are served by four chairlifts. At the base are rentals, ski school, nursery, cafeteria, and the 85-room EP lodge, a modern, shingle-roofed complex with pool, sauna, restaurant, and lounge with regularly scheduled entertainment. All rooms have private balcony, bath, and telephone. Some have fireplace. In summer, tennis programs are conducted on the eight Har-Tru courts and the Bolton Valley Writer's Workshop holds fiction, non-fiction, and poetry workshops here. Horses are available for riding on the 6000-acre grounds. ☐ **Mod to Lux** ☐ **S, R, T**

BROMLEY MOUNTAIN: The ski area here has a vertical drop of 1285 feet and the longest run is 2½ miles. Five double chairs and a J-bar serve 23 slopes with snowmaking, ski school, day lodge, food and drink facilities, and a nursery. In summer and for fall foliage a chairlift operates, and there is an Alpine Slide. Bromley Mountain is in Green Mountain National Forest, which has streams and ponds for fishing and the Long Trail for hiking.

Bromley Sun Lodge, Peru, VT 05152, (802) 824-6941 ☐ *Open all year* ☐ *On SR 11, 2 mi W of town at the ski area* ☐ This five-story modern lodge offers private balconies or patios in 50 motel-style units and operates on MAP in ski season. The lodge has an indoor pool, restaurant, and bar. ☐ **Mod**

Johnny Seesaw's, *Peru, VT 05152, (802) 824-5533* ☐ *Open Thanksgiving to early Apr and July through Oct* ☐ *1¾ mi W of town on SR 11* ☐ This rustic 6½-acre resort, which operates on MAP in ski season and EP in summer, is a favorite with families and college students. A half-mile from the slopes, it offers five economy bunk rooms and 18 no-frill rooms in the two-story inn, and some units share bath. More modern units are available in four chalets with fireplaces. The restaurant serves breakfast and dinner, and there is a bar. On the grounds are a swimming pool, tennis court, ball machine, and a separate playroom for children. You can also canoe, and a golf course and riding stables are nearby. ☐ **Inexp to Mod**

EAST DORSET: Erdmann's Eyrie, *East Dorset, VT 05253, (802) 362-1208* ☐ *Open mid-May to late-Oct and mid-Dec to mid-Apr* ☐ *Off US 7 in East Dorset* ☐ This small motor lodge is perched on a bluff overlooking Green Mountain Valley. Surrounding the lodge are 200 wooded acres, threaded with nature trails, some leading to a small waterfall, brook, and pond. Rooms have colonial-style pine furniture, private bath, TV. A small, four-bedroom farmhouse with fireplace and kitchen is also available for rent. Each morning a breakfast featuring blueberry pancakes is served in an old barn that has been converted to a dining room. The main lounge, with spectacular view and a fireplace, is in the motor lodge. On the grounds are a trout pond, swimming pool, and 200-yard driving range. Golf and tennis privileges are available at nearby country clubs, and three major ski areas are only a half-hour away. ☐ **Inexp** ☐ **S, G, T**

FAIRLEE: Bonnie Oaks Resort, *Box B, Fairlee, VT 05045, (802) 333-4302* ☐ *Open June–Oct* ☐ *2 mi W off I-91* ☐ Spread along the pretty wooded shores of Lake Morey, here are 75 rooms (many with balconies overlooking the lake) in rustic two-story lodges and three-dozen cottages (most with fireplace) on MAP and package plans. This large resort provides full facilities, including three New England meals a day in the lakeside restaurant, a snack bar (or box lunches), a bar for dancing and entertainment, a library, and indoor and outdoor game areas. In addition to the pool, there is a sandy beach for swimming, and canoes, sailboats, paddleboats, and other craft are available for fishing, waterskiing, and paddling about. Modest extra charges enable you to play on the tennis courts or the good nine-hole golf course. Special children's programs include supervised lake cruises and fishing. ☐ **Mod** ☐ **W, F, G**

Lake Morey Inn and Country Club, *Fairlee, VT 05045, (802) 333-4311* ☐ *Open all year* ☐ *I-91 at exit 15* ☐ This large, lakeside resort is best known for its 18-hole golf course, site of the annual Vermont State Open. The course is served by a pro shop, which has a dining room as well as showers, lockers, and cart and club rentals. A driving range, putting green, and 18-hole miniature golf course complete the golfing facilities. Active vacationers not particularly interested in golf will find solace on the four all-weather tennis courts or around the heated pool. The lodge has 1000 feet of lakefront and rents boats for sailing, waterskiing, and fishing. Lake Morey is considered good for lake bass, perch, pike, and pickerel, and the kitchen will prepare and serve your day's catch. All the MAP rooms overlook either the lake or golf course, and deluxe air-conditioned rooms and lakeside cottages with fireplace are available. The lodge also contains game room and TV room with wide screen TV. The menu of New England dishes in the dining room is complemented with occasional lakefront barbecues and clambakes. ☐ **Inexp to Mod** ☐ **G, T, W, F**

Rutledge Inn and Cottages, *Box 108, Fairlee, VT 05045, (802) 333-9722* ☐ *Open June–Labor Day* ☐ *W of town on Lake Morey Drive* ☐ Nestled among trees on a 1000-foot Lake Morey shoreline, this resort has a two-story inn with wide porch and four bedrooms plus two dozen cottages shaded by trees, all on

AP. The dining room, in colonial decor, makes its own breads and desserts and provides setups. The inn also has a library and recreation room. On the beach you can canoe, waterski, sail, or take out a boat for fishing. There is an all-weather tennis court and you can play golf at the Bonnie Oaks nine-hole course a quarter of a mile away. A hostess supervises activities for adults and children. ☐ **Mod** ☐ **F, W**

JEFFERSONVILLE: Smuggler's Notch Ski and Tennis Resort, *Jeffersonville, VT 05464, (802) 644-8851* ☐ *Open all year* ☐ *5 mi S on SR 108* ☐ More than 200 modern units, about half with kitchenettes–and some apartments with as many as five bedrooms–are offered in this complex set in luscious scenery. Some units have balconies, fireplaces, washers and driers, and other extras. Rates here offer a choice of EP, MAP, and package plans; the restaurant serves three meals daily and the bar has dancing and entertainment. There is supervised play for children, a recreation center for teens, tennis, an Alpine Slide, a swimming pool, and sauna. In winter the ski area offers four double chairs, rope tow, snowmaking, rentals, ski school, nursery, and day-lodge facilities. The vertical drop is over 2600 feet, and the longest run is 3½ miles. Rentals are also available for more than 30 miles of cross-country trails, and there is ice skating. ☐ **Mod** ☐ **T, S**

KILLINGTON: Killington ski area covers five mountains with more than 70 trails (up to five miles long) and over a dozen lifts distributed among the eight distinct sub-areas, yet skiers can easily start anywhere and ski to any of the slopes by taking the right combination of trails and lifts. Although the bowl catches an average annual snowfall of 25 feet, snowmaking ensures skiing from late October to late May. From the leisurely beginner slopes to the 62 per cent grades on Bear Mountain, Killington truly has a ski trail for everyone, and the U.S. Olympic team have trained here. Killington's ski school is famous and even trains instructors for other resorts. Its extensive programs include racing, rescue techniques, videotape diagnosis, and ski tuning seminars. Competitions and other events are frequent. Rentals, repairs, a nursery, and food and drink service are part of the ski complex, and the Killington community offers dozens of restaurants, bars, and night spots.

In summer, Killington is given over to hiking, sightseeing (the gondola and a chairlift operate), mountain climbing, pond and stream fishing, boating, swimming in the White River and at four nearby state parks, and cultural activities, such as the annual Festival of the Arts and the regularly scheduled musicals at the summer playhouse. Tennis is the biggest summertime sport, however. Killington School for Tennis offers two-day weekend and five-day mid-week sessions on the eight-court complex. The school instructors make extensive use of on-court videotaping and playback systems. Also offered are schools for tennis instructors, and canoeing and backpacking adventures. Five golf courses are also in the vicinity.

Accommodations here run the gamut–cheap bunk beds, moderately priced motels and country inns, and luxury chalets and condominiums. Almost half of the lodgings are available on MAP, at least in winter, and many offer two- to seven-day package plans that include lifts and instruction. Killington Lodging Bureau publishes an excellent directory (it classifies the most suitable lodging for families, couples, teenagers on their own, etc.) and provides free information and reservation service; contact them at 101 Killington Rd., Killington, VT 05751, (802) 422-3711. You may write the same address for the Skiers' Guide, the Summer Vacations brochure, or details on the winter or summer learning Vacations–for the last you may also phone (802) 422-3333.

Edgemont and Whiffletree Condominiums, *118 Killington Rd., Killington, VT 05751, (802) 422-3101* ☐ *Open all year* ☐ *From upper Snowshed parking lot, take Roaring Brook East Rd.; Edgemont Condominiums are right at fork, Whiffletree are left* ☐ These condominiums are only one-half mile from the Snowshed and Ramshead areas, an easy walk. A special trail allows guests to ski back to the lodgings in the evenings. Edgemont's units come with one to three bedrooms, fireplaces, living room with sofa bed, and fully equipped kitchens or kitchenettes. Whiffletree units are slightly more luxuriously furnished and varied. Room styles range from one- or two-bedroom units with one bath to two- to four-bedroom units with two baths. The three- and four-bedroom units are duplexes. All have furnished kitchens with dishwasher, full-size refrigerator, stove, all linens, fireplace, and wall-to-wall carpeting. ☐ **Mod to Lux**

Grey Bonnet Inn, *Killington, VT 05751, (802) 775-2537* ☐ *Open all year* ☐ *On SR 100, near jct US 4* ☐ This modern wood and shingle lodge with a rustic elegance is for those who want to get away from the crowd. But it is best for those with cars, since there is a small charge for trasnportation to the ski lifts four miles away. All rooms have two double beds, carpets, and TV, and a number of the rooms on the second floor have a balcony providing views of the Gifford Woods, the mountains, or the courtyard and pool. On the first floor guests will find an indoor whirlpool, saunas, game rooms, a card and party room with a rock garden, a library, and a piano lounge with nightly films. The lodge also has one all-weather tennis court. The candle-lit fireside dining room has a varied continental menu with salad bar. Rates are MAP (EP available only early winter and spring). ☐ **Mod**

Killington Village, *112 Killington Rd. Killington, VT 05751, (802) 422-3613* ☐ *Open all year* ☐ *Off Roaring Brook East Rd.* ☐ This is only a few hundred feet from the Snowshed and Ram's Head areas. The village comprises eight modern shingle and wood buildings, each with 12 private rooms, plus a Reception Center with Lakeside Lounge and Restaurant overlooking seven-acre Snowshed Pond. Rooms come in two types: Standard and fireplace rooms, each with double Murphy bed, a single bed that converts to a couch, and a private bath. Fireplace bedrooms also have fully equipped kitchenette-dining area. Various rooms can be combined to provide suites. In winter, MAP ski packages are available, with dining in the Angus Tavern Restaurant. Killington Village is headquarters for Killington's summer tennis program (see above) and there is rowing, canoeing, sailing, and fishing. Adjacent is the playhouse. □ **Mod to Lux** □ **T**

Summit Lodge, *101 Mountain Rd., Killington, VT 05751, (802) 422-3535* □ *Open all year* □ *Near jct US 4 and SR 100* □ About three miles from the slopes, this is a modern motor lodge countrified by fieldstone fireplaces, barnboard paneling, and roughhewn ceiling beams. All of the 42 rooms (available on MAP or EP) have private baths and some have a balcony or terrace. A typical breakfast offering would be homemade muffins, fresh fruit, crepes Romanoff, eggs Benedict, and apple or banana pancakes with maple syrup. Recent dinner menus have included sirloin Berey, sweet and sour pork chops, and chicken Kiev. The Summit boasts four clay and two all-weather tennis courts, heated outdoor pool, whirlpool, saunas, putting green, and game room. During the summer the lodge runs a tennis clinic that employs a ball-machine, videotaping, and a professional staff. In winter a complimentary shuttle bus ferries skiers to the slopes. □ **Lux** □ **T**

MT. SNOW: One expansive mountain with three faces, Mt. Snow has more than a dozen lifts serving more than 50 trails. Main Mountain has chiefly intermediate trails, some over two miles long. Sunbrook, on the south face, has leisurely intermediate trails, and the North Face has a half-dozen hair-raising expert slopes. The trails and lifts are interconnected and a third of the skiing area has snowmaking. At the base, the lodge has three cafeterias, three lounges, dining room, coffee shop, and a heated swimming pool, while the skiing center houses a rental shop, nursery, cafeteria, and lounge. At the ski school, private lessons and videotape sessions are available for any level, and there is a NASTAR program. An excellent Children's Learning Center provides programs for ages two through twelve. A cross-country center offers rentals, instruction, and guided tours. Ice skating, special events, and other activities are also available in winter.

In summer, the Mt. Snow Country Club and Public Golf Course offers a clubhouse with pro shop and restaurant, two practice greens, and instruction programs to benefit golfers at any level. The rolling 18-hole course, designed by Geoffrey Cornish up in the mountains, is cool all summer. More than 30 clay and all-weather riding stables, Snow Lake, and summer theater are also available in the valley. The gondolas operate to the 3600-foot summit of Mt. Snow, where markers along three-mile Thompson Nature Trail explain the mountain ecology. Mt. Snow is in Green Mountain National Forest, which provides ample hiking, lake and stream swimming and fishing for trout, pike, and perch. Canoe, sailboat, and other boat rentals are available in the vicinity, and cruises are given on Lake Whitingham.

The Mount Snow Lodging Bureau, 300 Mountain Rd., Mt. Snow, VT 05356, (802) 464-8501, will make reservations at more than 50 establishments in the area and will arrange two- to seven-day ski packages at virtually all. Most lodgings operate on MAP in winter. You can find bunks in dorms and farmhouses, motels, alpine lodges, modest and elegant country inns, and luxury chalets and condominiums.

Snow Lake Lodge, *West Dover, VT 05356, (802) 464-3333* □ *Open May–Oct and late Nov to mid-Apr* □ *300 Mountain Rd.* □ This lodge is one of the closest to the lifts, but that hardly matters because its own cable car ferries guests between the lobby and the slopes. The hotel has 117 doubles and 16 suites, all nicely paneled and furnished. The indoor Japanese dream pools are ideal for apres-ski soaks. Some of the restaurant's specialties are New England scrod, fondue, and various veal dishes. After dinner, there's often entertainment and dancing in the Spiral Cocktail Lounge. In summer, the lodge hosts the golf school, sailboats and paddle boats may be rented for use on Snow lake out front, and there are four tennis courts. The beautifully landscaped outdoor swimming pool is open all year. □ **Lux** □ **G, S**

NORTH HERO ISLAND: Shore Acres Resort Motel, *Rte. 2, North Hero Island, VT 05474, (802) 372-8722* □ *Open May–Oct* □ *½ mi S on US 2* □ On 50 acres of farmland on an island in Lake Champlain, this modest little resort offers only 20 paneled rooms with Vermont maple furnishings, but all have a lake view. The neat, paneled dining room has cocktails and features locally grown vegetables and home-baked bread. You can swim a few feet from your door on the half-mile shoreline, and a marina (half a mile away) has full services and boat rentals. Fishing is good for smallmouth and largemouth bass and for pike. The resort has lawn game areas and a driving range, and an 18-hole golf course is a 15-minute drive away. □ **Inexp** □ **F**

PITTSFIELD: Hawk, *Rte. 100 Pittsfield, VT 05762, (802) 746-5171 or (802) 451-4109* □ *Open all year* □ *½ mi W of Pittsfield General Store off SR 100* □ This may be the ultimate cabin-in-the-woods vacation; it is certainly the plushest. Hawk is a community of privately owned, custom-built vacation homes in four "colonies." Three colonies are on land bordering Green Mountain National Forest and the White River between Pittsfield and Stockbridge, and the fourth is across the state at Norwich. All 65 homes are sited for privacy and view. They vary from two to four bedrooms, and may have extras such as saunas or sleeping lofts, but all have equipped kitchens, stone fireplaces, and decks. Although designs vary, all exteriors (and some interiors) are of unpainted wood, giving these luxury homes a rustic look. If you wish privacy, you will be left alone. But if you wish service the staff is prepared to procure babysitters and limousines, do your shopping, and even come in to cook a special dinner, cater a party, or cook all your meals the length of your stay. Housekeeping services are available. Travel, tour, and secretarial services are provided on request. Annabelle's Restaurant, in a restored farmhouse, is near Stockbridge and serves dinner; the bar is open longer hours.

On the grounds is a barn housing squash courts, sauna, and exercise room. You can swim in two ponds or in the river. The White and Tweed rivers are stocked, and there is also fishing in streams and lakes in the national forest. The White River has fine canoeing, and liveries are nearby. Also on the premises are outdoor tennis courts, riding stables, and a TV lounge with game areas. A summer day-camp is run for kids aged five through 16. The staff will tell you where to play golf or rent bikes, and hiking trails extend into the forest, where you will also find the Long Trail. In winter there is cross-country skiing on the property and 11 miles south at Mountain Meadows in Killington. Killington is one of several nearby areas for downhill skiing. ☐ **Lux** ☐ **R, S, T, H, F**

QUECHEE: The Quechee Inn at Marshland Farm, *P.O. Box 457, Quechee, VT 05059, (802) 295-3133* ☐ *Open all year* ☐ *From I-89 take exit 1 to US 4 W; take first right on Clubhouse Rd., ½ mi to inn* ☐ The Quechee Inn was built in 1793 as the home of Joseph Marsh, the first Lt. Governor of Vermont. From 1793 until 1975, when it was renovated and converted into a country inn, the building served as a farmhouse. All of the rooms have Queen Anne period furniture or colonial antiques, including spindle-post beds and highboys. Front rooms overlook the river and pond across from the inn; the other rooms have a view of the meadows and hills. Each has private bath and cable color TV. The lounge has massive exposed beams, wing chairs, and fireplace.

The inn is nestled in the beautiful Ottauquechee River Valley on the outskirts of the restored village of Quechee and is part of the recreational community of the Quechee Lakes Club. Inn guests have available to them at nominal charge the two 18-hole golf courses, squash, racquetball and tennis courts, swimming pool, and sauna of the private club. On the lake there is swimming, boating, and fishing. Canoeing and sailing are also available. In winter, Quechee has downhill and cross-country ski areas, and within a dozen miles are Roundtop, Whaleback, and Suicide-Six ski areas. ☐ **Mod to Lux** ☐ **G, T, F, S, W**

SPRINGFIELD: Hartness House, *30 Orchard St., Springfield, VT 05156, (802) 885-2115* ☐ *Open all year* ☐ *from SR 143 take Summer Hill St. three blocks to Orchard St.* ☐ This unusual inn was built as a residence in 1903 by James Hartness, an inventor who patented 120 different machines, including the weird Turret Equatorial Telescope on the front lawn. The rambling inn has gables, ornate bay windows, and other Victorian favorites, and contains a dozen air-conditioned, modernized rooms. To one side are an additional 34 modern hotel units. Despite the comforts of the house, it did not suit Hartness, who found it noisy and distracting. First he tried to find solitude in a cabin he built behind the house, but a babbling brook was too annoying. Finally he built a series of soundproof underground rooms where he worked, thought, and housed his library. This retreat has now been converted into the cocktail lounge–which would have horrified Hartness who was a teetotaler–with dancing on Saturday nights. The inn has a dining room offering three good meals daily, and a comfortable lobby with fireplace. On the 30-acre grounds you'll find tennis and a swimming pool. Golf is available at the Crown Point Country Club's 18-holer three miles away. The inn operates on EP, but golf and skiing packages are offered. ☐ **Mod**

STOWE: Stowe Ski area–Mt. Mansfield is the name of the main peak (the highest point in Vermont, incidentally) and Spruce Peak the name of its sister summit, but Stowe is the name of the whole area. More than 36 miles of trails are serviced by a half dozen chairlifts and a gondola, and there is patrol and snowmaking. The ski school offers instruction for all levels, guided mountain experience, and NASTAR. Day care is available for children, with or without skiing lessons. The Mt. Mansfield Cross-Country Ski Center has instruction, rentals, and more than 100 miles of trail. Ice skating and snowmobiling are also available at Stowe. Three rental shops, and restaurants, bars, and cafeterias at base and slope lodges are part of the complex.

In summer, the toll road to Mt. Mansfield summit opens and the Mt. Mansfield gondola operates; both offer spectacular views of fall foliage. There is also an Alpine Slide. Mt. Mansfield State forest has a scenic drive through Smugglers' Notch, and you can hike the Long Trail through the forest and to the summit; you can also fish and boat on the reservoir on the southern part of the forest. Lake Elmore in Elmore State Park has swimming and boat rentals. Stowe has a historical museum, and the Stowe Area Assn., Box 1230, Stowe, Vt 05672, (802) 253-7321, offers tours of back roads. Stowe Country Club has an 18-hole course, practice courses, a clubhouse restaurant and bar, and pro shop rentals and instruction. Several hotels offer playing privileges and golf plans. Stables providing trail rides are also in the area.

Edson Hill Manor, *Rte. 1, Stowe, VT 05672, (802) 253-7371* ☐ *Open mid-June to late-Oct and mid-Dec to early-Apr* ☐ *Take SR 108 3.3 mi N of Stowe village and Edson Hill Rd. 1.7 mi to resort* ☐ Its steep shingle roofs, brick chimneys, and small dormer windows give Edson Hill Manor the look of an English country home. The inside is as cozy and quietly elegant as the outside. Most of the walls are pine-paneled, many ceilings are beamed, and fireplaces in the public rooms have Delft tiles. Dinner is by candlelight, and there is a bar and wine cellar. MAP is offered in winter and EP in summer, and there are ski and riding packages as well. All but three of the 16 rooms and suites in the manor and annex have private bath, several have fireplaces, and two have kitchenettes. The manor is set on a beautiful hillside surrounded by mountains; some areas are landscaped, but most of the 400-acres has been left wild. The pool is surrounded by a flagstone terrace, lawns, and old trees. Under the same ownership since 1953, this resort's specialty is English riding instruction in the ring and on the miles of trail through the woods around the resort. British instructors give group and private lessons and trail rides at hourly rates. Also available are picnic rides and side-saddle lessons. Wintertime activity centers on the cross-country ski school. Staffed with professional instructors, the school has a fully equipped shop with rental packages for children and adults that include trail fees. Guided tours are available on request. The 60 kilometers of marked trails start at the doorstep. Mt. Mansfield ski area is only one mile away and the golf club a little over four miles away. ☐ **Mod to Lux** ☐ **S, R, F**

Golden Eagle, *Box 1110, Stowe, VT 05672, (802) 253-4811* ☐ *Open all year* ☐ *½ mi N on SR 108* ☐ On 14 pretty acres of lawns and trees, this complex offers an amazing variety of double rooms, one- and two-room efficiencies, and two-bedroom apartments in a motel, modern lodges, and 19th-century buildings. In addition, a two-bedroom cottage screened by trees is off to one side; and secluded beside a brook a mile away from the complex is a three-bedroom chalet. Most of the 60 units have kitchens or kitchenettes, many have fireplaces, some have balconies, patios, or decks, and a few have whirlpools. A few rooms sharing bath are on the second floor of the Partridge Inn, the attractive early-American-style restaurant and bar. In the center of the complex are an outdoor pool, a tennis court, game areas, and a small stocked pond. Also available are a sauna, a game room, and a coffee shop. This is a friendly, homey resort, very good for families (many units sleep up to six people). EP, MAP, and packages are available. ☐ **Inexp to Lux**

The Inn at the Mountain and the Town Houses, *Mt. Mansfield Co., Stowe, VT 05672 (802) 253-7311* ☐ *Open early Dec to mid-Apr and mid-June to mid-Oct* ☐ *Off SR 108 6 mi NW of Stowe* ☐ So close to the slopes that you can ski down to the lift, this contemporary inn, formerly the Toll House Inn, incorporates the original tollhouse structure, with its barnboard paneling and three stone fireplaces. It has 26 double rooms and eight suites, each with a window wall and a balcony affording a view of the surrounding valley and slopes, all on MAP and ski plan. The rustic restaurant serves both European cuisine and New England dishes, and there is a cozy tavern. In summer, lunch is offered on the terrace. The nearby Town Houses are two-story units with a choice of one, two, or three bedrooms; these units have a fireplace and fully equipped modern kitchen, and operate on EP or ski plan. In summer, guests have privileges at the golf course, six clay tennis courts, a recreation room, an outdoor pool, and access to hiking trails at their disposal. ☐ **Lux** ☐ **S, T, G**

Stowe Flake/Nordic Inn, *Box B, Stowe, VT 05672, (802) 253-7355* ☐ *Open all year* ☐ This complex consists of an older inn with 19 economy rooms plus 50 modern rooms and suites in an alpine-style motel. Some units have steam baths. The restaurant is open all day in summer, when you may register EP, but in the winter MAP season it serves only breakfast and dinner. The cocktail lounge is lively year round. There are two pools (you can lunch poolside in summer), sauna, game room, outdoor game areas, and indoor and outdoor tennis. On the grounds are a putting green and chip and putt course, and golf privileges are available. ☐ **Mod to Lux** ☐ **T, G, S**

Stowehof Inn, *Edson Hill Rd., Stowe, VT 05672, (802) 253-8500* □ *Open late Nov to late Apr and early June to mid-Oct* □ *Take SR 108 3.3 mi N of Stowe village and Edson Hill Rd. ½ mi to resort* □ This lodge, a whimsical bit of architecture with a soaring, prow-like grass roof, is a Tyrolean-Norwegian inn from a storybook. The huge timbers of the entryway are supported by still larger tree trunks. The front hall is floored with wide, hand-planed boards, and intricately carved or brightly painted Bavarian clocks adorn the mantels. The guest rooms and suites are all different, though each has a private bath and a view from its own balcony or patio. Appointments include Chippendale desks, Queen Anne chairs, and hand-hewn beams. The "demi-suites" come with Murphy beds, and some have kitchenettes and fireplaces. Town houses with two or three bedrooms are also available. The restaurant leans heavily toward continental dishes, with such specialties as homemade lahvosh, sauteed veal piccata, Salzburger nockerl, and Dover sole amandine. On the resort's 18 acres of birch and pine forests are four all-weather tennis courts and a heated pool. Golf privileges are available. Rates here may be EP, MAP, or on ski and tennis plans. □ **Lux** □ **T, S**

Topnotch at Stowe, *Mt. Mansfield Rd., Stowe, VT 05672, (802) 253-8585* □ *Open all year* □ *Off SR 108, 6 mi NW of Stowe* □ Here you are not greeted by a desk clerk at the front desk but by a concierge in a wood-beamed library with a Louise Nevelson sculpture, and the aim is to provide the grace and charm of a European inn and still remain relaxed and informal. The 30-by-30-foot living room, where tea and pastry are served at 4 pm, has a moose head mounted on the wall of exposed stone, an armoir with antique toys, 12-foot-high windows, and interesting furnishings. The two lodging wings of the inn provide 72 individually decorated rooms and suites all with picture-window views, touches such as plants or dried flowers–and your bed is turned down each night. Town houses provide one or two bedrooms, living room, and porch. A three-bedroom chalet with fireplace and dining balcony is also available. Even if you stay here on EP rather than on MAP or on a golf, tennis, ski, or other plan, you will want to try the candlelit restaurant, which has been featured in *Gourmet* magazine. You may find on the menu iced fruit soups, avocado bisque, Nova Scotia salmon, chateaubriand, rack of lamb rosemary, or sherry trifle. You may breakfast in your room or in the dining room, where the hearty fare includes lamb chops or steak in addition to the usual eggs and cereals. In good weather, a lunch of quiche or salad may be on the patio. The cozy bar has an annex for games, dancing, or entertainment.

On the 120-acre grounds are an outdoor heated pool, areas for croquet and other outdoor games, an indoor game room, and a tennis center with 10 all-weather, one clay, and four indoor courts, instruction with videotape replay and other modern aids, and rentals and repair. The spa has sauna, massage, an exercise room, and whirlpool. The stables provide English and Western instruction. The golf course is two miles away, and the resort will arrange boating, canoeing, and bicycling, and give you tips on where to fish. In winter the ski touring center opens, with instruction, rentals, and 40 miles of trails. As its name implies, this resort is topnotch. ☐ **Lux** ☐ **R, S, G, T**

Trapp Family Lodge, *Trapp Hill Rd., Stowe, VT 05672, (802) 253-8511* ☐ *Open Memorial Day weekend to mid-Apr* ☐ *2 mi N off SR 108, then 2 mi on Trapp Hill Rd.* ☐ These three alpine lodges are owned by the Trapp family of "Sound of Music" fame. They offer 50 rooms (some share bath) on MAP and ski plans. Some units have fireplaces and balconies. On the grounds are a swimming pool, a game room with TV, outdoor game areas, and hiking trails. Entertainment is offered in the bar, and special activities are scheduled. In winter, cross-country skiing equipment is provided. ☐ **Lux** ☐ **S**

STRATTON MOUNTAIN: This broad mountain has more than 59 miles of groomed trails served by seven double chairs and other lifts. Five snowmaking systems cover 80 acres. At the base are rental and repair shops, food, drink, and other services, and there are summit facilities. The ski school takes children from age three and sponsors NASTAR races, seminars, and special events. At the Stratton Mountain Country Club is the ski-touring center with 20 miles of machine-groomed trails, rentals, instruction, guided tours, and dining room. In summer, the John Newcombe Tennis Center offers programs of professional instruction with videotape and other modern aids. Facilities include eight Har-Tru courts, four indoor courts, and pro shop. The Country Club's 18-hole golf course was designed by Geoffery Cornish, and clubhouse services include a pro shop, restaurant and bar, and cart rental. The golf academy has a special 22-acre training center designed to simulate and isolate every type of golf shot. From mid-September to mid-October, Stratton's Arts Festival presents exhibitions by potters, sculptors, and craftsmen as well as performances by musicians and dancers. Film and photographic workshops are also offered. Also available are stables offering lessons and trails rides and 17-acre Stratton Lake for swimming, fishing, and sailing. There is also a Wilderness Pursuit Center plus a health spa with complete programs. Accommodations here are open from late November to April and from about mid-May to September or October and are available on a variety of weekend and other package plans. The inns may be contacted directly at Stratton Mountain, VT 05155, or through a central reservations number–(802) 297-2200 or toll-free (800) 451-4261.

Birkenhaus, *(802) 297-2700* ☐ Set in a grove of white birch from which it gets its name, this 20-room modern lodge is in Austrian decor, has a dining room with continental food and home baking, a bar, beauty shop, family room, and swimming pool. All rooms have balconies. Bunkrooms are available. ☐ **Mod**

Liftline Lodge, *(802) 297-2600* ☐ Wood-paneled gables, yellow shutters, and white stucco arches give this the look of an Austrian inn. It has 75 rooms and boasts a health spa with California hot tub, sauna, swimming pool, and special summer fitness program. Also on the premises are a library, piano bar, game room, boutique, pebbled courtyard, and red-clay tennis court. ☐ **Mod**

Stratton Mountain Inn, *(802) 297-2500* ☐ This 100-room, three-story motor lodge on a little rise overlooks its swimming pool and four tennis courts. Each room has a balcony. Downstairs there's a huge stone fireplace, elevated lounge, and a sun terrace dining room. ☐ **Mod**

SUGARBUSH VALLEY: Six miles northwest of Warren, in Green Mountain National Forest, the Sugarbush community spreads out across the valley at the base of two ski complexes connected by free shuttlebus. The areas have 70 trails served by 13 lifts, snowmaking, rentals, good ski school, NASTAR, and a nursery that accepts infants and children of all ages. The day lodges have fireplaces and sun decks and provide food and drink; in the valley are two dozen restaurants and bars, and many shops. If you join Sugarbush Sports Center (day memberships available) you'll find racquetball/handball courts, squash, 3 indoor and 6 outdoor tennis courts, indoor/outdoor pool, saunas, whirlpool, massage, and exercise room. Sugarbush also has riding stables. Many special events are sponsored throughout the year. The village has hotels, inns, town houses, apartments, and private home rentals. Sugarbush Tour Bureau—Warren, VT 05674, toll-free (800) 451-5030 or (802) 583-2381—will make reservations. Most lodgings are available on fly and ski packages of one day or longer. These may include lift tickets and optional meal plans, rentals, lessons, or airport transportation.

Sugarbush Inn, *Warren, VT 05674, (802) 583-2301* ☐ *Open all year* ☐ *2 mi W of SR 100 on Sugarbush Valley Access Rd.* ☐ This long and large two-story structure with colonial touches offers 70 inn rooms—some with fireplace, kitchenette, private patio or balcony—as well as suites and condominium units. In the main building are the formal Colonial Room for continental dishes, the informal Onion Patch for steak and seafood, and the bar. EP and package plans are available. Specials here include the 18-hole Robert Trent Jones golf course, 16 Har-Tru and red-clay tennis courts, two indoor courts, a sauna, and a heated pool. From May to October, Sugarbush hosts a John Gardiner Tennis Clinic. An extra-special is soaring instruction available at the nearby airport. ☐ **Lux** ☐ **T, G, S**

VERGENNES: Basin Harbor Club, *Vergennes, VT 05491, (802) 475-2311* ☐ *Open mid-June to mid-Oct* ☐ *7 mi W of Vergennes on private rd.* ☐ Basin Harbor is a little cove of blue-green on the wooded shore of Lake Champlain. Circling the cove are white, colonial-style buildings—the old Homestead and newer lodges with 37 rooms; 65 cottages; and a recreation hall. Most of the cottages are colonial clapboards. Of varying shapes and ages, they range from comfortable to lavish, and come in sizes from studio to three bedroom, and may face the water or be hidden among old trees. Your unit may come with refrigerator, private balcony, patio, or fireplace. The AP rates include ample breakfasts and excellent dinners in the lakeside dining room, where you choose from varied menus. Late risers may have a continental breakfast in the lounge. Lunches are poolside buffets. You may have quiet drinks at the pool or lodge deck, or with dancing to a trio in the cocktail lounge. Dress code is required evenings. On the 700 acres stretching back from the shore is an excellent 18-hole, 6000-yard golf course, with gently rolling fairways and views of the lake and the tiers of Adirondack Mountains beyond it. Tennis facilities include two Har-Tru and three Plexipave courts. Instruction is available in both sports from resident pros. In addition to the swimming pool, sandy beaches are on both sides of the harbor. The hotel runs daily cruises on Lake Champlain, but you can also rent a canoe, motorboat, rowboat, or Sunfish from the harbor master, or make arrangements for watersking. There are badminton, croquet, and shuffleboard courts, and short nature trails for which the club prints guides. In high season a full children's program and teenage activities are offered, and adult entertainment may include movies in the Town Hall. The club has a 3200-foot airstrip. ☐ **Mod to Lux** ☐ **G, T, W, F**

WILMINGTON: On-the-Rocks Lodge, *Wilmington, VT 05363, (802) 464-8364* ☐ *Opens June to mid-Oct and Dec–Apr* ☐ *4.2 mi N on SR 100, then follow signs* ☐ Run by an interesting couple, Margie and Orla Larsen (he was Mt. Snow's first ski school director and is an inventor), this is a modern inn with an old-fashioned country look. All of the 15 double rooms have private bath, designer fabrics and wallpapers, plus many personal touches, and some have fireplaces. You may book here on EP, MAP, or packages, and the fare includes homemade soups, entrees such as filet of sole stuffed with crabmeat, homemade desserts, and a good wine list. Lunch is on a rose-bordered patio with beautiful views. The lodge has handsome lounge with a huge stone fireplace, paisley-covered sofas, and picture windows overlooking the gardens, plus a charming bar, and two game rooms. The 20-acre grounds are landscaped with gardens near the lodge and surrounded by evergreen forests. Included in the rates are use of a trout stream, landscaped outdoor pool, two clay tennis courts (instruction and equipment available), and a 30-mile network of trails for birdwatching and hiking in summer and for cross-country skiing in winter. Within walking distance is a par-3, 18-hole golf course, and three miles away is the Geoffrey Cornish 18-holer at Mt. Snow, with full clubhouse facilities. Wilderness Lake (5 miles) has boat rental, sailing, and canoeing. Mt. Snow and Haystack ski areas are four miles away. ☐ **Lux** ☐ **F, S, H, T**

WOODSTOCK: The Woodstock Inn, *Woodstock, VT 05091, (802) 457-1100* ☐ *Open all year* ☐ *Take US 4 on the village green* ☐ One of the Rockresorts developed by Laurance S. Rockefeller, this 121-room stately white manse is on the green of the charming village. Much of the appeal derives from the small-town atmosphere of Woodstock itself–postcard-neat houses along the streets, no trace of commercialism; forested hills roll off into the distance, and only a few feet from the inn, a covered bridge spans the Ottauquechee River. Although the building is modern, many old-fashioned touches—brass hinges, weathered timbers, hand-cut stone—suggest a colonial inn. Your room will have a handmade bed quilt, oak furniture custom-made in Vermont, and a vanity of Vermont marble; 20 rooms have a steam bath. The two suites have a refrigerator and steam bath. The main dining room serves continental food and fresh fish in elegant candlelit surroundings (dress code), and the informal coffee shop has a full family menu. The Pine Room lounge has an immense fieldstone fireplace. A buffet brunch is served on Sunday, and in summer there is outdoor dining at the Garden Grille.

The beautiful 18-hole golf course, spreading behind the inn, was designed by Robert Trent Jones. A golf pro, pro shop, two putting greens, and electric carts are on hand. Tennis buffs have their own pro and shop and 10 courts—clay and all-weather—to play on. Also here are nature trails, Stanley Steamer rides, bicycle rentals for lovely marked paths, two paddle tennis courts, and a large outdoor pool. Riding stables are five miles away and offer lessons and many miles of marked trails. In March and April maple sugaring demonstrations are given at the inn's sugarhouse. In winter there are horse-drawn sleigh rides, and the ski touring center opens. This has rentals, instructions, a lounge, guided tours and picnic trips, and 47 miles of groomed and marked trails for all levels that include the beautiful Skyline Trail, a self-guiding nature trail, a three-mile snowshoe trail, and a racing loop. Suicide Six is the downhill ski area, with a vertical drop of 650 feet. The J-bar lift is free, and chairlifts also service the 18 slopes. There are snowmaking, ski school, rentals, and a base lodge. Woodstock Tom is a smaller downhill facility a mile away. Available in the vicinity are fishing, hiking, mountain climbing, hunting and game bird shooting, lake swimming and boating, summer theater, auctions, and country fairs, and the inn sponsors many year-round special events. Truly outstanding, this inn offers something for everyone, from senior citizens to romantic couples to active families. A travel desk, babysitting, and other services are, of course, provided. ☐ **Mod to Lux** ☐ **G, T, S**

Virginia

BRYCE MOUNTAIN RESORT: Bryce Resort, *Basye, VA 22810, (703) 856-2121* ☐ *Open all year* ☐ *11 mi W of Mt. Jackson on SR 263* ☐ An all-in-one place to escape everyday life, this vacation community is tucked away in a beautiful mountain valley. In buildings highlighted by warm woods and massive stone fireplaces, it offers 33 units in large condominiums; two-level, two-bedroom units in town houses; and two- to six-bedroom chalets. All units have kitchens, fireplaces, and either a balcony or a deck. Bryce Resort is set up to provide as much or as little privacy as you want. If you want to socialize you can have some or all of your meals at the rustic, wood-beamed Copper Kettle restaurant looking out on the ski slopes, or join the evening gathering at the Copper Kettle Lounge for dancing, musical entertainment, backgammon, and electronic games. Bryce has a 2500-foot lighted landing strip, and an FAA-approved flying school. In winter, Bryce's four ski slopes have snowmaking, rentals, ski school, racing, and lodge facilities. In summer, the same downhill trails are trimmed and groomed for grass skiing. Also here are a swimming pool, wading pool, tennis (lighted, indoor), horseback riding and instruction, golf on a 6175-yard Ed Ault-designed course that winds and twists along the valley floor, putting green, bicycling, hiking, canoeing, sailing, and fishing. ☐ **Mod** ☐ **T, S, G, F, W**

CHARLOTTESVILLE: The Boar's Head Inn, *P.O. 5185, Charlottesville, VA 29905, (804) 296-2181* ☐ *Open all year* ☐ *4 mi W off US 250 in Ednam Forest* ☐ In Shakespeare's time the head of a boar symbolized hospitality, and this inn, named for London's Boar's Head Tavern, displays a mounted boar's head in the lobby. The aim here is to re-create the atmosphere of a Virginia country estate, and the inn, nestled in a lush valley 15 miles from Monticello, incorporates beams and paneling from a local grist mill built in 1834 to rival Thomas Jefferson's mill. The inn's charm has drawn notables as diverse as Elizabeth Taylor, Chief Justice Earl Warren, James Dickey, and Walter Cronkite. The inn has 92 double rooms and seven two-room suites (five with fireplace). Some units in the three-story colonial-style main building have chestnut woodwork, old beams, or alcove beds modeled on those Jefferson built at Monticello. Rooms in the newer but matching wings overlook the fairways, a pond, and the distant mountains. All units are furnished in early American decor and most have private balconies or patios. The Old Mill Dining Room is built of heartwood pine timbers from the old mill and has a fireplace. Here you dine by candlelight (dress code for dinner) on duckling a l'orange, capon, crabmeat, or flambe specialties all served by a staff in colonial dress. The less formal Downstairs Hearth Room feature steaks and weekend buffets. The lounge, also beamed and paneled, has live music and dancing nightly.

The inn has a swimming pool, a lawn lounging area with poolside service, a putting green, croquet and other outdoor game areas, a labeled garden, two outdoor tennis courts, and two ponds with ducks, geese, and swans where you may fish. For an extra charge you may use the Sports Club facilities. This complex, set beside a pond, has a restaurant for luncheon and dinner, a 75-foot pool, saunas, and exercise rooms. Indoors are three lighted Grasstex tennis courts and outdoors are 10 clay courts. The four squash courts include a doubles court. There are also three lighted platform tennis courts. The shop offers pro services and stocks supplies for all the sports. Horseback riding is available five miles away. An 18-hole golf course is 15 miles away, although fairways stretch to the inn. The Boar's Head has complete services from a florist to babysitting. At Thanksgiving there's a foot hunt with basset hounds, and at Christmas the medieval festival has stagecoach rides, costumed troubadors, and a boar's head brought into the dining room to the sound of trumpet calls. □ **Mod** □ **T, G**

CHINCOTEAGUE: Chincoteague Island is a fishing and tourist community connected by causeway to beautiful Assateague Island National Seashore. The seashore, a 37-mile-long barrier island, has lovely beaches, dunes, and forests which may be seen on nature trails and by auto tour. Assateague is a wildlife sanctuary, famous for birds, small Sika deer, and wild ponies that are rounded up in an annual festival in late July. In addition to swimming and fishing, there are boat cruises and interpretive programs. Assateague has no accommodations, but Chincoteague has many motels, seafood restaurants, a full-service marina, and an Oyster Museum.

Driftwood Motor Lodge, *P.O. Box 92, Chincoteague, VA 23336, (804) 336-6557* □ *Open all year* □ *On Beach Road at Assateague bridge* □ This three-story motel offers 28 rooms with private balconies and patios, some of which overlook marshes and the water. It has a pool, but you must go to the center of town (about a mile) for a restaurant. □ **Mod**

The Refuge Motor Inn, *P.O. Box 173, Chincoteague, VA 23336, (804) 336-5511* □ *Open all year* □ *On Beach Road at Assateague bridge* □ This 68-room, two-story motel has private balconies and patios, some overlooking marshes and water. On the grounds are a pool and picnic areas with grills. The motel rents bicycles and arranges tours of Assateague as well as boat tours. □ **Mod**

HOT SPRINGS: The Homestead, *Hot Springs, VA 24445, (703) 839-5500* □ *Open all year* □ *On US 220, 20 mi N of Covington* □ The epitome of Southern hospitality, with superb service, this is one of America's grand old resorts. The first hotel was built at the springs in 1764, but the present handsome brick building with white trim was constructed after a devastating fire in 1901. You approach along a curved landscaped drive to a white-columned entrance under an imposing tower topped with a cupola. To either side are lower lodging wings. The inn provides almost 700 units, all with private bath and about half with air conditioning. The main building and its wings have singles, twins, and parlor suites. The south wing, an ultramodern structure in brick that angles around a lawn and large outdoor pool, provides many rooms and duplex suites with recessed private balconies. Sparkling white cottages, set rather closely together behind the inn and the road, are the most expensive units. All guest units are expertly decorated, and the cottages are charming. Public rooms are immense, with high ceiling, and impeccably furnished. The Homestead is an AP resort and famous for its food, which ranges from cornbread to sophisticated dishes, all served on china and silver serving dishes by armies of waiters and captains. Tea is still served to music, and dining areas include the glass-walled grill overlooking gardens as well as outdoor terraces.

The 17,000 acres of lush, mountain-ringed greenery include three 18-hole golf courses, with some of the best golf in the nation. There are 15 tennis courts. Three swimming pools include one indoors and another at the quaint complex where the spa still functions, updated with saunas and steam rooms. Also on the grounds are skeet and trapshooting, fishing in Cascades Stream, bowling alleys, riding stables, surrey and buckboard rides, and a supervised playground and playroom for children. In winter (about Nov to Mar) the ski area provides a double chairlift, surface tows, a ski school, snowmaking, a restaurant and bar, a cafeteria, and a nursery. An olympic-size skating rink is also open. The Homestead has a 5600-foot airstrip, and of course provides all the extras from dancing and entertainment to chauffeur service. ☐ **Lux** ☐ **G, T, R, F, S**

The Cascades Inn, *Hot Springs, VA 24445, (703) 839-5355* ☐ *Open all year* ☐ *Off US 220, 3 mi from the Homestead* ☐ Operated by the Homestead, this MAP inn at the head of Cascades Gorge was originally established more than a hundred years ago, when people came to Bath Country for the mineral baths. Today 50 units are available in the inn with a wide colonnaded porch, a matching white motel, and in little white cottages with large porches. The dining room and glassed-in lounge have an unpretentious country air, and the lovely grounds have an outdoor pool and three miles of trout stream. Guests may use all of the facilities of the Homestead. The mountainous Cascades Golf Course adjoins the inn, and the nearby Lower Cascades, with its own golf pro, was designed by Robert Trent Jones. ☐ **Mod** ☐ **F, G**

IRVINGTON: The Tides Inn and Lodge, *Irvington, VA 22480, (804) 438-5000* ☐ *Inn open mid-Mar to early Jan, Lodge open mid-Mar–early Nov* ☐ *S off SR 3* ☐ The Virginia tradition of gracious living makes a superior resort like the

Tides not only possible but also probable. Located on a height, on a beautifully landscaped peninsula between the Rappahannock River and Carter's Creek, the Tides Inn provides 110 rooms and suites in white, balcony-lined buildings, separated from the waterfront by a fringe of trees and an impeccable lawn. All units are air-conditioned and most have private balconies or patios. The Tides Lodge is across the creek and is reached by a free ferry that shuttles across every half hour. Here there are an additional 50 rooms in buildings of a more rustic design set among trees at the edges of fairways. The lodge also rents two-bedroom cottages and privately owned condominiums on EP. Both complexes are run on AP and have restaurants with garden and water views and cocktail service. Guests may dine at either complex (dress code evenings). The inn also has the Chesapeake Club, a lovely room open for a nominal extra fee, where your bottles will be stored and where dances and special parties are held. Lunch is usually also served at the golf club and at the inn's Summer House, a pavilion secluded by trees at the tip of the peninsula; the terrace overlooks the olympic-size saltwater pool. Included in the rates are cruises aboard a yacht, bicycles, use of indoor and outdoor game areas, fishing in the stocked ponds on the golf courses, and the use of Sailfish, canoes, or other boats without motors (motorized craft and guided fishing trips are available at extra cost at the dock).

The 45 holes of fine golf here include the 6400-yard Tartan Course at the lodge and the 6950-yard Golden Eagle. The latter is two miles away (free transportation is provided) and has club facilities. At the inn there is also a nine-hole, par-three course. Tennis rentals are available for the four courts at the inn and the three (one lighted) at the lodge. The inn also provides supervised play for children in summer. ☐ **Mod to Lux** ☐ **G, T, W, F**

MOUNTAIN LAKE: Mountain Lake Hotel, *Mountain Lake, VA 24136, (703) 626-7121* ☐ *Opens early June to early Oct* ☐ *NW of Blacksburg via US 460, then 7 mi N on SR 700* ☐ On 2000 acres beside a lake high in the Blue Ridge Mountains, this AP resort has about 100 rooms and suites in a traditional hotel plus cottages. The restaurant has mountain ham and other local specialties, and wine, beer, and setups are available. The lobby has a TV set and a fireplace, and there is a recreation room. The lake, surrounded by heavily forested mountain slopes, has a swimming beach and boats. As no motors are allowed, this is a quiet retreat for fishing and lazing. Horseback riding, archery, tennis on lighted courts, and outdoor game areas are also provided. In the evenings movies, bingo, or other entertainment may be scheduled, but in general this resort, surrounded by thousands of wilderness acres of Jefferson National Forest, is for those looking for scenery rather than action. ☐ **Mod** ☐ **R, F, W**

NEW MARKET: Shenvalee Lodge, *P.O. Box 430, New Market, VA 22844, (703) 740-3181* ☐ *Open all year* ☐ *15 mi W of Skyline Drive via US 211, then 1½ mi S on US 11* ☐ This unpretentious 52-room resort, consisting of a colonial-style lodge and one-story motel annexes, spreads over 300 acres of fairways in the Shenandoah Valley. Most guest rooms have small porches or balconies overlooking the swimming pool, golf courses, and the mountains beyond. Rates are EP, and the restaurant specializes in evening buffets of sugar-cured ham and other regional fare. Also available are a snack bar and tavern. The 18-hole golf course surrounding the lodge and motel is par-70 and has a driving range and a pro shop. Also on the grounds are an outdoor swimming pool, three tennis courts, lawn games, and a stocked fishing pond. Nearby are New Market Battlefield Park, the Endless, Shenandoah, and Luray caverns, and the Bryce Mountain Ski Area. ☐ **Inexp to Mod** ☐ **G**

VIRGINIA STATE PARKS: Eight of Virginia's state parks offer housekeeping cabins that are remarkable buys for families. Four of the parks are in the mountains and focus on high-altitude lakes–Claytor Lake, Douthat, Fairy Stone, and Hungry Mother. Staunton River Park is on a peninsula that juts into Buggs Island Lake, part of an enormous Kerr dam project on the North Carolina border. Goodwin Lake–Prince Edward park is in the Midlands and has two lakes surrounded by pines. Westmoreland is on the Potomac River. Seashore State Park, near Virginia Beach, has a beach on Chesapeake Bay and a beautiful boardwalk nature trail through a cypress swamp. Cabins come in several sizes and are of two types. Some are log cabins–with rough beams inside and out, and covered porches–that were traditional in the Appalachian Mountinas. Furnishings are also traditional, including mountain rocking chairs. Other cabins are less appealing in design but more modern in appointments, constructed of concrete block, with screened porches. Douthat also has a magnificent mountaintop log-and-stone lodge that sleeps eight and is so rambling that you won't get in each other's way. In most of the parks the cabins are well spaced and screened with vegetation for privacy. All of the parks offer swimming, fishing, and boating. Information and reservation forms are available from Divison of State Parks, 1201 State Office Building, Capitol Square, Richmond, VA 23219, (804) 786-5045. You may also reserve (with a small fee) by calling Ticketron or writing their Virginia State Parks Reservation System, Box 62284, Virginia Beach, VA 23462.

WINTERGREEN: Wintergreen Resort, *Wintergreen, VA 22938, (804) 361-2220* ☐ *Open all year* ☐ *S of Waynesboro on Rte. 664* ☐ This resort community offers some 200 condominium and home rentals in attractive modern buildings in the Blue Ridge Mountains. Treeloft homes, set among trees on mountain slopes, offer 360-degree views and sleep two to six people. Condominiums, also surrounded by trees, are near the Mountain Village shops and Sports Center and sleep two to six people in studios and one- to three-bedroom units. Rodes Farm Houses, set in the grassy valley, have Franklin stoves and sleep up to six. Secluded in the woods away from the village are individually designed homes with three or four bedrooms that will sleep up to eight people. All units have equipped kitchens or kitchenettes, most have balconies or decks and fabulous mountain views, and some have fireplaces. The village has a market, but if you want to eat out you can try the varied menu at the Copper Mine Restaurant at the Sports Center. This is warm with wood beams and furnishings, and serves lunch, dinner, and drinks. Rodes Farm Inn, a restored 19th-century farmhouse with long wooden tables and open fireplaces, offers country ham, crispy fried chicken, homemade biscuits, and similar mountain fare. Three meals a day are available here.

Rates here are EP or on package plans that may include a choice of sports facilities. The 18-hole golf course is on a 3851-foot-high plateau called Devil's Knob. Tennis courts and swimming pools are located on the mountain slopes and in the valley. Rodes Farm offers stables, lessons, jumping, and miles of bridle trails in meadows and woodland. In winter, there is snowmaking for six slopes (two lighted) served by one double and two triple chairlifts. The ski school offers rentals and instruction. For children aged 7 to 13 a program of organized activity is run in summer. In ski season a program is run for children aged two to eight. In all seasons the Sports Center offers sports instruction. ☐ **Mod to Lux** ☐ **T, G, R, F, S**

Washington

CHELAN: This stunning lake cuts through 8000-foot peaks of the Cascades and is one of the most scenic areas of the nation. The lake offers good swimming, fishing, sailing, and boating from several public recreation areas along the shores. At the north end is Stehekin, gateway to the wilderness of the Cascades, with a lodge. A passenger ferry makes frequent trips spring through fall (daily in summer) the 55 miles from Chelan to Stehekin. Chelan has an 18-hole public golf course with pro shop and pro services, public tennis courts, and a riding stable. Trout fishing is good in the Wenatchee and Methrow River, and ice fishing is popular at Alta Lake. Echo Valley ski area, north of Entiat, has surface tows and a ski school, and cross-country skiing is popular in the area. For information, contact the Chamber of Commerce, Box 216, Chelan, WA 98816.

Campbell's Lodge, *Box 278, Chelan, WA 98816, (509) 682-2561* ☐ *Open all year* ☐ *On US 97* ☐ On Lake Chelan, this motel offers 83 motel rooms and suites, and 17 cottages (some with kitchenettes). Many units have balconies and lake views. Also on the grounds are a restaurant, bar, sandy beach, a dock, two swimming pools, and a whirlpool. ☐ **Mod** ☐ **W, F**

Cannon's, *Box 1107, Chelan, WA 98816, (509) 682-2932* ☐ *Open all year* ☐ *On SR 150* ☐ A good base for families exploring Lake Chelan and the surrounding Wenatchee National Forest, this motel offers 21 one- and two-bedroom suites, each with lake view, kitchenette, color TV, and air conditioning. On the four-acre grounds, with spacious lawns and terraced flower beds, are picnic area with umbrellas, tables, and grills. Along the sandy private beach is the dock with boat and waterskiing rentals. A heated outdoor pool permits swimming when the lake is too cold. The motel has a play area and can provide a list of babysitters. Restaurants are nearby. ☐ **Mod** ☐ **W, F**

CLE ELUM: Hidden Valley Ranch, *Cle Elum, WA 98922, (509) 674-2422* ☐ *Open all year* ☐ *From exit 85 (Cle Elum) on I-90, go N 8½ mi on SR 970 to Hidden Valley Rd.* ☐ When the original homesteaders were setting up their angora sheep ranch in 1895, they probably never dreamed that within 100 years their ranch would be better known for its oyster-beef Wellington than its wool. Small, rustic but genteel, this guest ranch offers apartments (accommodating up to 6 persons), frame and log cabins (some with kitchen), horseback riding on 40 miles of wilderness trails in the Cascades, and fine food. One especially nice cabin is the Hilltop (for two only), with fireplace, private deck, and Japanese soaking tub. Other guest have the ranch's heated pool to soak in.

Besides the beef Wellington, the gourmet dinner menu includes Provimi veal cordon bleu, fillet of sole with shrimp and crab, and baked king crab legs. A reservation must be made at least three days in advance for these dinners. Package plans are available that include all meals (with family-style dinners less elegant than those mentioned above), and some horseback riding. Also offered are exhilarating photographic trips down the Yakima River. The valley itself is good for wildlife photography, with more than 100 species of birds as well as coyote, bobcat, deer, and elk making their homes there. □ **Mod to Lux** □ **R, H**

COPALIS BEACH: Iron Springs Resort, *P.O. Box 207, Copalis Beach, WA 98535, (206) 276-4230* □ *Open all year* □ *3 mi N on SR 109* □ This little colony of 20 houskeeping cottages is scattered in 100-acre spruce woods on the beach. Cottages accommodate from two to ten people and have fireplaces, equipped kitchens, linens, and decks or outdoor sitting areas. Swimming, piper boarding, surfing, beachcombing, surf fishing, crabbing, and especially digging for razor clams are the activities here. □ **Inexp to Mod** □ **F**

CRYSTAL MOUNTAIN SKI AREA: Located near the northeastern corner of Mt. Rainier Park off SR 410, this is Washington's highest ski area, offering one of the longest seasons in the state. One triple-chair, five double-chairs, one T-bar, and ten rope tows ferry skiers up 2400 feet. The 4500 acres of trails and slopes includes runs of all difficulty and even ski jumps. One mile-long run is lit for night skiing. The Summit House Restaurant and Bar has a superb view of Mt. Rainier. At the base are rentals, repairs, ski shop, cafeteria, restaurant, bar, and nursery. Accommodations in this major resort area range from budget rooms to luxury apartments. For more information, contact: Crystal Mountain Ski Area, Crystal Mountain, WA 98022, (206) 663-2265.

LONG BEACH: This resort is on a narrow peninsula that separates Willapa Bay (a wildlife refuge) from the Pacific. Clamming and beachcombing are popular on the 28-mile-long beach. The peninsula was once a strategic area guarding the mouth of the Columbia River, and state historic parks commemorate Fort Canby and Fort Columbia with interpretive exhibits.

Chautauqua Lodge, *Box 757, Long Beach, WA 98631, (206) 642-2244* □ *Open all year* □ *N and W of town off SR 103* □ This attractive three-story motel is on the beach and offers 120 double rooms and suites; half the units have fireplaces and kitchenettes, and many have private patios and balconies with water views. The indoor pool has a whirlpool and sauna. The nearby restaurant opens seasonally in the wee hours of the morning for fishermen and birdwatchers, and has a bar. ☐ **Mod** ☐ **W, F**

Surfside Inn, *P.O. Box 281, Rte. 1, Long Beach, WA 98631, (206) 655-5211* ☐ *Open all year* ☐ *15 mi N on SR 103* ☐ On the beach, this three-story motel offers 39 comfortable double rooms and suites, some with kitchenettes, with private patios or balconies. The restaurant specializes in seafood and the bar has dancing and entertainment. The inn has golf with pro services, a putting green, and a driving range. ☐ **Mod**

MOCLIPS: Ocean Crest, *Moclips, WA 98562, (206) 276-4465* ☐ *Open all year* ☐ *1 mi S of Moclips on SR 109* ☐ Set along a surfside bluff, this complex of modern motor lodges and restaurants commands a sweeping view of beach and ocean. Even the pool house, with small heated pool and sauna, has panoramic view. Most of these contemporary doubles and suites have wood paneling, large picture window and balcony, fireplace, and kitchenette or refrigerator. All have cable color TV. The prints and other stylistic accents decorating the lodge are reminiscent of Northwest Indian patterns; a fine collection of authentic Indian baskets is on display. The dining rooms, open for dinner only, have steak and chicken dishes, but are best known for their seafood—clam chowder, razor clams, and blueback salmon. The lounge has ocean view, fireplace, and wide wine list. The resort also has a gift shop and private path to the beach. ☐ **Mod**

MOUNT RANIER NATIONAL PARK: This scenic mammoth, where mountaineers have trained for Everest and whose broad peak dominates the park's skyline, is a study in superlatives. It is the highest (base to peak) mountain and has the longest and largest glaciers in continental United States. The park terrain ranges from the wind-swept icy summit to tranquil wildflower meadows on the western slopes. Waterfalls, lakes, streams, and ancient forests abound. The park headquarters are at Ashford, and visitor centers are at Longmire, Paradise, Ohanapecosh, and Sunrise. The 90-mile-long Wonderland hiking trail (permit required) encircles the mountain. Connected to the park's roads, it can be used as a day trail, but shelters are spaced every 10 miles or so for those who want to hike larger portions of the circuit. Guided climbs to the summit are available. Paradise Valley, accessible in the winter from the east via highway 706, has a small ski area with rope tows. No fishing license is required in the park. Park wildlife includes deer and bear, but most tourists' favorites are the agile mountain goats. For more information contact Mt. Rainier National Park, Ashford, WA 98304, (206) 569-2211.

Paradise Inn, *Mt. Rainer Park, Paradise, WA, (206) 569-2706* ☐ *Open mid-June to early Sept* ☐ *On SR 706, 20 mi E of Nisqually* ☐ This venerable, rustic lodge, built in 1916, has the classic roughhewn cedar beams, cut-stone fireplaces, and Indian rugs one has come to expect of Northwestern backcountry hotels. The spacious lobby and dining room provide spectacular views of Mt. Rainier through tall windows. The dining room serves a variety of hearty dishes, with fresh salmon a house specialty. Almost all of the 100 or so rooms have private bath, and all have telephones and are simply furnished in contemporary style. The Visitor Center is next door to the inn.

Only the snack bar here is open fall through spring, but National Inn in Longmire offers 14 inexpensively priced rooms all year round. That inn also has a country store, a cafeteria, and a gift shop. Reservations for either inn can be made through Mt. Ranier National Park Hospitality Service, 4820 S. Washington, Tacoma, WA 98409, (206) 475-6260. ☐ **Inexp to Mod** ☐ **H, F**

MOUNT SPOKANE SKI AREA: This ski area has some of the most extensive night skiing anywhere. Three of the five chairlifts here are lit, providing over 8200 feet of lighted trails. Altogether, 31 trails of all levels of difficulty lace the slopes. At the base, both of the day lodges have ski rentals on the first floor and cafeteria on the second. One also has a day-care center, the other a repair shop. The ski school holds private, group, and children's lessons. An ice skating rink and a network of cross-country trails are also in the area. For more information, contact: Mt. Spokane Skiing Corp., N. 7322 Division St., Spokane, WA 99208, (509) 484-3908 or 238-6281.

Snowblaze Condominiums, *Rte. 1, Box 416-B, Mead, WA 99021, (509) 238-4630* ☐ *Open all year* ☐ *From Spokane, take US 2 N past Mead to SR 206 NE to end* ☐ These condominiums range from standard hotel rooms up to deluxe two-bedroom units with two baths, kitchen, and fireplace. Efficiencies, studios with loft and fireplace, and one-bedroom suites are also available. Kitchens are fully equipped with utensils and cookware. All units are individually decorated–some contemporary, some rustic, some early American. Tennis and a heated pool are open in summer, and ski packages are offered in winter. Next door is Grambo's, a steak and seafood restaurant and lounge that has music, singing waiters, and a Sunday brunch that features trout. ☐ **Mod** ☐ **S**

NORTH CASCADES NATIONAL PARK: Fur trader Alexander Ross, who recorded his exploration of the North Cascades in his 1814 diary, was the first white man known to have crossed this craggy mountain range. In the 1850s, gold was discovered, but the area was so inaccessible the boom never got very big. It wasn't until 1960, when highway 20 was completed, that the range's majestic beauty was revealed to the public. But even this modern-day highway is often closed by winter snows.

Twenty-four-mile-long Ross Lake, an impoundment of the Skagit River behind Ross Dam, is a major stop in the North Cascades. The National Recreation Area here, formed by Ross lake and George and Diablo lakes below it, is a favorite of fishermen because of the excellent trout catches. Surrounding the recreation area, the glaciated mountains and the alpine meadows of the half-million-acre park provide superb hiking, much of it requiring only amateur skill. Many campgrounds are scattered around the area. Hunting is permitted only in the recreation area, not in the park. For more information, write North Cascades National Park, Sedro Woolley, WA 98284.

Ross Lake Resort, *Rockport, WA 98283, Newhalem 4735 (via Mt. Vernon operator)* ☐ *Open mid-June–Oct* ☐ *From SR 20, turn N over Diablo Dam to City Light boat landing. Boats leave the dock at 8:30 am (7:30 Sat.) and at 3:00 pm* ☐ This resort is unusual in several ways. You can't get there by car; you must take the passenger boat from Diablo Lake to Ross Lake. And, because Ross Lake is contained by steep mountain walls and has little level shoreline, the resort is built on huge rafts. It has no restaurant and no liquor or food sales, but it provides hunters, fishermen, and sightseers with a fine base camp in this superb wilderness area. The modern one-bedroom cabins have private bath, kitchen, fireplace of gas heat, and four or five beds. The older one-room cabins share a community bath and have a wood stove for cooking and heating and a refrigerator. The resort rents boats, motors, and canoes, and has refrigeration, gas, tackle, and fishing licenses. ☐ **Inexp** ☐ **F, H, W**

Skagit Motel, *1977 Highway 20, Sedro Woolley, WA 98284* ☐ *Open all year* ☐ *On SR 20, 3 mi E of I-5 at Burlington* ☐ For those who would rather not be in deep wilderness and want to enjoy the scenery from a car, this motel would be a good jumping-off point for a motor tour through the North Cascades. Among its 47 rooms and suites are eight with kitchenette (half of these with oven). All rooms have color TV and some have air conditioning. A restaurant is next door. ☐ **Inexp to Mod**

OCEAN SHORES: This peninsula, dividing Gray's Harbor from the Pacific, has developed over the last quarter century into a resort community complete with fine 18-hole municipal golf course, canals, and lakes, excellent marina, and 2500-foot lighted landing strip. Freshwater and saltwater fishing guides and charters are available, and clam digging and crabbing are very popular on the beaches.

The Canterbury Inn, *P.O. Box 306, Ocean Shores, WA 98569, (206) 289-3317* ☐ *Open all year* ☐ *On Ocean Shores Blvd.* ☐ Almost all of the contemporary rooms and suites in this handsome motor lodge have sliding glass doors that open onto a beachfront private patio. All have cable color TV, direct-dial telephone, and kitchen (most with oven). Many have a fireplace. Downstairs are a small indoor heated pool, pleasant social room with fireplace, and recreation room. The resort will provide a list of babysitters. A restaurant and lounge are a block away. ☐ **Mod** ☐ **W**

The Grey Gull, *P.O. Box 674, Ocean Shores, WA 98569, (206) 289-3381* ☐ *Open all year* ☐ *On Ocean Shores Blvd.* ☐ Laid out like a V with the open end facing the ocean and a swimming pool set between its two arms, this attractive, weathered-wood motor lodge has many rooms with ocean views from their recessed private balconies. The efficiencies and one- to three-bedroom suites typically feature walls of exposed brick and ceilings of exposed beams. All have fireplace, color TV, and kitchen with dishwasher (except efficiencies). The lodge has a sauna and will furnish a list of babysitters. Restaurants are nearby. ☐ **Mod to Lux** ☐ **W**

The Polynesian, *P.O. Box 998, Ocean Shores, WA 98569, (206) 289-3361* ☐ *Open all year* ☐ *On Ocean Shores Blvd.* ☐ Although some of the roof lines are reminiscent of Polynesian long houses, the exterior siding of this modern motor lodge is log and stone, a little more in keeping with most people's expectations of a Northwestern lodge. Most of the 70 units are one- or two-bedroom suites with color TV, fireplace, and kitchen (some with dishwasher). A three-bedroom penthouse with sauna, three baths, and fireplace is available for up to six persons. Many of the rooms have a balcony facing the ocean. The restaurant, housed in an octagonal building set in the middle of the parking lot, has beautiful wood paneling, high ceiling, varied menu of continental dishes, and lounge with frequent entertainment. The lodge also boasts an indoor pool and recreation rooms. ☐ **Mod** ☐ **W**

PORT LUDLOW: The Admiralty Resort, *Box 75, Port Ludlow, WA 98365, (206) 437-2222* ☐ *Open all year* ☐ *Follow signs 8 mi N to resort* ☐ Among the picturesque little Puget Sound towns, set up by shipping and lumber companies in the 19th-century, Port Gamble and Port Ludlow were homes for the Puget Mill Company, which built the original Admiralty Hotel for its company officials and their guests. This old wood building burned down long ago, and little remains of the original town. But the company, under its present name of Pope and Taulbot, has created a 3000-acre modern resort and residential community on the site. On landscaped grounds stretching back from Puget Sound are modern one- and two-story lodges housing more than 200 brightly furnished double rooms and suites, many of which look out over the sound through wide expanses of glass. The suites include two-level units with kitchen and living areas downstairs plus sleeping lofts and parlor suites with one to four bedrooms. Some suites have stone or brick fireplaces; all have private patios or balconies. The handsome Harbormaster restaurant specializes in good seafood and dramatic views. The lounge has entertainment nightly in summer and weekends in winter. In summer the P & T Deli makes special sandwich treats.

Port Ludlow Golf Course, designed by Robert Muir Graves, is regarded as one of the finest in the Northwest. Almost 7000 yards of fairways wind through fir forest by the side of Admiralty Inlet, numerous creeks and lagoons. The hilltop clubhouse has rentals, pro services, putting green, and driving range. There are also seven tennis courts. At the L-shaped olympic-size outdoor pool is a modern beach club with indoor game areas, a billiard room, and a deck overlooking the bay. A squash, a racquetball court, and a youth activities center are nearby. The full-service marina provides moorage for 300 boats, rents sailboats and power boats, and arranges salmon fishing charters.

Next to the marina is a saltwater lagoon for swimming. Skin diving, clamming, and beachcombing are popular. Bicycle paths run throughout the property, with rental bikes available. If the constant sight of the Olympic Mountains doesn't draw you all the way to the park for a hike, you will at least want to wander the trails in Port Ludlow. Parents will find supervised programs for children and teenagers, and babysitting can be arranged. ☐ **Mod to Lux** ☐ **G, T, W, F**

QUINAULT: Rain Forest Resort, *Quinault, WA 98575, (206) 288-2535* ☐ *Open all year* ☐ *South Shore Rd., 3½ mi NE of US 101* ☐ Composed of a lakefront motel and a string of cottages, a restaurant, dock area, and campground, this informal resort provides a base for exploring Lake Quinault and the magical rain forest at the southwestern end of Olympic National Park. The housekeeping cottages, in varying sizes accommodating from two to six persons, are equipped with all appliances, linens, and utensils. The larger cottages have carpeted bedrooms. The motel rooms are also carpeted and have a wood paneling and TV. The restaurant, open for dinner, specializes in Quinault salmon, and the Jack of Nines lounge has cocktails. Also on the grounds are swimming beach, general store, laundromat, wine and beer shop, and service station.

The lodge marina rents rowboats and boats with outboards. The lake and lower river, within the boundaries of the Quinault Indian reservation, are subject to tribal fishing regulation. The upper river is outside the reservation and public fishing laws pertain. Winter and spring, steelhead averaging 12 to 15 pounds can be taken from the river, and spring through autumn, Dolly Varden, rainbow, searun cutthroat, and summer steelhead inhabit the lake and river. The primevally verdant rain forests just east of the resort in the Enchanted Valley are the product of damp prevailing westerlies condensing on the cool western slopes of the Olympic Range. Mist hangs in the air and trees are exotically draped with moss and lichen. ☐ **Inexp** ☐ **F, W**

SAN JUAN ISLANDS AND ANACORTES: These islands have been the secret of Northwesterners for years. Many Washington and British Columbia residents first get a look at them on the ferry ride between Sidney and Anacortes and share the fantasy of jumping ship and running to live on one of the these idyllic hideaways. There are 172—enough to get lost on forever. The ferries from Anacortes serve four islands—San Juan, Orcas, Lopez, and Shaw. For schedules, contact: Ferry Terminal, Pier 52, Marion St., Seattle, WA 98104, (206) 464-6400. You may take your car, and these four islands have paved or gravel roads. Friday Harbor, on San Juan Island, is the major city, complete with two airports, nine-hole golf course, marina services, several motels, laundromat, drug store, and grocery.

Both the developed and undeveloped islands are dotted with state parks—51 in all, varying in size from less than an acre to Orcas Island's 4934-acre Moran Park, with hiking and nature trails, freshwater swimming and fishing, and

2400-foot mountaintop lookout. Another park, on San Juan Island, is a monument to the important role these islands once played in history. Vancouver Strait, in the early 1800s, was the loosely defined border between the province of British Columbia and the United States. English and American settlers lived on San Juan in an uneasy truce until the death of an English pig, shot while rooting in an American potato patch, sparked a feud that sent both sides running for their guns. London and Washington were horrified to learn that commanders at local outposts had already called out soldiers and warships. At the verge of stumbling into war, the governments agreed to set up garrisons at opposite ends of the island and find a neutral party to settle the issue. In 1872, Kaiser Wilhelm I of Germany ruled in favor of the United States. The National Park Service now maintains the "Pig War" garrisons as a historical park.

Before catching the ferry in Anacortes, you might make a side trip to Deception Pass State Park (9 miles south of town on highway 20) to enjoy its clamming, fishing, boating, swimming, and especially its stunning shoreline scenery.

The Islander Lopez, *Fisherman Bay, Lopez, WA 98261, (206) 468-2233* □ *Open all year* □ The Islander is a two-level motel–a pretty one, with dark board-and-batten siding–set against a backdrop of sharp fir-forested hills by the side of quiet Fisherman Bay on the west side of Lopez Island. There is a method to its modesty: An effort has been made to keep the building unobtrusive and let the surroundings speak for themselves. The motif is Polynesian, and the choice is not altogether inappropriate: the San Juans are isolated and overgrown enough to suggest a Hawaiian retreat. The dining room and bar have woven-mat ceilings with bamboo and log crossbeams. Specialties include salmon (pan fried, or baked with wine, dill, and garlic) and home-baked breads and pastries. The dining room will cook your catch, or you can cook it yourself at the barbecue pit on the beach. There is dancing in the lounge. There's little trace of Polynesia in the 26 paneled guest rooms and suites, all of which have picture-window views and decks overlooking the bay; many units have kitchenettes.

The hotel offers moorage for 40 to 50 boats with gas, electric outlets, and fresh water. Rentals and charters are available. The bay was once a smuggler's cove, and searching for buried treasure in the caves and on the beaches of the island is one good way to take in the scenery. Also on the grounds are a nice enclosed swimming pool, a whirlpool, lawn games, and bicycles. Tennis and golf (nine holes) are available on the island. □ **Mod** □ **W, F**

Roche Harbor Boatel, *Roche Harbor, San Juan Island, WA 98250, (206) 378-2155* □ *Open Apr–Nov* □ *At NW tip of island at end of Roche Harbor Rd.* □ This is a self-sufficient complex consisting of the Hotel de Haro; two-bedroom housekeeping cabins with fireplaces; one- and two-bedroom condominium units with kitchens, fireplaces (some), and patios; a large heated pool; tennis courts; a restaurant; a customs office; a nearby airstrip; a grocery; and a complete marina with moorage for 200 boats. The ivy-covered Hotel de Haro, set in semiformal flower gardens, is a three-story cupola-capped wooden structue built in the 1880s by John McMillan, pioneer mining magnate. Timbers from the Hudson's Bay trading post originally on the site still support the hotel. Furnished with some antiques, its simple share-bath rooms have a creaky charm. The Presidential suite, occupied by Teddy Roosevelt in 1906, has a fireplace, private bath, and private veranda. John McMillan's own house, overlooking Speiden Channel, became the restaurant. It has a lounge and a varied menu that emphasizes fresh seafood.

Besides the usual island pleasures of boating, fishing, and hunting the innumerable rabbits, the resort has its own rather bizarre attraction: the McMillan family mausoleum. Set deep in the woods, seven classic columns guard a limestone dining table and chairs, symbols of the family's communion in the afterlife. ☐ **Inexp to Mod** ☐ **F, W**

Rosario Resort Hotel, *Orca Island, Eastsound, WA 98245, (206) 376-2222* ☐ *Open all year* ☐ Robert Moran built Rosario in 1905, on thousands of wooded acres on Orcas Island, the largest of the San Juans. Most of the land is now Moran State Park, but the enormous mansion he built for himself by the water's edge still stands at the center of the remaining 1300-acre resort–one of the loveliest in the Northwest. This graceful white building, with an arched and pillared porch running along its length and width, stained-glass windows, and a many-gabled copper roof, houses five inexpensive guest rooms and a five-tiered dining room that offers extraordinary views of Cascade Bay and the woods. More than 100 luxurious double rooms and suites are offered in modern lodges scattered on the landscaped grounds. Most of these units have balconies (some with spectacular views of the bay), many have kitchenettes, and some have fireplaces. Many units will sleep four adults, and the largest will sleep up to eight. At dockside is a boatel with cheaper rooms and suites. The bay-view dining room serves three meals daily, with seafood the specialty, and has dinner dancing. Adjoining is a cocktail lounge that has frequent entertainment. On the water is Discovery House, with steaks broiled in a sunken barbecue pit, a salad bar, and a dance combo.

At the well-equipped marina you can dock your boat or rent one to explore the wooded coastline of Cascade Bay or for waterskiing. Fishing is good for cod, perch, and cutthroat and salmon, and charter trips are available. Of the three swimming pools, one is for the whole family, one with cocktail service is for adults, and the small indoor pool has sauna and whirlpool. You can also swim along the two miles of beachfront or at the mountain lakes in the adjacent park. Rosario also offers tennis, scuba-diving, a 360-foot canoeing lagoon, and great walks–from strolls along the resort paths and gardens to more ambitious hikes in the woods. Golf is available at the nearby Orcas Island course. Services here include supervised play for children, half a dozen shops, a game room, self-service laundry, courtesy buses, and an organ-pipe-lined Music

Room, the site of organ concerts and teas. This is a relaxed and informal resort, a good compromise between the elegance of a grand hotel and the quiet seclusion and casualness of a cabin in the woods. ☐ **Inexp to Mod** ☐ **F, W, T**

Ship Harbor Inn, *Anacortes, WA 98221, (206) 293-5177* ☐ *Open all year* ☐ *On SR 20, 3½ mi W of Anacortes* ☐ This modern motor lodge and rustic cottages are only a short walk from the ferry terminal. All units have cable color TV and most have private patio or balcony. The Crow's Nest cottage has a fireplace, as do some units in the main lodge, and some two-room cottages have kitchenettes. Trails from the lodge lead through a wildlife sanctuary to the beach. Restaurants are closeby, and the lodge has picnic tables and grills as well as playground. ☐ **Inexp to Mod**

SEKIU: Coho Resort, *Sekiu, WA 98381, (206) 963-2233* ☐ *Open all year* ☐ *1 mi E on SR 112* ☐ On Juan de Fuca Strait, this tiny, unpretentious resort offers only 12 motel rooms with shower, 10 with kitchenette, on the waterfront. The adjacent restaurant has a salad bar and specializes in salmon. Although the water is usually too cold for swimming, fishing is great–you can rent boats and motorboats, or arrange for charters for salmon and bottom fish. Steelhead are in the nearby Hoko River. In addition to beachcombing and watching seals and porpoises (and whales in season), the Makah Indian digs and museum are of interest. Jeep service is available to Lake Ozette at the Pacific coastline section of Olympic National Park. ☐ **Inexp** ☐ **F**

UNION: Alderbrook Inn, *Union, WA 98592, (206) 898-2200* ☐ *Open all year* ☐ *5 mi E of Union on SR 106* ☐ The Alderbrook is on 525 wooded acres near the horseshoe turn of Hood Canal, at the foot of the spectacular Olympic Mountains. The canal is a favorite of boaters and fishermen; the Alderbrook catches the attention of other sports enthusiasts too, with a good 18-hole golf course (6300 yards, par 72), and four lighted tennis courts. You can stay in either the inn itself–a modern cedar building with distinctive "matchstick" balconies running along each of its three stories–or in one of 22 housekeeping cottages in the adjacent woods. Thirty-six of the inn rooms have private balconies, and many have views of the canal; 47 rooms look inward onto an open-air courtyard. The cottages, each two bedrooms, have electric kitchens, fireplaces, and rustic furnishings. If you arrive at the inn by water, you'll find 1200 feet of moorage on their floating dock. You can rent or charter a boat through the inn and go out after salmon, or sail, waterski, or skin dive in the canal. The glass-enclosed swimming pavilion has a large heated pool, therapy pool, and saunas. You can go beachcombing or hike the nature trails in the nearby woods. Fresh seafood is, naturally, the specialty of the house at the Alderbrook dining room, and it's served buffet-style Friday nights; there's also a Champagne Brunch on Sundays. The Waterwheel Room offers cocktails and dancing every night. ☐ **Mod** ☐ **G, T, W, F**

WHITE PASS SKI AREA: Although this ski area is only 14 miles southeast of Mt. Rainier, you may not be able to combine a visit to the park with a ski vacation here because snow often closes the road into the park during winter. But even at this distance, the mountain spectacularly dominates the horizon. From the base area at 4500 feet, tow chairlifts carry skiers up to 6000-foot Pigtail Peak and eight trails, from novice to expert difficulty. The longest trail is two miles. Another chairlift and a Poma and a rope tow serve the broad open slope on the lower part of the mountain (lit for night skiing Thursday, Friday, and Saturday). At the base is a day lodge where lessons and rentals are offered daily. Across US 12 is a service station, store, and the Village Inn—a modern motor lodge with a restaurant and a heated pool. For more information, contact: White Pass Co., P.O. Box 354, Yakima, WA 98907, (509) 453-8731.

WINTHROP: Sun Mountain Lodge, *Box 1000, Winthrop, WA 98862, (509) 996-2211* □ *Open Dec–Oct* □ *From Pateros, 40 mi N on SR 153* □ Perched on a mesa high above Methow Valley, this ranch-style aerie offers magnificent panoramic views of pine-covered hills and snowcapped mountains. The main lodge is rustic, with peeled logs, exposed beams, and huge stone fireplaces. It houses the dining room, lounges (one with TV), a gift shop, and a cocktail lounge with a deck hanging over the mountainside. Two matching two-story guest lodges offer 48 paneled double rooms and two suites with fireplaces. All units have a patio or deck with a view either of the Cascades or of Patterson Lake and the valley. The dining room specialties include mountain trout, poached Columbia River salmon, and broiled Cornish game hen; fruits and vegetables are freshly picked at local farms, and bread is home-baked. In addition there are breakfast trail rides, pit barbecues, a summer chuckwagon snack bar, and the lodge will pack a picnic lunch.

On the 1000-acre grounds are a heated swimming pool, a whirlpool, four asphalt tennis courts, a playground and outdoor game areas, and nature trails for wild flower or fossil hunting or watching the abundant wildlife. Patterson Lake has a private beach, boat rentals, waterskiing, and fishing, and stream fishing is excellent for cutthroat, rainbow, and wastern brook trout. Within a dozen miles are three rivers and 12 lakes. Available at extra charges are guided trail rides and pack trips, scenic and whitewater raft trips, and nine-hole golf at the nearby Bear Creek course. In fall, hunting trips are organized. In winter, there is downhill skiing at nearby Loup Loup, with a ski shop for rentals, or snowmobiling, ice fishing, sleigh rides, and cross-country skiing. □ **Mod** □ **W, F, R, S, H**

West Virginia

BERKELEY SPRINGS: Coolfont Recreation, *Berkeley Springs, WV 25411, (304) 424-1232* ☐ *Open all year* ☐ *3 mi W off SR 9* ☐ In a beautiful rural mountain area near the historic colonial spa of Berkeley Springs (still operating but not modernized), Coolfont offers unpretentious country living in chalets and cabins with one to five bedrooms and kitchens. Most units have fireplaces, porches, balconies, or patios. On the 1000-acre grounds are a restaurant, bar, package store, and other facilities. The pretty mountain lake has boat rental, cruises, a sandy beach, and fishing. Also available are tennis, bicycles, outdoor game areas, and mountain craft instruction. Evenings there are hootenannies and other rousing mountain entertainments. Golf is available nearby. In winter there is cross-country skiing and tobogganing. This area is especially beautiful in spring and summer when wildflowers are profuse. ☐ **Inexp** ☐ **W, F**

MORGANTOWN: Lakeview Inn, *Box 88A, Rte. 6, Morgantown, WV 26505, (304) 292-6311* ☐ *Open all year* ☐ *7 mi N off SR 857* ☐ This comfortable country inn offers some 200 air-conditioned rooms and suites on 400 hilly acres. Many units have refrigerators or private patios. The restaurant serves local and continental fare, and there is a private club with dancing and entertainment. On the grounds are an 18-hole golf course with pro services, a putting green, lighted tennis courts, indoor and outdoor pools, and indoor and outdoor game areas. Also available are boating and waterskiing. The health club has saunas and exercise rooms. In winter there is cross-country skiing. Nearby you can arrange raft trips on the Cheat, New, and Gauley rivers, visit glass works, or see West Virginia University. ☐ **Mod** ☐ **W, F, T, S, F**

SNOWSHOE: Snowshoe, *Box 10, Snowshoe, WV 26209, (304) 799-6762* ☐ *Open all year* ☐ *42 mi S of Elkins on US 219* ☐ In beautiful Monongahela National Forest, in one of the state's most scenic areas, this ultramodern resort community is built on slopes over 4000 feet high, giving most lodgings magnificent views. The cheapest units are double rooms and bunk rooms (sleeping four) at Spruce Lodge and the larger rooms that sleep two to four in Timberline Lodge. But there is also a wide range of more luxurious units with fireplaces in some 200 condominiums that vary from efficiencies to three bedrooms; in three-bedroom chalets and four-bedroom town houses; and in private homes that have between two and four bedrooms. All are available on daily rates or package plans, and most have decks, balconies, or patios. The resort center has restaurant, cafeteria, snacks, bars, and entertainment lounges, ski shop, game room, and other facilities.

Three lifts service the ski slopes, which have trails at all levels–of which the expert Cup Run is the most famous. The ski school has a full schedule of activities, including NASTAR, and the BRR Rabbit Center has programs for ages two to eight and eight to twelve. There are also miles of cross-country trails. In summer the tennis center opens seven all-weather courts with pro services and aids such as ball machines. In addition to adult tennis clinics, programs in a variety of sports (tennis, soccer, canoeing, etc.) are offered ages nine through 15. Canoeing and rafting on the Greenbrier and other rivers is stunning and includes whitewater. The trout fishing and hiking in this area are also spectacular. The Cranberry Glades boardwalk through unique alpine vegetation is also nearby. ☐ **Inexp to Mod** ☐ **S, F, T, W**

WEST VIRGINIA STATE PARKS: For self-sufficient couples or families who want the outdoors, privacy, and low prices, West Virginia's state parks are hard to beat. The parks are wild and beautiful, and the lodgings vary from modern motel-style units to charming stone and log housekeeping cottages. Sometimes (as in Watoga State Park) your cottage may be so far from any other, and so carefully sited, that it's like having your own wilderness retreat. Cabins come in three styles: **Economy** (open Memorial Day to Labor Day weekends) are one-room combining kitchenette, double bunk beds, living area, and bath with shower. These sleep four people. **Standard** (open late April to late October) are log cabins with modern kitchen and baths, and open fireplaces that sleep up to six people. **Deluxe** cabins (may be open all year or closed part of winter) are of modern design with paneled walls, open fireplace, modern kitchen and baths, and heat. All come with cooking and eating utensils and linens, but without maid service. Cots and cribs are available for small charges. In summer, cabins are rented weekly only. Recreational facilities are often seasonal. Most parks have naturalist programs and other free activities. Further information may be obtained through West Virginia Department of Natural Resources, Division of Parks and Recreation, 1800 Washington St. E, Charleston, WV 25305. Reservations should be made through the park or–from certain states–through the toll-free central reservations office in Charleston: From WV, (800) 642-9058. From IN, OH, KY PA, MD, VA, NJ, DE, NC, SC, or Washington, DC, (800) 624-8632.

Blackwater Falls State Park, *Davis, WV 26260 (304) 259-5216* ☐ *Open all year* ☐ *Off SR 32, 5 mi S of Davis* ☐ At the falls, the amber water of the Blackwater River drops five stories to Blackwater Canyon, a deep gorge strewn with boulders. Smaller waterfalls, the Falls of Elakala, are only 100 yards from the main lodge. A footbridge over the falls leads to one of the park's hiking trails. On the canyon rim, the modern 55-room lodge has an excellent view, a large restaurant, gift shop, lounge, and recreation room. Also on the South Rim are 25 Deluxe cabins. Pendleton Lake has a sandy swimming beach and a large slide into the water. Horseback riding and tennis are also available. In winter, there is cross-country skiing. Downhill skiing is nearby at Canaan Valley (see below). ☐ **Inexp** ☐ **H, W, R, S**

Cacapon State Park, *Berkeley Springs, WV 25411, (304) 258-1022* ☐ *Open all year* ☐ *Off US 522, 10 mi S of Berkeley Springs* ☐ Here you have a choice of lodgings. The attractive two-story, red-roofed stone lodge has 50 rooms (some large enough to sleep four) overlooking fairways and mountains. The lodge houses a congenial restaurant and a recreation room. The Old Inn, open only in the summer, is a rustic alternative; it offers 11 low-ceilinged rooms, most of which share baths. Also available are 6 Economy, 12 Standard, and 11 Deluxe cabins. This is beautiful country–6115 forested acres near the Presidential retreat at Camp David–with hiking, camping, picnicking, and fishing–plus an 18-hole golf course designed by Robert Trent Jones and laid out on the gently rolling plain at the foot of Cacapon Mountain. Cacapon Lake has a sandy beach, fishing, and rental of rowboats and paddle boats. Also available are tennis, horseback riding, and lawn sports. ☐ **Inexp** ☐ **G, H, F, W**

Canaan Valley State Resort Park, *Davis, WV 26260, (304) 866-4121* ☐ *Open all year* ☐ *10 mi S of David on SR 32* ☐ The story goes that when the discoverer of this Appalachian mountain valley got his first look at it he couldn't resist quoting the Bible: "Behold! The land of Canaan!" The woods and hills and streams of this 6000-acre state park are undeniably beautiful. The altitude, (3000 feet above sea level) makes for mild summers and early snow in the winter. There's an 18-hole, 6300-yard, par-72 golf course, with pro shop, snack bar, and driving range. In addition, there are six lighted tennis courts and a heated outdoor swimming pool. The park's trails offer challenges for every level of hiker. Backpackers can head for the Otter Creek or Dolly Sods Wilderness Area. The rivers and streams are stocked with bass and trout, and the area is fine for hunting (deer, bear, and wild turkey). In winter there's ice skating in the outdoor rink and cross-country skiing: a ski-touring center operates out of the golf pro shop. Downhill skiing is at the Canaan Valley Ski Area, with two double chairs and one Poma servicing about 20 trails (some lighted), a lodge with cafeteria and pub, ski school, rentals, snowmaking, and nursery. Grass skiing is popular in summer. The attractive, 250-room modern lodge sits on a grassy knoll. It consists of a main building and five smaller ones, all two-story and connected to the main building by covered wooden walkways. Each of these buildings offers a different view of the park. The main building houses the Aspen dining room, a snack bar, lounge, game room, and a gift shop featuring West Virginia crafts. The 15 Deluxe cabins are clustered in the woods just south of the lodge. ☐ **Inexp** ☐ **G, T, S**

Hawks Nest State Park, *Ansted, WV 25812, (304) 658-5212* ☐ *Open all year* ☐ *1 mi W of Ansted off US 60* ☐ This beautiful park has a mountaintop lodge with 20 doubles and five two-room suites (one with fireplace) overlooking scenic New River Gorge. Most rooms offer balconies, many with sweeping views of the gorge or the lake, and all have private bath. Also here are a restaurant and mountainside pool. The lures are beautiful mountain scenery and whitewater rafting (children must be 14 or over) by professional outfitters on the New and Gauley rivers (reservations required). Seasonal aerial tramway takes you 600 feet down the mountain to lakeside activities on the canyon floor–picnicking, fishing, rental boats, a historical museum, and tennis courts. Isolated Economy and Standard cabins hidden in the woods are available at nearby Babcock State Park, where a restaurant and swimming pool are open in summer. (See also Twin Falls State Park.) ☐ **Inexp** ☐ **W, F**

Lost River State Park, *Mathias, WV 26812, (304) 897-5372* ☐ *Open Apr–Dec* ☐ *From SR 259, turn W at Mathias; follow road 4 mi to park* ☐ This mountainside preserve of 3712 heavily wooded acres was once the family summer retreat–called Howard's Lick–of the Virginia Lees. It was Lighthorse Harry Lee himself who built the original cabin on the site, and that cabin is preserved as a museum. Charles Carter Lee added cabins and a hotel. By the time the property was acquired by the Carr family, around the turn of this century, it was a popular and fashionable spa, renamed Lee White Sulphur Springs. The hotel burned down in 1910. The state park now on the site is one of the most rugged and beautiful in West Virginia. All of the accommodations are in cabins–15 Standard and 9 Deluxe–and there are a restaurant, snack bar, two tennis courts, a swimming pool, areas for lawn games, and riding stables. You can ride or hike to Cranny Crow, the park's premier lookout point. A short drive away is Spruce Knob National Recreation Area, where the windy road will bring you to the highest point in West Virginia. ☐ **Inexp** ☐ **H, R**

Tygart Lake State Park, *Grafton, WV 26354, (304) 265-2320* ☐ *Open Apr–Oct* ☐ *From Clarksburg, take US 50 E to Grafton; cross Tygart River to eastern shore; follow road S to park* ☐ The modern 20-room lodge here overlooks 13-mile-long Tygart Lake and has a restaurant and private club. Ten Deluxe cabins are also available. On the lake is a private beach, a marina with rental canoes, waterskiing, sailboats, scuba diving, and fishing equipment. A public golf course is four miles away. ☐ **Inexp** ☐ **W, F**

Twin Falls State Park, *P.O. Box 236, Mullens, WV 25882, (304) 294-4000* ☐ *Open all year* ☐ *Take SR 16 from Beckley to SR 54; at Maben take SR 97 to park* ☐ From the 20 rooms and suites, all with picture windows and most with balconies or patios, this hillside lodge offers wonderful views. Separated for privacy in the woods are 13 Deluxe cabins. The lodge has a restaurant, gift shop, and lounge. If golf is your game, there's a good nine-hole course with pro shop and clubhouse. The 3776-acre park has a large swimming pool, tennis courts, and an archery range. Scenic trails lead to nowhere in particular, but if you must have a destination, head for the falls for which the park was named, or to one of the overlooks, or to the pioneer homestead on Bower's River, restored to its early 19th-century original condition. Nearby are Babcock, Grandview (with regional dramas in summer), and Hawks Nest state parks, the Beckley Exhibition Coal Mine, and whitewater rafting. ☐ **Inexp** ☐ **H**

Pipestem State Resort Park, *Pipestem, WV 25979, (304) 466-1800* ☐ *Open all year* ☐ *On SR 20, 15 mi S of Hinton* ☐ Set in stunning 4000-acre mountain scenery, Pipestem Lodge is an attractive, modern, seven-story building with appealing public areas, restaurant, and terrace. The 113 rooms are comfortably furnished, and many offer glass-wall views and private balconies. In addition, 25 Deluxe cottages (with two, three, or four bedrooms) are set in secluded woods. Guests may use the lodge facilities–shops, game room, sauna, massage, exercise room, heated indoor pool, outdoor olympic-size pool, wading pool, and sitter service. Finally, there is 30-room Mountain Creek Lodge, set by a mountain stream at the bottom of Bluestone River Gorge and accessible only by a scenic aerial tramway that operates April through October. This lodge offers its own restaurant, lounge, and giftshop. The park offers an 18-hole championship golf course, clubhouse with the Rhododendron Room supper club and pro shop, plus a nine-hole, par-three course set in the woods. A park visitor center features a restaurant with panoramic views of Bluestone Gorge, a country store, and Appalachian handcrafts. The recreation center has a cafeteria and snack bar, plus a game room. Park activities include the scenic tramway ride into the gorge, horseback riding and pony rides, hayrides, archery, tennis, mini-golf, shuffleboard, outdoor plays and musicals, hiking, fishing, boating and waterskiing on Lake Bluestone, picnicking, playgrounds, game courts, rental bicycles, and more. ☐ **Inexp** ☐ **G, H, F, W**

Watoga State Park, *Star Route 1, Box 252, Marlinton, WV 24954, (304) 799-4087* ☐ *Open all year* ☐ *1 mi N, then SE on Park Rd.* ☐ This beautiful 10,000-acre mountain park has 8 Deluxe and 25 Standard cabins, all widely separated in scenic wooded areas, plus a restaurant and grocery store. Watoga Lake has fishing and boating, and the park has a swimming pool, tennis, horseback riding, outdoor game areas, and an arboretum with labeled trails. Nearby is the wonderful Cranberry Mountain Visitor Center, with a boardwalk and guided tours of the unique cranberry glades vegetation. Also nearby are Pearl S. Buck's birthplace, now a museum, Droop Mountain Battlefield, and many public recreation areas in state and national forests. ☐ **Inexp** ☐ **W, R**

WHEELING: Wilson Lodge, *Oglebay Park, Wheeling, WV 26003, (304) 242-3000* ☐ *Open all year* ☐ *3 mi NE on SR 88* ☐ In a 1500-acre municipal park, this lodge offers beautiful rustic accommodations in a 125-room lodge (some units have fireplaces), plus 30 cabins with one to four bedrooms, a kitchen, and private outdoor sitting areas. In addition to the restaurant, there are a private club and snack bars. On the grounds are an olympic-size swimming pool, a wading pool, and indoor and outdoor game areas. The park is a center for dog shows, horse shows, outdoor dramas and concerts, and other events. The nature center has guided walks and lectures, and there are a zoo, gardens, an arboretum, and a planetarium. A sports center provides full services for the 18-hole and par-3 golf courses and for tennis. Schenk Lake has boating and fishing. In winter there is downhill skiing with snowmaking and rentals. ☐ **Mod to Lux** ☐ **G, T, W, F**

WHITE SULPHUR SPRINGS: The Greenbrier, *White Sulphur Springs, WV 24986, (800) 536-1110* ☐ *Open all year* ☐ *Off I-64 in White Sulphur Springs* ☐ The Greenbrier is dazzling–6500 acres of beautiful parklike grounds at the foot of the Blue Ridge Mountains. The hotel itself is mammoth, and the facilities are apparently infinite: once you've finished playing one golf course, there's another, and another. What holds it all together is the dignity of the Georgian architecture, the elegance of the formal gardens and huge public rooms, of crystal chandeliers, hand-carved wall ornaments, colonnades, and the atmosphere, everywhere, of the Old, Old South.

White Sulphur Springs began to be popular in 1778, after Mrs. Amanda Anderson was dumped into a hollowed-out tree trunk filled with water from the springs. She jumped out instantly, cured of crippling rheumatism. By 1858 a summer community large enough to support a hotel had grown up around the springs. The first was the Grand Central, known popularly as the Old White. It was a smash; the Civil War came and went, leaving the building, and its popularity, unaltered. It even survived the influx of rich Yankees who poured into White Sulphur Springs in private railroad cars when the Chesapeake and Ohio Railroad extended its line into the town. In 1910 the C & O bought the resort and began to design an addition, which is now the central wing of the Greenbrier. When the Old White was torn down in 1922 the Greenbrier succeeded it as the area's uncontested premier resort.

There are three golf courses: the Lakeside (par 70, 6289 yards); the Old White (par 70, 6368 yards) and the Greenbrier, redesigned for the 1979 Ryder Cup Matches by Jack Nicklaus (par 72, 6721 yards). All three courses begin and end at the Club House, which contains not only the pro shop and locker rooms, but also the informal terrace dining room, site of the Greenbrier's gorgeous lunch buffet. The tennis facilities are equally elaborate. Five Har-Tru courts are near the golf clubhouse, ten more are up the hill aways, and five Dynaturf courts are in a lavish indoor tennis center with viewer's gallery, pro shop, and videotape facilities. The two swimming pools, both enormous, are a kidney-shaped one by the golf club and a spectacular rectangular indoor pool with a colonnaded terrace. The hotel also has a modern eight-lane bowling center, platform tennis courts, and a stable for the use of guests who'd like to explore the 200 miles of mountain trails on horseback (and for those who'd just like to take a leisurely carriage ride around the grounds). The Greenbrier Gun Club–you are driven to it in a limousine–is a completely equipped trapshooting and skeet-shooting center, with a pro shop and instruction.

The Greenbrier Spa, in the north wing of the hotel, offers mineral baths, massage, saunas, whirlpool, and Scotch sprays. Before or after your treatment you can earn your aerobic points in the swimming pool, exercise room, or by walking, jogging, or bicycling on the measured trails the hotel has mapped out around the grounds. One of the most unusual services the spa offers is the Morning After Preservation–a steam bath, sauna, and massage designed to aid recovery from overindulgence the night before. The Greenbrier is famous for its cuisine–there's a cooking school on the premises–and breakfast and dinner are included in the room rates. Lunch can take a variety of forms: the buffet at the Golf Club or the Old White Club, or a snack at the recently added coffee shop and ice-cream parlor. In addition, there's tea every afternoon in the main lobby, served to the strains of Viennese waltzes played by the Greeenbrier's own chamber music group. The Old White Club and Tavern serves late dinners, and offers cocktails, dancing, and entertainment. There's also music for disco and contemporary dancing, well into the night. Separate meals, by the way, are served for children, as part of the optional children's program, which is in operation Mondays through Saturdays from June 15 to Labor Day. Parents can sign their kids up for all or part of the day's activities.

The rooms are comfortable, spacious and all you'd expect from a super-luxury hotel. Each is decorated differently, but always tastefully, with color-coordinated furniture, draperies, and wall to wall carpeting. Housekeeping and service are impeccable. Maids visit the rooms three times a day. The last time is to turn down your covers. If you're feeling flush you might want to rent the split-level Presidential Suite for a little elbow room. It has a living room, den, office, two kitchens, and seven bedrooms, each with its own bath. The Valley View Estate House is another good choice for big spenders: you'll have to put up with only one kitchen and five bedrooms, but a couple of hundred dollars are shaved off the daily rate, and a car is included. Cottage rentals are also available, at slightly more reasonable tariffs.

The Greenbrier prides itself on having at least one employee for every guest, and this personalized service, along with the scaling-down effect of the careful interior design work done since the war, gives it comfort that matches its splendor. It is a charming giant. ☐ **Lux** ☐ **G, T, R**

Wisconsin

CABLE: Eagle Knob Lodge, *Lake Owen Rd., Cable, WI 54821, (715) 798-3297* □ *Open late May to early Sept* □ *County M 2 mi E to Lake Owen Rd. N to resort* □ This comfortable resort on Lake Owen has 4 lodge units and 11 one- to four-bedroom cottages; some feature fireplace, color TV, or balcony. A sandy beach provides a swimming area, and guests can rent rowboats, sailboats, canoes, ski boats, motors, and guide service at the dock area. Also on the grounds are a recreation room, tennis, and a playground. The dining room, open to the public by reservation, serves three meals a day and has beer and wine. □ **Mod** □ **W, F**

Telemark Lodge, *Cable, WI 54821, (715) 798-3811* □ *Open all year* □ *From US 63 in Cable, take County M 3 mi E to lodge* □ Although the lodge has a sleek, modern design, a great deal of native stone and varnished or exposed wood was used to give it a countrified, hunting-lodge atmosphere. Especially impressive is the lobby's stone fireplace, a towering 55-foot-high, 155-ton pile of rock. Since the two long lodging wings stretch away from the large, central lobby, most of the 200 studio rooms have uninterrupted views of the woods and golf course. Each room has a queen-size bed and queen-size sofa bed, a kitchenette without oven, two vanities, color TV, and private balcony. More luxurious are the 28 two- to four-bedroom townhouses in a wooded setting not far from the lodge. All have fireplace, sundeck, dishwasher, washer, and drier, two baths, and living room.

On the 2000 acres surrounding the lodge is the ski area, Mount Telemark. Though the vertical drop is only a modest 375 feet, the "mount" has 10 trails—the longest 2600 feet—and a base lodge with snack bar, rentals, repairs, nursery, and so on. The excellent cross-country ski area has been used as a training area by the U.S. Olympic team. Movies, ice skating, and sleigh rides are also offered in winter. In summer, pro shops and instruction are available on the 18-hole golf course and four tennis courts. Also on the property are a riding stable, 3700-foot airstrip, indoor and outdoor pools, health club with sauna and whirlpool, and a shopping mall. The resort rents canoes and has pick-up service for those who want to go down the Namekagon River, part of the St. Croix National Scenic Riverway. The river and the streams that feed it are also good for trout fishing. Adjoining Chequamegon National Forest has hiking trails and roads for bicycling.

The lodge has two dining rooms and a coffee shop, providing everything from a quick lunch to a gourmet dinner with fine wines. The lounge is known for the jazz greats who have appeared there—Harry James, Buddy Rich, and Stan Kenton. ☐ **Mod to Lux** ☐ **S, T, G, R**

DELAVAN: Lake Lawn Lodge, Box 527, Delavan, WI 53115, (414) 728-5511 ☐ *Open all year* ☐ *On SR 50, near jct SR 15* ☐ This cluster of interconnected hotel and lodges sits on lawns at the edge of Lake Delavan. The motel-style lodges are modern-rustic with much glass and wood, and provide 364 rooms and 22 suites. Many units have balconies or patios facing shaded lawns or the swimming pool courtyard, and some face the lake. Some units have sleeping lofts or fireplaces, and in summer 18 kitchenettes are available. Rustic appointments also characterize the public rooms—a lakeside sun lounge, the dining room with continental menu, the Frontier restaurant with cheerful gingham tablecloths and large windows over the lake, the coffee shop, and the cozy lakeview bars. Evidently, this site was once sacred to the Indians, for about 40 burial mounds dot the property. Arrowheads, spear points, and other artifacts are displayed in the dining room.

Most of the 278 acres is devoted to the 18-hole golf course surrounding the lodging complex. The clubhouse has a pro in attendance and rents clubs and carts. Tennis players have seven lighted courts, and an array of miniature golf and other outdoor game areas as well as pool tables and other indoor games is provided. Two indoor pools, one surrounded with glass walls, complement a health club with steambaths, saunas, and whirlpool. On the lake are another swimming area and three docks with canoes, motorboats, waterskiing, waterbikes, and other boats. Seasonal horseback riding, supervised play for children, dancing on outdoor terraces, and evening entertainment are also provided. In winter there is tobogganing, and ice skating on the pond, and you can rent snowmobiles. The lodge also has a number of shops and a 4400-foot airstrip. ☐ **Mod to Lux** ☐ **T, G, R, W**

DOOR COUNTY: The Door County peninsula, stretching about 40 miles from Sturgeon Bay to Gills Rock, cuts into Lake Michigan, forming Green Bay on its northwestern side. Six miles off the tip of the peninsula is Washington Island. The county gets its name from the strait separating the island from the peninsula that French explorers labeled *Portes des Morts* or "Death's Door." Door County has miles of scenic shoreline with bluffs, dunes, and swimming beaches, as well as picturesque fishing ports that provide fishing piers, full-service marinas, and fishing charters. Scenic boat cruises leave from Sturgeon Bay and other ports, and ferries to Washington Island dock at Ellison Bay. Settled largely by Scandinavian fisherfolk, Door County retains many reminders of its heritage—shipbuilding and historic museums plus a pioneer farm at Sturgeon Bay, a lighthouse museum at Peninsula State Park, an Indian museum at Egg Harbor, a maritime museum at Ellison Bay, a Norwegian chapel at Baileys Harbor, and pioneer Moravian displays at Ephraim. Washington Island, settled by Icelanders, has a museum, and Rock Island State Park (reached on cruises from Washington Island or Gills Rock) has interesting Icelandic-style buildings. Fish boils, Scandinavian festivals, parades, fireworks, and other events are held June through October, and there is summer theater and a music festival at Fish Creek.

At Sturgeon Bay there are caves in limestone bluffs at Cave Point State Park, and the 1100-acre Potawatomi State Park has fishing, boating, an observation tower, rope tows for downhill skiing, and cross-country trails. Kangaroo Lake at Baileys Harbor has inland swimming, fishing, and boating, and there is a flower preserve nearby. Newport State Park at Ellison Bay also has swimming and cross-country ski trails. The 3700-acre Peninsula State Park, between Fish Creek and Ephraim, offers nine miles of sandy beaches and bluffs along Eagle Harbor. In summer this park has fishing, swimming, boating, and riding plus an 18-hole golf course. In winter there are trails for cross-country skiing. Just south of Fish Creek is the Omnibus Nordic Ski Center with full facilities for cross-country skiing on trails that cover 300 acres and are lighted at night. A double chair, a T-bar, and rope tows are provided for skiing. All year there is hang gliding here.

Alpine Resort, *Egg Harbor, WI 54209, (414) 868-3000* ☐ *Open Late May to mid-Oct* ☐ *¾ mi S off County G* ☐ This pleasant family resort on 400 acres on Green Bay offers 60 units in a three-story alpine lodge plus 30 cottages with one or two bedrooms and screeened porches. Rates in high season are AP (EP off season), and the lounge provides beer and setups. The clubhouse provides 27 holes of golf, putting green, tennis, and rentals. There is a sandy swimming beach, and you can rent bikes as well as boats for fishing and excursions. Stables are nearby. Hayrides, bonfires, dancing, and evening entertainment are also offered. ☐ **Inexp to Mod** ☐ **W, F, G, T**

Bay Shore Inn, *4205 Bay Shore Dr, Sturgeon Bay, WI 54235, (414) 743-4551* ☐ *Open late May to late Oct* ☐ *3 mi N on Bay Shore Dr.* ☐ On a wooded knoll on Sturgeon Bay, this family resort has seven air-conditioned rooms with shared baths in the old white-frame homestead, 10 motel rooms with showers, and six A-frame cottages with wooden decks and glass fronts that are nestled in the woods. Most units have a bay view. The dining room in the main lodge has early American decor, and the comfortable lobby has a fireplace. Home-grown vegetables, home-baked breads, and homemade desserts are featured, and there's a weekly barbecue. On the spacious, wooded grounds are a recreation building, tennis court, archery, playground, and game areas. At the waterfront is a small beach, a slide, free rowboats and sailboats, a dock, a waterskiing boat, and a 22-foot sloop for cruises. EP is offered all season and MAP in high season. ☐ **Mod to Lux** ☐ **F, W**

Cliff Dwellers, *3540 Duluth, Sturgeon Bay, WI 54235, (414) 743-4260* ☐ *Open Apr–Nov* ☐ *From SR 57/42 just before crossing bay, take Duluth Ave.*

(County C) N 2 mi to resort ☐ Overlooking Sturgeon Bay in front and a pleasant wooded hillside and moss-draped ledges out back, this appealing motor lodge offers tasteful and extra-comfortable accommodations close to Potawatomi State Park. The 16 rooms in the lodge have floor-to-ceiling windows, recessed private balconies with bay views, carpeting, dual vanity baths, and color TV. The 14 individually decorated chalets suites have kitchenettes, one to two-and-a-half bedrooms, color TV, and maid service; some have screened porch or fireplace. The 12-acre grounds are nicely landscaped with fieldstone paths, flower beds, and manicured lawns. The small, kidney-shaped, heated pool is surrounded by a sundeck and is on the water's edge. Rowboats, fishing boats, charters, and motors are available at the lodge's dock. Although the Cliff Dwellers has no restaurant, continental breakfast and hot sandwiches are available at shaded tables on the lawns or in your room. Honeymoon and anniversary packages are offered. ☐ **Mod to Lux** ☐ **W, F**

Country House, *715 Highland Rd., Sister Bay, WI 54234, (414) 854-4551* ☐ *Open all year* ☐ *½ mi S on SR 42* ☐ On five heavily wooded acres on the bay, this two-story motel has 16 air-conditioned rooms with balconies. You can swim and fish in the bay, and Peninsula State Park is five miles south. ☐ **Mod** ☐ **F**

Edgewater Motel, *Ephraim, WI 54211, (414) 854-2734* ☐ *Open early May–Nov; a few units all year* ☐ On a beach overlooking Green Bay, this two-story motel has 34 air-conditioned rooms and suites, a restaurant, and picnic tables and grills. ☐ **Inexp to Mod** ☐ **F**

Ephraim Shores, *P.O. Box 166, Ephraim, WI 54211, (414) 854-2371* ☐ *Open late May to late Oct, Dec 27 to New Year's, and weekends in Nov and Feb* ☐ *On SR 42* ☐ In town, this modern, two-story shingled motel offers 46 paneled rooms, some of which sleep four people; some units face Eagle Harbor, some have kitchenette, some have private balconies. The Harborside restaurant features Norwegian meatballs, broiled whitefish, Moravian sugarcake, and carmel rolls. The game room, with pinball and pool table, indoor swimming pool, and whirlpool are in large rooms. ☐ **Mod**

Glidden Lodge, *Rte. 3, Sturgeon Bay, WI 54235, (414) 743-4944* ☐ *Open late May-late Oct* ☐ *8½ mi N on SR 57 to Valmy, then 2 mi E on County T, then follow signs* ☐ Secluded among towering pines on the Lake Michigan shore, this resort offers 18 pine-paneled rooms with lake view in a lodge and motel as well as six nicely spaced wooden cottages with two or three bedrooms and kitchens. You may book here on EP or MAP; the dining room serves family-style breakfast and dinners, and there is a bar. There is a wide sandy swimming beach and a dock for fishing, charters, boat rental, and waterskiing. Also on the grounds are patios and lawns with lounging chairs, tennis, a putting green, a playground, and rental bikes. Golf and riding are available nearby. This is a good buy for families. ☐ **Inexp to Mod** ☐ **F, W**

Gordon Lodge, *Bailey's Harbor, WI 54202, (414) 839-2331* ☐ *Open early May–late Oct* ☐ *7 mi N via SR 57, County Q* ☐ On North Bay, with a beautiful view of the lake, this is one of the nicest resorts in Door County. The lodge has 44 unusually large, attractive, and comfortable units, half in the main lodge and half in one- and two-bedroom cottages. Many units have fireplaces or private patios, and all are air conditioned and have TV. Rates are EP early in the season and MAP after the restaurant opens in June. The restaurant is open for lunch as well. The food is very good here. The cocktail lounge has seasonal dancing and entertainment. On the spacious wooded grounds bordering the sandy swimming beach are a pool with poolside service, putting green, tennis, playground, and game areas. At the marina you can fish, go waterskiing, or rent boats or waterbikes. Fishing charters are available, and stables are nearby. ☐ **Mod** ☐ **F, W**

Helm's Four Seasons Motel, *Sister Bay, WI 54234, (414) 854-2356* ☐ *Open all year* ☐ *Off SR 42, N of town* ☐ Overlooking the bay, this motel offers 22 air-conditioned rooms, all with private balconies or patios and floor-to-ceiling windows giving lawn and water views. Continental breakfast is included in the rates, and a restaurant is within walking distance. From spring to fall six housekeeping cottages are also open. In addition to the private swimming beach, which has a sundeck, there is fishing and charter boat service. In winter the motel rents snowmobiles and there is good cross-country skiing around the bay. ☐ **Mod** ☐ **F**

Hotel du Nord and Laura Maxwell Inn, *P.O. Box 68, Sister Bay, WI 54234, (414) 854-4221* ☐ *Open early May–March* ☐ *1 mi N on SR 42* ☐ Although run by the same people and right across the street from each other, these two lodgings are quite different in style. The Laura Maxwell, a modern, two-story brick motel, is set on the water's edge, giving all 16 rooms a view of Green Bay. The lounge has a fireplace and in front, by the seawall, are lounging chairs and a sun deck on the water. Next to the inn are two beach villas with refrigerator and private deck overlooking the bay. The shingled Hotel du Nord has a much more rustic feel than the inn, and the high-ceilinged, open-beam lobby is dominated by a huge stone fireplace. Off the lobby are the Back Room, the charming luncheon area, and the main dining room with a steak and seafood menu. Cocktails and wines are available. The hotel has rooms of varying sizes that will sleep up to four people. The adjacent lodge has doubles reached from common balconies that ring both levels. Three nearby cottages have fireplace, refrigerator, living room, and one or two bedrooms. All units are heated and have private bath and maid service; all units except those in the main hotel have air conditioning and TV. At the waterfront there is swimming and a dock with complimentary rowboats. ☐ **Inexp to Mod** ☐ **W**

Leathem Smith Lodge, *Sturgeon Bay, WI 54235, (414) 743-5555* ☐ *Open all year* ☐ *1640 Memorial Dr.* ☐ Set on the bay, this lodge offers 56 units that vary from motel-style doubles to suites with sunken living room; most units have a private balcony or patio. Six air-conditioned cottages, three set in a Japanese garden, have between one and three bedrooms. Four candlelit dining rooms specialize in local fish dishes, and there is an outdoor patio lounge as well as an indoor cocktail bar with dancing and entertainment. On the grounds are a huge swimming pool, with a sunning apron and a game room to one side, plus tennis courts, a playground, shuffleboard courts, and a driving range. For an extra fee you can play the nine-hole golf course. Bike rental is available. At the waterfront is a 300-foot pier and yacht basin where waterskiing, charter boats, and fishing guides are available. ☐ **Mod to Lux** ☐ **F, W, T**

The Mariner Motel and Cottages, *Egg Harbor, WI 54209, (414) 868-3131* ☐ *Open May to late Oct* ☐ *On County G, 4 mi W of jct SR 42* ☐ Isolated in 60 acres of heavy woods on Green Bay, this motor inn offers 22 doubles in a one-story motel overlooking lawns and the swimming pool. In a newer two-story shingled lodge are eight spacious air-conditioned units with private decks overlooking the bay; four of these have kitchenettes. Four two- and three-bedroom housekeeping cottages are secluded under trees near the shore and are rented by the week without maid service. Bicycles, picnic tables, grills, lake swimming, and fishing are available on the grounds. Tennis, golf, a municipal fishing pier, boating, riding, and a restaurant are nearby. ☐ **Mod** ☐ **F**

Maxwelton Braes Resort, *Baileys Harbor, WI 54202, (414) 839-2321* ☐ *Open early June to late Oct* ☐ *1½ mi S on SR 57* ☐ On the lovely grounds bordering the 18-hole golf course is a handsome lodge with 16 rooms. And scattered in the woods are another 28 units in one- and two-bedroom cottages with refrigerators. Rates are MAP, and the dining room does not serve lunch. The bar has seasonal dancing and entertainment. On the grounds are a heated pool and game areas. Tennis on lighted courts, and the golf course, are extra. ☐ **Inexp to Mod** ☐ **G, T**

The Shallows Resort Motel, *Egg Harbor, WI 54209, (414) 868-3458* ☐ *Open Apr–Oct* ☐ *2 mi S on County G* ☐ On Green Bay, this modern, comfortable complex consists of 18 attractive rooms and suites in a two-story wooden lodge with private sundecks facing the water. Rates include continental breakfast. Also available are four kitchenette cottages under trees. On the grounds are a swimming pool, beach, picnic tables, and grills. Tennis, bikes and boats are included in the rates, and sailing lessons are available. Within driving distance are stables, golf, and restaurants. ☐ **Mod** ☐ **F, W**

EAGLE RIVER: This little river town is the hub of activity for the famed Eagle chain of 28 lakes that are navigable for more than 20 miles. Canoeing and houseboating are popular, and the fishing is great. Ten miles east of town is million-acre Nicolet National Forest, with public recreation sites on additional trout streams, rivers, and lakes. In winter the emphasis is on cross-country skiing and snowmobiling, and there are two downhill areas. Newcomers will find it profitable to stop in at the Trees for Tomorrow center in town on the riverbank. An environmental camp, it offers a self-guiding ecology trail and special programs on subjects such as wilderness survival, canoeing, and cross-country skiing.

Chanticleer Inn, *Rte. 3, Eagle River, WI 54521, (715) 479-4486* ☐ *Open all year* ☐ *1½ mi E off SR 70* ☐ On a wooded bend of Voyageur Lake, one of the 28 in the Eagle chain Chanticleer's cheapest accommodations are 21 rooms with refrigerators in a two-story motel and six wooden cottages with fireplaces, kitchens, wall to wall carpeting, and one to three bedrooms. More expensive are some 50 three floor town house units with kitchen, dining area, and fireplace on the ground floor, and studios and bedrooms upstairs; the complete unit sleeps up to ten people but may be subdivided. Most units have paneled and have balconies, but most are screened behind trees and lack lake views. There is also a cocktail lounge.

On the lake are two sandy swimming beaches and docks with canoes, pontoons, sailboats, waterskiing, and fishing boats (guides are available too). In addition to the usual lawn games and recreation room, archery and two tennis courts (one lighted) are provided, and there is a nine-hole golf course. In winter you can go cross-country skiing, ice skating, tobogganing, or snowmobiling. You have a choice of booking on meal plans that also include golf and other sports, of paying for everything separately, or of packages. ☐ **Inexp to Mod** ☐ **W, F, G, S, H**

Earle Waters, *Box 730, Eagle River, WI 54521, (715) 479-4411* ☐ *Open all year* ☐ *6 mi E off SR 70* ☐ Beautifully set on lakeside on the Eagle chain of lakes, this resort offers 100 comfortable units of varying sizes, some with refrigerators, fireplaces, private balconies, or patios, and lake views. Most units are rented on AP rates and in addition to a dining room there are outdoor barbecues, poolside service, a lounge with dancing, and entertainment planned by a social director. On the spacious grounds are a swimming pool, putting green, tennis courts, and outdoor game areas. At the lake, you can swim, fish, or rent canoes, sailboats, or other boats, or take waterskiing lessons. Golf can be arranged. Other services include bicycle rental and children's programs. In winter you can rent a snowmobile. ☐ **Mod to Lux** ☐ **W, F, T**

White Eagle Motel, *Eagle River, WI 54521, (715) 479-4426* ☐ *Open all year* ☐ *½ mi W on SR 70, 17* ☐ Strung out along the shore of Eagle River is a strip of one-story motel rooms plus three cottages with kitchenette that will sleep up to six people. The best thing is the view—the swimming pool and fishing dock overlook a beautiful bend in the river, overhung with trees. You'll have to walk or drive to a restaurant, but on the premises are picnic tables and grills. Canoes and other boats are available. ☐ **Inexp to Mod** ☐ **W, F**

FONTANA: The Abbey, *Fontana, WI 53125, (414) 275-6811* ☐ *Open all year* ☐ *Off SR 67* ☐ This established resort on the western shore of Lake Geneva, has a glass-walled A-frame main lodge flanked by 2-story wings containing 300 air-conditioned rooms, each with air conditioning, color TV, and private patio or balcony. Some units have wetbars or steam baths. La Rondell is the family dining area; La Tour de Bois is for gourmet dining. The Lower Deck and La Rondelle's lounge offer dancing and live entertainment in season. AP, MAP, and packages are offered.

The Health Club facilities are excellent–indoor pool, sauna, whirlpool, massage, and exercise room. In summer, you will find plenty to do around the outdoor pool with service terrace, beach, and marina, where you can rent boats for fishing and waterskiing. Apart from bowling, tennis, outdoor games, and bicycling, you can play the nearby Abbey Springs 18-hole golf course. In winter, the resort has snowmobile rentals, tobogganing, ice skating, iceboating, and ice fishing. Supervised play for children, shops, and other services are provided. ☐ **Lux** ☐ **W, F, G**

GREEN LAKE: Heidel House Resort, *P.O. Box 537B, Green Lake, WI 54941, (414) 294-3344* ☐ *Open all year* ☐ *On SR 23 2 mi E of Green Lake* ☐ This resort is perhaps best known for its restaurant, housed in a gracious old mansion on the shores of Green Lake. Highlights of the authentic German and Swiss menu include veal Oscar, seafood, and fondue, a specialty of the Fondue Chalet dining room. The Rathskellar offers dancing and entertainment. Guests are accommodated in 78 doubles and five suites in a modern, wood-sided motor lodge or in three elegant cottages, including the 10,000-square-foot Grey Rock. Rooms with kitchenette or fireplace are available; all have color TV and air conditioning.

Spring-fed Green Lake, 7½ miles long and 2½ miles wide, is Wisconsin's deepest lake. The resort's yacht, "Yacht's of Fun," has breakfast, lunch, and dinner cruises. Besides canoes, rowboats, and skiing boats, fishing boats may be rented or chartered to go after the lake's trout, pan fish, or pike. The well-known Lawsonia Links championship golf course is across the street and two other courses are within seven miles. The resort has its own tennis court. The Green Lake Center has 1100 acres of cross-country skiing terrain and marked trails. The heated indoor pool and sauna are great places to relax sore muscles. ☐ **Mod** ☐ **W, F, S**

HAYWARD: Ross Teal Lake Lodge, *Hayward, WI 54843, (715) 462-3631* ☐ *Open early May through Oct* ☐ *21 mi E off SR 77, then 1 mi N* ☐ Widely spaced along the shores of Teal Lake, these two dozen nice cottages have one to four bedrooms. Most have livingrooms and screened porches, half have kitchenettes, and a few have fireplaces. AP and EP are available, the dining room serves three meals, and cocktails are available. On the lake is a swimming beach and dock for boat rentals. Fishing guides can be arranged and your catch will be frozen. (That the fishing is fine here is shown by the fact the National Fresh Water Fishing Hall of Fame is in town–part of the complex is a 4-story high muskie, and there are mounted and live fish displays.) Guns are provided for the rifle range and trapshooting, and there is seasonal hunting. ☐ **Mod** ☐ **W, F**

LAC DU FLAMBEAU: Dillman's Sand Lake Lodge, *Lac Du Flambeau, WI 54538, (715) 588-3143* ☐ *Open mid-May to late Oct and New Year's week* ☐ *3 mi N on County D* ☐ This hamlet, in the center of the Lac de Flambeau Indian Reservation, was named for the Chippewa practice of attracting fish at night with flaming torches. The lodge complex is beautifully designed and landscaped so that all facilities are widely spaced and screened by thick woodlands on the 250-acre peninsula on Big White Sand Lake. In the modern lakeside lodge and annex are 11 paneled comfortable doubles, studio rooms, and suites. The annex has a lounge with refrigerator and TV. Scattered along the peninsula shores are 18-wooden cottages that vary from bedroom/living room combinations to one- to three-bedroom units with separate living rooms; virtually all have lake views, and may have fireplaces, patios, screened porches, or other extras. Rates are EP, MAP, or family plans. The waterview dining room serves three meals daily. Breakfast includes blueberry and pecan pancakes. A lunch will be packed if you want to stay out all day. Dinner entrees include walleye pike, and there is Sunday buffet. In summer steak cookouts are featured. Cocktails, wines, and beer are available.

Extensive on-premises facilities include three beaches for swimming and scuba, one with gradual slope for children and another with a steep drop-off for divers and waterskiers; three tennis courts in quiet wooded area; archery range and free equipment; outdoor game areas; free waterskiing rides and lessons; game room with TV; driving range; and hiking trails with a good chance of seeing wildlife. At the full-service marina you can launch your own boat free or rent canoes, sailboats, kayaks, or fishing boats. In addition to stocking by the state, the tribe maintains its own hatchery and stocks waters in the reservation. Fishing guides and supplies are available and your catch will be filleted and frozen at no charge. From May to September the lodge sponsors workshops in the fine arts, photography, and nature study. Trap and shooting ranges are on the property, and in fall there is hunting. In winter trails are marked and groomed for cross-county skiers and snowmobiles, and the lodge provides instructors, guides, and rentals. Off the premises but nearby you can rent horses, golf at the Minocqua and Trout lake courses, take canoe trips and seaplane rides. Indian powwows are frequent in Lac Du Flambeau. This resort, which has been operated by the same family for about half a century, provides many services–supervised children's activities in summer, bicycle rental, get-acquainted cocktail parties and other events, free newspapers, lectures on history, airport pickup, and continental breakfast delivered to your cabin if you are a late sleeper. ☐ **Mod to Lux** ☐ **W, F, S, H, T**

LAKE GENEVA: Interlaken, *Lake Geneva, WI 53147, (414) 248-9121* ☐ *Open all year* ☐ *3 mi W of Lake Geneva on SR 50* ☐ In addition to the 144 contemporary double rooms in the three-story lodge, this EP resort offers 150 villa suites, each with kitchen, fireplace in the living room, and patio overlooking Lake Como. Set amid 90 acres of woods, gardens, and landscaped grounds are 5 all-weather tennis courts, 1 indoor and 2 outdoor pools, health club with whirlpool and sauna, and recreation room. Boat, waterskiing, and snowmobile rentals are available. Interlaken's dining and drinking spots frequently have dancing and entertainment. □ **Mod** □ **T, W**

Playboy Lake Geneva, *Lake Geneva, WI 53147, (414) 248-8811* □ *Open all year* □ *At jct of US 12 and SR 50* □ This long, low-rise, 350-unit hotel stretches along a hill overlooking the beautiful fairways, willow trees, and water hazards of one of the golf courses. The redwood balconies and window walls of the rooms give the building a nice balance between sleekness and rusticity, a proper pose for this sophisticate in the woods. Many units have kitchenettes. The lobby has impressive multi-level terraces of aggregate concrete overgrown with greenery. As you might expect from a Playboy resort, Lake Geneva has many dining, drinking, and entertainment areas. The big showroom regularly presents "name" stars as well as musical groups, Broadway revues, and other acts. The Sidewalk Cafe, the VIP Room, and Playboy Club dining areas–serving everything from the hearty to exotic–are all overseen by a bevy of bunnies. There's even an ice cream parlor for the kids. After dinner and a show, there's a disco that goes until the wee hours, and to recuperate from all that late night activity, there's a health club complete with indoor pool, sauna, massage, and exercise rooms. Guest rooms here are ultramodern, with stripes on the drapes and bedspreads and chrome plating on the furniture. The VIP suites have patio, wetbar, and fireplace.

The two 18-hole golf courses, the Brute and the Briar Patch, were designed by Harris, Fazio, and Dye, with Nicklaus and Sanders as consultants. Other warm weather facilities on the 1400 acres include eight tennis courts, racquetball courts, large outdoor pool, horseback riding, skeet and trapshooting, archery, and swimming, boating, and sailing on the small private lake. In winter, the resort opens its own ski area with three double chairs and rope tows, an attractive chalet with snacks and drinks before two massive fireplaces, lessons, and rentals. Also available are snowmobile rental, ice skating, sleighing, and cross-country trails. Social and recreation directors supervise adult and children's activities, and there is a day care center. The many activities include plane rides. Rates are EP. □ **Lux** □ **T, G, S**

LAND O' LAKES: Forest Lodge, *Star Route, Land O'Lakes 54540, (715) 385-2128* □ *Open mid May through Mid-Oct* □ *20 mi W on County B* □ Strung along the pretty wooded shore of High Lake, this AP resort has 14 one- to four-bedroom cottages with living rooms, refrigerators, screened porches, fireplaces, and private docks. The dining room will pack a lunch if you want to be out all day. There is a recreation room and sometimes movies or other entertainment is provided. You may swim at the sandy beach or rent a pontoon, sailboat, paddleboat, motorboat, or go waterskiing. Fishing Guides are available. ☐ **Mod to Lux** ☐ **F, W**

Sunrise Lodge, *Star Route, Land O'Lakes, WI 54540, (715) 547-3684* ☐ *Open mid-May through Oct* ☐ *2 mi S on US 45, 3 mi E on County E, then 1 mi N* ☐ This tiny resort is on beautiful Lac Vieux Sesert, which extends into Michigan, surrounded by national forest. It has only 15 one- to three-bedroom frame cottages, some paneled in knotty pine, and some more modern. All have separate living rooms and refrigerators, and four have kitchenette. Half a dozen of these units are on the lakefront and the rest are set back in the woods. The one-story white-shingled main lodge contains the recreation room and dining room. Depending on the cottage and the season, you can check in here on plans that include three meals and fish cleaning services, or with optional meals and fish cleaning. This family-oriented resort does not have a bar or scheduled activities, but there are informal get-togethers at the Sunday evening cookouts, berrying, or a mine tour, and babysitting, organized play, and other services will be supplied on request. On the lake is a private beach, the dock has canoes and fishing boat rental, and guides are available. Other facilities include tennis, badminton, other game areas, a jogging trail, and a playground. ☐ **Mod** ☐ **F, W**

MINOCQUA: This town, gateway to the Northwoods lake district, calls itself the "Island City" because it is on an island in Lake Kawaquesaga. At the edge of Northern Highlands–American Legion State Forest, it has the same fine muskie fishing and watersports. The town has a zoo and a wildlife park, both with animal contact areas.

Driftwood Lodge, *Rte. 2, Minocqua, WI 54548, (715) 356-6080* ☐ *Open mid-May to late Oct* ☐ *5 mi NW via US 51, then 4 mi S on Mercer Lake Rd.* ☐ On Mercer Lake, these 11 attractive and comfortable cottages are offered on MAP. Each has a living room, refrigerator, and one to four bedrooms, and some have fireplaces. The restaurant serves breakfast and dinner and features a weekly outdoor barbecue, and there is a bar. The recreation room has dancing. In addition to archery and outdoor games, there are bicycles, an outdoor pool, and a swimming beach. Rentals, waterskiing, and guides are available at the dock. ☐ **Inexp** ☐ **F, W**

Jansen's Squirrel Lake Lodge, *Box 205, Rte. 1, Minocqua WI 54548, (715) 356-3442* ☐ *Open May–Oct* ☐ *8 mi NW on US 51, SR 70, then follow signs 5 mi S* ☐ On Squirrel Lake, this AP resort has eight basic one-bedroom cottages with refrigerators. The three-meal-a-day restaurant serves beer. A recreation room, playground, and outdoor game areas supplement the lakeside swimming beach and dock with rentals and guides. ☐ **Mod to Lux** ☐ **F, W**

Northern Resort, *Box 239, Rte. 2, Minocqua, WI 54548, (715) 356-6150* ☐ *Open late May to late Sept or Oct* ☐ *2½ mi NW off US 51 at 8293 Northern Rd.*

☐ Set on a 16-acre wooded peninsula between lakes Minocqua and Kawaquesaga, this resort has 20 rooms (two share bath) in an early American-style white-frame lodge. In the woods are a half-dozen nice cottages with early American furnishings (including some antiques) two to four bedrooms, and screened porches. Four cottages have kitchenettes. These are EP, as only snacks are available here and you must drive to a restaurant. Apart from archery, outdoor game areas, bicycles, and a swimming beach with rafts, there is a marina with boat rental, waterskiing instruction, and guides. ☐ **Inexp to Mod** ☐ **F, W**

MISHICOT: *Fox Hills Resort, Box 129, Mishicot, WI 54228, (414) 755-2376* ☐ *Open all year* ☐ *11 mi N of Manitowoc on County B* ☐ This modern motel-style resort has 154 comfortable air-conditioned rooms and suites with private patios and balconies. The restaurant offers steak and seafood and the bar has dancing and entertainment. On the premises are a health club with whirlpool and sauna, an indoor pool, and a recreation room. Outdoors are a putting green, badminton and tennis courts, and 27 holes of golf with clubhouse services. In winter there is cross-country skiing here, and a few miles away is Hidden Valley downhill ski area. This has a vertical drop of 200 feet, chairlift and rope tows, snowmaking, ski school, rentals, and food and beverage service. ☐ **Mod to Lux** ☐ **G, T, S**

NORTHERN HIGHLANDS–AMERICAN LEGION STATE FOREST: This beautiful area of hundreds of lakes rimmed with evergreen forest is one of Wisconsin's prime vacation meccas. Boulder Junction alone claims 194 lakes within a nine-mile radius, and Manitowish Waters has a chain of ten lakes you can navigate without portage. Because state ownership has prevented development, many lakes have pristine shorelines and beautiful beaches for swimmers. The crystal-clear waters are popular with skin and scuba divers. Canoeing and boating are naturally big here, and you can take guided trips or stay overnight at many campsites along the water trails. Whitewater trips are given on the Flambeau River west of Manitowish. The fishing is famous. Boulder Junction, which calls itself the "Muskie Capital of the World," holds an annual jamboree in mid-August where muskie stuffed with wild rice and baked over barbecue pits are served between parades, sky diving exhibitions, and other events. Manitowish Waters also has a muskie festival (Sept) and waterskiing shows in July and August. The state fish hatchery, with aquarium displays, is at Woodruff. For landlubbers there are miles of trail used by hikers and horseback riders, and in winter these are groomed by the state and by local clubs for cross-country skiing, snowmobiling, and snowshoing. Ice skating and ice fishing are also popular. The Plum Lake Golf Course at Sayner is a good 18-holer, and there are other courses at Trout Lake and Minocqua. Several towns have riding stables and public tennis courts. In fall there are colorama sightseeing tours and festivals, hunting, and cranberry harvests. The towns here are no more than hamlets, and the resorts are unpretentious, but for outdoors people this is marvelous vacation country.

Arbor Vitae Motel, *P.O. Woodruff, WI 54568, (715) 356-3393* ☐ *Open all year* ☐ *1½ mi N on US 51* ☐ This little one- and two-story motel overlooks lawn and trees on the lake shore and has 20 rooms. It has its own sand beach and a dock for boat rental, fishing, and waterskiing. They will freeze your catch free. In winter there is ice skating and cross-country skiing, and you can rent snowmobiles. Trails start at the door. ☐ **Inexp** ☐ **W, F, S**

Cardinal's Manitowish Lodge, *Woodruff, WI 54568, (715) 385-2710* ☐ *Open mid-June to early Sept* ☐ *8½ mi N off US 51* ☐ Here 20 nice cottages–with fireplaces and screened porches and with one to four bedrooms–are strung along the lakefront under trees. These are offered on AP, with three meals served in the dining room (or, if you want to be out all day they will pack you a lunch). On the water is a swimming beach, and dock rentals include pontoons, sailboats, and waterskiing boats. Also on the grounds are outdoor game areas, tennis, and skeet and trapshooting. ☐ **Mod to Lux** ☐ **F, W**

Deer Park Lodge, *Manitowish Waters, WI 54545, winter, (305) 733-1749; Feb to Sept: (800) 327-1012 or 826-7215* ☐ *Open June–Sept* ☐ *3 mi SE off US 51 on County K* ☐ Huddled on a peninsula in Lake Manitowish, this resort's motel-like facilities include 10 rooms in the main lodge set among pines, 18 rooms and suites in a two-story motel-style unit built right over the water, and 40 little frame houses with one or two bedrooms that are winter on the lakefront or back from the shore among the pines. Except for five EP cottages with kitchenettes, rates are MAP, but lunch is also available and the bars have nightly piano music, dancing, and entertainment. This resort caters to parents by offering a free playschool eight hours daily, seven days a week, plus wading pool, playground, indoor game room, and children's menus. Also included in the rates are an outdoor pool with a glass wind screen, four tennis courts, nine riding horses, waterskiing, instruction, and outdoor game areas. Golf (eight miles away) is also free in July and August. At the docks are a variety of boats, and fishing guides are available. Deer Park also has skeet and trapshooting and bicycle rental. ☐ **Lux** ☐ **F, W, G, T**

Esch's Serenity Bay, *Rte. 1, St. Germain, WI 54558, (715) 479-8866* ☐ *Open all year* ☐ *1 mi E, take Halberstadt Rd. N* ☐ This lakeside cottage colony is a fine choice for a family. Besides a good view of Little St. Germain Lake, each of the EP log cabins has a fully equipped kitchen, knotty-pine paneling, carpeting, and two to six bedrooms. Some have patio deck, screened-in porch, or fireplace. Surrounded by woods, the nine acres of grounds include picnic and play areas, and tennis courts. Canoes, sailboats, ski boats, and motors may be rented, and in season, a fishing boat is supplied free with each cabin. The 600-foot-long sandy beach has a gentle slope good for children. In winter snowmobiles are available for rent. ☐ **Mod** ☐ **W, F**

Froelich's *Sayner Lodge, Sayner, WI 54560, WI 54560, (715) 542-3261* □ *Open early May to early Oct* □ *1 mi NE to Lake St., then follow signs* □ On Plum Lake, the resort's main lodge has a library, recreation room, and a dozen rooms, but most lodgings are in the 27 cottages, which have living rooms and screened porches, (some also have fireplaces). All units are attractive and are available EP or AP. The restaurant, with background music and a lake view, is excellent. It has a varied menu that ranges from duckling and prime rib to Italian and French specialties. The bar has dancing and entertainment. On the grounds are the usual lawn games, lighted tennis, a heated outdoor pool, a wading pool, a playground, and a pitch-and-putt course. The adjoining golf club may be played on half or all 18 holes and has clubhouse services. At the dock you can rent all manner of boats and arrange for a fishing guide. This resort is unusually comfortable and better endowed than most in the area. □ **Lux** □ **F, W, G, T**

Lofty Pines Lodge, *St. Germain, WI 54558, (715) 542-3100* □ *Open all year* □ *5 mi NW via SR 155* □ *Big St. Germain Drive* □ On Big St. Germain Lake, this former estate which has been a family-owned resort for a quarter of a century, offers an assortment of 10 older and contemporary one- and two-story lodges, some overlooking the lake, some set back in the pines. Most units have between one and three bedrooms, kitchenettes, fireplaces, living rooms, and some have extras such as lofts, patios, or steam bath. While some are contemporary, others have braided rugs or hardwood floors and antique furnishings. Rates are EP and the closest restaurant is a half-mile away. Facilities include a recreation room, outdoor games, picnic tables and grills, a playground, and a sand beach. At the dock are canoe, sailboat, and other rentals, with instruction in waterskiing, sailing, and canoeing. Behind the lodge is Plum Creek, excellent for trout. Lost Creek is a good canoe trail. Birds and wildflowers are prominent on nature trails. In winter you can rent a snowmobile (800 miles of trails) or cross-country skiing equipment (10 miles of trail) or go ice fishing. □ **Inexp to Lux** □ **W, F, S**

Miller's Resort of the Woods, *Boulder Junction, WI 54512, (715) 385-2137* □ *Open early May–Nov* □ *½ mi N off County M* □ This pleasant lakeside resort offers 18 cottages with one to four bedrooms, a living room, and most have refrigerator; some have fireplaces and screened porches. AP, EP, and package plans are offered, and the restaurant serves three meals daily. The recreation lodge has TV and games, and outdoor game areas include tennis and a playground. There is a sandy beach and sun deck. The dock has launching facilities and rents pontoons, sailboats, paddleboats, motorboats, and boats for waterskiing. Fishing guides are available. □ **Mod** □ **W, F**

Motel in the Woods, *P.O. Box 257, Boulder Junction, WI 54512, (715) 385-2182* □ *Open mid-May to mid-Oct* □ *8 mi SE on County M, K* □ On the 10-acre lakefront, on a chain of three lakes, this rustic complex has four rooms with large windows in the two-story main lodge and 10 frame cottages strung out along the white birches on the shore so that each has a water view. All cottages have refrigerators, and some have screened porches or fireplaces. EP and MAP are offered, the dining room serves lunch and dinner, and cookouts and box lunches are available. Facilities include a recreation room, outdoor game areas, nature trails, swimming beach, and boat rentals and fishing guides. □ **Mod** □ **W, F**

Wildcat Lodge, *Box 138, Boulder Junction, WI 54512, (715) 385-2421* □ *Open all year* □ *4 mi N on County M* □ On 120 wooded acres on Wildcat Lake, part of a chain of three lakes, this resort has 10 rustic cabins with paneled interiors, heat, separate living room, refrigerator, and large picture windows overlooking the water. These are available on MAP in season and EP off season. Four modern EP cabins have full kitchens, fireplaces, four bedrooms, and sliding glass doors that open onto decks overlooking the lake. The main lodge has a game room and an attractive dining room with a fireplace. Breakfast features homemade buttermilk pancakes with wild blueberries or raspberries picked locally, and dinner is also served. They will also pack a lunch. There is a beach for swimming and the dock has free sailboats and rentals of canoes and other types of boats. Guided fishing trips are available, and a fishing clinic is provided weekly. On the grounds are a tennis court, game areas, playground, and two nature trails with good birdwatching (including bald eagle and osprey). In winter the resort's 20 miles of cross-country trails connect to those in state lands. In town is a free warming hut and ice skating rink. Services include babysitting and airport pickup. □ **Mod** □ **F, W, S**

RHINELANDER: Headquarters for Nicolet National Forest, and set at the confluence of the Wisconsin and Pelican rivers, this city boasts over 200 lakes within a twelve-mile radius. In addition to fishing, hunting, and watersports, it has a logging museum and a self-guiding 11-mile auto tour of a demonstration forest on Consolidated Papers' lands. The latter makes its lands available also for cross-country skiing. Camp 10 Ski Area, ten miles south, has T-bars, rope tows, ski school, and rentals for downhill (350-foot vertical drop) and cross-country skiing.

Holiday Acres, *Box 460, Rhinelander, WI 54501, (715) 369-1500* □ *Open early May to mid-Mar* □ *4 mi E on SR 8, then 2 mi N on Lake George Rd.* □ Spreading along 1000 beautiful wooded acres on the shore on Lake Thompson, this resort (EP) has 28 air-conditioned rooms in a modern lodge and 36 modern frame cottages. Most units have picture windows with a lake view and more than half have fireplaces. The cabins have kitchens or kitchenettes, one to four bedrooms, and come with a boat; some have screened porches. Six of the cottages are a mile away from the main complex, at a campground area with an outdoor pool. Maid service for cottages is available at an extra charge. At the main complex is a handsome restaurant of roughhewn cedar and open-beam ceiling with an indoor stream, waterfalls, a stone fireplace, and combo for dancing. The cuisine is good and includes walleye pike, steaks, seafood, and a wine list. A Friday night buffet and Sunday brunch are featured. There is also a coffee shop and a cocktail lounge. On the premises are an indoor pool, recreation room, outdoor barbecue, lawn games, three tennis courts, hiking trails, horseback riding, and a putting green. On the 6000-foot shoreline are a sandy beach, canoeing, boating, sailing, and fishing (muskie, pike, pan fish). In winter there is cross-country skiing, with equipment available, and snowmobiling. Trails start on the resort property. Several golf courses are 6 to 30 miles away. ☐ **Mod to Lux** ☐ **W, F, T, R**

Miller's Shorewood Vista Resort, *Rte. 6, Rhinelander, WI 54501, (715) 362-4818* ☐ *Open mid-June to mid-Sept* ☐ *4 mi E on SR 8, then ½ mi N on Lake George Rd.* ☐ On Lake George, this is a cluster of a motel with 11 units and 33 cottages with one to four bedrooms each (some with fireplace), all operated AP. In addition to regular dining room meals, there are buffets and snacks. The bar has seasonal entertainment. There is a ballroom with dancing and other events frequently scheduled. Facilities include lake swimming, tennis, bicycling, and canoeing, boating, and waterskiing. ☐ **Inexp** ☐ **W, F**

SPRING GREEN: The Spring Green restaurant at Wintergreen Ski Area complex was designed by native son Frank Lloyd Wright, who also built this summer home, Taliesin East, nearby. The Frank Lloyd Wright Foundation (in Hillside School on SR 23) has continued Wright's work locally in the design of the Spring Bank and the cocktail lounge of the Dutch Kitchen (127 East Jefferson St.). Another master of design, Robert Trent Jones, laid out the Springs, the fine 18-hole golf course. A tourist attraction (9 mi S on SR 23) is House on the Rock, an imaginative house perched 450 feet above Wyoming Valley. At Tower Hill State Park, a shot tower has been reconstructed. Spring Green also has tennis courts, canoe rentals, riding stables, and Wisconsin River fishing. The ski area has nine trails, the longest 4100 feet, from a wooded bowl facing the river. Two chairlifts and one J-bar service the runs. At the top of the bowl, which has a vertical drop of 400 feet, is the lodge with ski school, rentals, cafeteria, and lounge. In winter the golf course's pro shop is transformed into a cross-country skiing center (rentals, ski school, and food service) for 28 kilometers of trails that loop over 300 rolling acres of the golf course and river shoreline. For more information contact: Wintergreen, P.O. Box 467, Spring Green, WI 53588.

Hayloft Lodge, *Spring Green, WI 53588, (608) 588-2568* ☐ *Open all year* ☐ *On US 14, 1 mi W of jct SR 23* ☐ The extra-attractive motor lodge and restaurant resembling Scandinavian round barns, were designed by a Wright student. Doubles have color TV and telephone. In addition to the 27 air-conditioned doubles, the lodge offers honeymoon and executive suites, and an apartment with fireplace, kitchenette, and three bedrooms. The Rock Restaurant, serving steakhouse fare, frequently schedules dancing and live entertainment. The lodge also has a recreation room, wooden-domed indoor pool, whirlpool, and sauna. Mid-week package plans with Wintergreen are available. ☐ **Mod** ☐ **S**

THREE LAKES: Van Kirk's Lake Breeze Resort, *Three Lakes, WI 54562, (715) 546-3616* ☐ *Open all year* ☐ *1 mi SE on US 45, SR 3 D, then 1 mi NE on County X* ☐ Set on Town Line Lake in the Northwoods, at the edge of Nicolet National Forest, this cottage colony has more than a dozen comfortable cottages on AP. All cottages have screened porches. Half a dozen have kitchenettes and are available off-season on EP. Rates include your three meals, setups, your own boat, and waterskiing. Motorboats and fishing guides are available at extra cost. Facilities also include tennis and lawn games, and in winter snowmobile and cross-country ski trails start on the property. The forest has more than 80 miles of trails. Sheltered Valley Ski Area, nine miles north, has a Poma lift, rope tows, ski school, rentals, and a day lodge with restaurant. ☐ **Inexp** ☐ **W, F, S**

WISCONSIN DELLS: The scenic Dells of the Wisconsin River are the state's premier tourist magnet. The best way to see the Dells is on boat tours, some of which have stopovers so you can explore the terrain on foot paths. Land and water tours are also given on Wisconsin "Ducks" (amphibious crafts). Other attractions include a narrow-gauge railway, Storybook Gardens with children's rides and shows, Biblical Gardens, a deer park, water-ski shows, Indian dances, and historic sites, Fort Dells, a replica of a frontier town with rides, attractions, and a sightseeing gondola ride up a 335-foot tower. Just twelve miles south is Baraboo, once the home of the Ringling Brothers circus, with live circus acts, parades, and calliope concerts at the World Circus Museum, plus a railway museum. Swimming, fishing, and boating are available at Mirror Lake State Park south of Wisconsin Dells and at Devil's Lake State Park south of Baraboo; the latter is especially beautiful and interesting, with unusual cliffs and rock formations, Indian mounds, exhibition on the Ice Age, and naturalist programs on glaciation. In winter, downhill and cross-country ski complexes are open at Christmas Mountain and Sky Line north of Wisconsin Dells and at Devils Head south of Baraboo.

Birchcliff Lodge, *River Rd. Wisconsin Dells, WI 53965, (608) 254-7515* ☐ *Open May to mid-Oct* ☐ *On River Rd., 1½ mi N of Wisconsin Dells* ☐ This mini-resort has one- and two-bedroom units in air-conditioned rustic lodges and log cabins, all nicely maintained and furnished with early American reproductions in maple. All units are heated, most are carpeted, and some have a fireplace. The lodge serves breakfast but offers only sandwiches and snacks the rest of the day. Facilities on the pretty 28 acres include a landscaped freeform pool, recreation hall, pitch-and-putt course, archery, an asphalt tennis court, a playground, and bicycle rental. Within a mile you can find golf, fishing, horseback riding, and restaurants. ☐ **Inexp**

Chula Vista, *Wisconsin Dells, WI 53965, (608) 254-8366* ☐ *Open Apr–Nov* ☐ *3 mi N of Wisconsin Dells on River Rd.* ☐ Chula Vista has 130 double rooms and suites in 14 one- and two-story motel-style building nicely set among virgin pines on 44 acres of landscaped grounds beside the Wisconsin River. Riverside executive suites are the choicest lodgings, but all units are air-conditioned and have wall to wall carpeting and climate control. Four dining rooms allow you to choose formal and informal meals. The cuisine is well-regarded locally and includes prime rib, seafood, local fish, and daily specials. On Sundays there is a champagne brunch. In addition to a cocktail lounge there is a night club with dancing, dinner show, and other entertainment. Scattered about on the grounds are areas for badminton and lawn games, a kidney-shaped pool, miniature golf, and tennis courts. Also available are a recreation room and TV lounge. Babysitters, a playground, and a wading pool are provided a for children. Adjacent is a nine-hole golf course, and two riding stables are nearby. Along the river is a walking trail from which you can fish. ☐ **Mod to Lux** ☐ **G, F, T**

Dells Mound Lake Resort, *Box 218, Rte. 1, Wisconsin Dells, WI 53965, (608) 253-2222* ☐ *Open May–Oct* ☐ *3 mi S on US 12, SR 23, ½ mi E on Hiawatha Dr.* ☐ Here a dozen modest housekeeping cottages sleeping up to four or six people overlook Lake Delton. Most are two-room units and all are air-conditioned and pine paneled. Facilities include swimming beach, free boats, motorboat rental, outdoor game areas, and fishing. Golf, tennis, riding, and good restaurants are available within a mile. ☐ **Inexp** ☐ **W, F**

Wyoming

BUFFALO: Paradise Guest Ranch, *Box 790, Buffalo, WY 82834, (307) 684-5233* ☐ *Open all year* ☐ *From Buffalo take US 16 12 mi W; turn onto Hunter Creek Rd. and follow signs to ranch* ☐ Paradise Guest Ranch, one of the oldest dude ranches in the country, occupies a stunning 180-acre valley site in Big Horn National Forest. The gentle slopes offer gorgeous views of the Big Horn mountains and pine and spruce forest, but the location is convenient as well as picturesque: there is easy access to both the wilderness and the nearby town of Buffalo.

There are nine log cabins, which sleep from two to eight people. Most have living rooms with fireplaces, and porches; some have kitchens; all have private baths. (There are also a few single rooms which share a bath.) Home-cooked meals, which are included in the room rates, are mostly served in a central dining room, but there are also cookouts. The clubhouse is a gathering place for BYO cocktails, and for the night's entertainment, which is often square dancing. Spectator sports center on the corral: roping, bronc breaking, horse-shoeing, and some competitive games, including a children's rodeo. The ranch runs trail rides and pack trips. The pack trip to the high camp in the Cloud Peak Wilderness Area gives guests access to great trout fishing in mountain streams, as well as hunting for deer and elk. There's also fishing for brook trout in the stream which runs through the ranch itself. Riding instruction is available, or experienced riders can take their pick of the ranch's 40 head of gentle horses for trail rides just about any time of the day or night. In addition, the ranch offers licensed guide service for mountain climbing, and in the winter the location is convenient for ice fishing, snowmobiling and even down-hill skiing on the Meadowlark run (vertical drop 600 feet) 30 miles away. ☐ **Mod** ☐ **R, F, H**

CODY: Bill Cody's Ranch Inn, *P.O. Box 1390, Cody, WY 82414, (307) 587-2097* ☐ *Open May–Sept and mid-Dec to mid-Mar* ☐ *25 mi W of Cody on US 14/16/20* ☐ Halfway between Yellowstone National Park and the bustling tourist town of Cody, Wild Bill's grandson runs this civilized seven-acre dude ranch in Shoshone National Forest. The 11 double rooms, two singles, and three suites are housed in a cluster of log cabins, elegantly decked out in Western style with custom-made furniture and lamps, pine paneling, and either carpeting or polished hardwood floors. Every room has a private bath and individually controlled heat. Room rates include three meals a day (with a minimum stay of four days); the dining room's specialties are chili, and–for breakfast–sourdough hotcakes. Unlike many ranches, this inn has flexible seating times for meals–two hours each are allowed for breakfast and dinner.

During the summer and fall, your schedule will undoubtedly include horseback riding on one of the 30 horses, and cowboy guides lead all-day picnic and cookout rides every day of the week. The management will find horse, tack, and trail to suit your skills. There is no shortage of challenging territory for the experienced rider. Shoshone National Forest has trails covering a variety of terrain from the creekside lawns and gardens of the ranch to the rugged mountaintops of the Absarokas. Whitewater and float trips on the Shoshone River are also popular summertime activities. Trout fishing is a major attraction in the fall. In the winter, the Ranch Inn becomes a base for skiers at Sleeping Giant Ski Area, about 20 miles away, which has one T-Bar and one rope tow, and a vertical drop of 450 feet. There's also cross-country skiing along the miles of forest trails and snowmobiling through Sylvan Pass into Yellowstone. Package plans for both winter and summer seasons are available. ☐ **Mod to Lux** ☐ **R, S, F**

Blackwater Lodge Motel, *Box 1162, Cody, WY 82414, (307) 587-3709* ☐ *Open early May to mid-Oct* ☐ *37 mi W of Cody on US 14/16/20* ☐ The Blackwater Lodge is a good base for Yellowstone explorers; it's just 15 miles from the park's east entrance. There are 16 log cabins, with log or knotty-pine-paneled interiors and private baths; most are carpeted and one has an impressive floor-to-ceiling stone fireplace. The dining room specialty is barbecued ribs. Drinks and bottles-to-go are available from the cedar-paneled bar and lounge. If you choose not to spend all your time at the park or visiting Cody, there's plenty to do around the 22-acre site. Blackwater Creek, which runs through the grounds, is fine for fishing, and there are badminton, volley ball, and horseshoe courts on the premises. The recreation lounge has a Ping-Pong table and cable color TV. The management can arrange trail rides into the high country of the Absaroka Mountains; these should be of special interest to big-game hunters and photographers, as well as fishermen in search of the mountain lakes' golden trout. ☐ **Inexp to Mod** ☐ **F, R**

Castle Rock Ranch, *Southfork Rte. 2, Cody, WY 82414, (307) 587-2076* ☐ *Open all year* ☐ *On SR 291 SW of Cody* ☐ This working ranch in the beautiful valley of the South Fork sits on grassland surrounded by snowcapped peaks. The few cozy log cabins come with private bath and will sleep up to four people; they vary from one room to a two-bedroom unit with living room and fireplace. AP rates include horseback riding, river fishing, a pool, a game room, cookouts, outdoor game areas, and evening activities. The ranch has two miles of private river frontage on the South Fork, but also arranges day trips to Yellowstone and day-long floats on Clarks' Fork. Also for extra charges the ranch will take you on weeklong wilderness pack trips (some especially for teenagers), trips to Yellowstone, and big-game hunting. With advance notice, they will also arrange for babysitters. ☐ **Inexp to Mod** ☐ **R, F**

Elephant Head Lodge, *Wapiti, WY 82450, (307) 587-3980* ☐ *Open early-May to early-Oct* ☐ *11 mi E of E entrance to Yellowstone, on US 14, 16, 20* ☐ Built in 1910 by Buffalo Bill's niece and her husband, this lodge was named for a strangely shaped rock on the Shoshone River. The lodge, on 4½ riverside acres, offers EP or AP, and the dining room specializes in trout. Cookouts are arranged on request. Ten log cabins (three built in 1910) with handmade furniture and private bath sleep two to five people each. Some have a fireplace. The river, 300 yards from the lodge, has excellent trout fishing. Horseback trips include guided wilderness all-day fishing trips, and photography trips for elk, moose, sheep, and deer on Mesa Creek, Fishhawk Creek, and Gunbarrel Creek. Singalongs around campfires are held evenings. ☐ **Mod** ☐ **R, F**

Goff Creek Lodge and Motel, *Box 167, Cody, WY 82414, (307) 587-3753* ☐ *Open mid-May to mid-Oct* ☐ *41 mi W of Cody on US 14/16/20* ☐ This cluster of log cabins is only 10 miles from the east gate of Yellowstone National Park. You can sign up for a full, meals-included plan or just for lodgings, and since horseback riding is available at extra cost, it's possible to customize your vacation here to suit your needs. Whatever options you choose, you'll stay in one of the 14 comfortable log cabin units grouped in a clearing in the Wapiti Valley of Shoshone National Forest by the side of Goff Creek. Most of the cabins are single rooms for one to three people; there are a few one- and two-bedroom units. Featured are knotty-pine interiors, exposed-beam ceilings, and ranch-style oak furniture; the lamps are custom-made converted cowboy boots. All of the rooms have private baths. The family-style dining room serves home-cooked meals; the kitchen also provides chuck-wagon dinners and box lunches on trail rides. In addition to activities on horseback, the lodge offers excellent fishing for cutthroat trout in the Shoshone River, good hiking and rock hounding on the forest trails, and hunting for deer, elk, moose and mountain sheep in the fall. At night there's square dancing. ☐ **Inexp to Mod** ☐ **R, F, H**

Grizzly Ranch, *North Fork Route, Cody, WY 82414, (307) 587-3966* ☐ *Open June to Labor Day* ☐ *26 mi W of Cody, just off US 14, 20* ☐ Set in the high range country of Wapiti Valley against the steep slopes of Grizzly Ridge, this 100-acre ranch has a sweeping view of the Absaroka Mountains at the far end of the valley. Capacity is only 15 guests. You are housed in a cluster of modern log cabins among a stand of shade trees along a creek, a short distance from the main house. The dining room serves family-style meals and provides set ups. The AP rates include two guided rides a day, breakfast rides, cookouts, and all-day rides, as well as a two-hour float trip and the Cody Night Rodeo.

Rock and fossil hunting, tennis, horseshoes, fishing on the nearby Shoshone River, hiking, birdwatching, and evening square dances are among the other activities. For an extra charge five-, seven-, and ten-day pack trips can be arranged. Fall hunting trips are arranged after the regular season. ☐ **Mod** ☐ **R, F**

Hidden Valley Ranch, *Southfork, Cody WY 82414, (307) 587-5090* ☐ *Open Apr to late Nov* ☐ *From Cody take South Fork Rd. 14 mi, turn right across river and follow signs* ☐ This 4000-acre working cattle ranch, at a 6000-foot elevation in the arid mountain and range country east of the Absaroka range, takes only 32 guests. Built in 1974, the ranch buildings are neat and rustic, constructed from unpainted timber, with spacious interiors finished with wood paneling and rough beams. You stay in log cabins with private baths, living rooms, and picture windows. Family-style meals in the pine-paneled main lodge include ranch-grown beef as well as home-baked breads and pastries. There is a bar. You may stay here on EP, and rent a horse by the hour, or on AP, which includes unlimited riding and all ranch activities and facilities. You are welcome to accompany ranch hands on their rounds, and in spring there is branding and a ranch rodeo. Guided horseback trips are conducted daily, and you can take overnight pack trips into the mountains. Instruction is available for beginners. Saddle sores can be immersed in the heated swimming pool. The ranch maintains a stocked trout pond, and the Shoshone River is nearby. Hidden Valley provides boats and motors for fishing on nearby Yellowstone and Buffalo Bill lakes, and in fall outfits big-game hunts. ☐ **Mod** ☐ **R, F**

Valley Ranch, *South Fork Route, Cody, WY 82414, (307) 587-4661* ☐ *Open May to late Oct* ☐ *40 mi SW of Cody on South Fork Route* ☐ With its clay-chinked cabins of roughhewn logs, this working ranch looks like it dates from pioneer days, which it does. Valley was homesteaded in 1892 and established as a guest ranch in 1915. The original sod-roof homestead cabins are still standing for the benefit of artists and photographers (the ranch has a darkroom), but the 35 guest cabins are less spartan—with large beds, baths, wood stoves, and fresh flower arrangements. Cabins without bath are available at a reduced rate. The AP rates include all ranch activities, horseback riding, and meals of ranch-grown meats and vegetables.

Valley raises cattle, pigs, lambs, ducks, chicken, and turkeys, and you are welcome to help with ranch chores or in the garden. Surrounded by the high peaks of the Absaroka range and Washakie Wilderness, the area abounds in wildlife, including deer, elk, moose, and bighorn sheep. Dr. Oak Thorne, the owner, is a biologist who enjoys arranging nature walks and fossil hunts. The South Fork of the Shoshone River, a noted trout stream, flows through two miles of ranch property, and you can swim or tube on it. There's also fishing in the ranch lake. And of course there's horseback riding. All-day or half-day trips are available, and family pack trips or fishing trips to the high country can be arranged. Trips are also available to nearby museums and national parks. At various times during the season there are special workshops in oil painting, photography, and stained-glass making. ☐ **Mod** ☐ **R, F**

DRIGGS: Grand Targhee Resort, *Alta, Wyoming, via Driggs, ID 83422, (307) 353-2304* ☐ *mid-Nov to late Apr and early July to Labor Day* ☐ *From SR 33 in Driggs, take turn-off for resort 12 mi E* ☐ With an annual snowfall of 500 inches, Grand Targhee is one of the few ski resorts that can happily do without snowmaking. Although the slopes are technically in Wyoming, the only way you can get to them is through Driggs, Idaho—unless you snowshoe over the Tetons. The ski area is formed by two bowls, serviced by three double chairs and one rope tow. The vertical drop is 220 feet, and the longest trail is 2½ miles.

At the base of the area are three lodges, none more than 50 yards from the lifts or other facilities. The 15-rooms at Targhee Lodge and the 48-rooms at Teewinot Lodge have twin queen beds, color TV, and telphone. Sioux Lodge offers 32 studios, studios with with loft bedrooms, and 2-bedroom apartments. All units here have double bunks, sofa bed, kitchenette, fireplace, and TV. Guests at all three lodges use the outdoor pool (heated for wintertime swimming). The complex of buildings at the base includes a steak house, cafeteria, bar, sports shop, rentals, nursery, and game room. The resort also has cross-country rentals and trails, and the ski school offers private, semi-private, and group lessons for both children and adults. The instructors give special lessons in skiing deep powder, the stuff that Targhee is made of. Two Laykold tennis courts are open in summer, and the chairlift ferries guests up to 10,200 feet to study the view. A soaring center is 12 miles away. ☐ **Mod to Lux** ☐ **S**

JACKSON HOLE: Jackson Hole offers some of the most spectacular mountain scenery in the world. The 19th-century mountain men referred to any valley surrounded by high mountains as a "hole" and this one was named for Davey Jackson, the first trapper to come to the valley. Although Jackson Hole today is visited by more than four million tourists a year, Grand Teton National Park has preserved the rugged wilderness beauty of Davey Jackson's valley. Bordered by Yellowstone National Park on the north, and surrounded by 3½-million-acre Bridger–Teton National Forest, the 40-mile-long valley is ringed with mountains and presided over by 13,770-foot Grand Teton. The jagged peaks rise abruptly 7000 feet from the wide valley floor, without the foothills generally associated with mountain ranges. The sheer granite faces make this a mountaineer's paradise, and instruction and guides are available. Grand Teton National Park maintains more than 200 miles of trails for the casual hiker as well as the experienced mountaineer. Jackson Hole is also a paradise for outdoor people and photographers. The valley teems with wildlife, including the 8000 elk that winter at the National Elk Refuge. In winter you may visit the refuge on horsedrawn sleigh rides. A number of outfitters provide summer pack trips into the mountains and organize big-game hunts. The Snake River, full of cutthroat trout and rough with rapids, rushes through the valley, and fishing, scenic, and whitewater float trips are available. Jackson Lake, 20 miles long, has watersports. Also available are covered-wagon rides, helicopter tours, sightseeing chairlifts, glider instruction and rides, polo, summer music festival, Western melodramas, and special events. For golfers, the Jackson Hole Golf and Tennis Club has an 18-hole golf course designed by Robert Trent Jones, as well as fine tennis. Jackson Hole is one of the most famous ski areas in the country, with its wide-open bowls, deep powder, and 4000-foot vertical drop. Excellent skiing can also be had at Snow King Mountain, just outside the town of Jackson, and Grand Targhee on the west side of the Tetons. Full facilities and services are available for downhill and cross-country.

The town of Jackson has its own unique charms. Understatement is not one of them. This is an exuberant monument to the Old West, with stagecoaches circling the town square, elk-antler fences, raised board sidewalks, silver-dollar-studded saloons, saddled barstools, and a weekly rodeo. Jackson boasts three theaters, weekly square dances, and a number of art galleries and museums dedicated to the area's frontier past. Teton Village is a more recent resort community at the base of the ski slopes. For a full list of accommodations, which include housekeeping cabins, motels, hotels, condominiums, inns, ranches, and resorts, contact Jackson Hole Chamber of Commerce, 532 N Cache St., Jackson, WY 83001, (307) 733-3316.

Best Western Alpenhof, *Box 288, Teton Village, Jackson Hole, WY (307) 733-3462* □ *Open mid-Dec to mid-Apr, and mid-May to mid-Oct* □ *12 mi NW of Jackson on SR 390* □ This comfortable inn offers 40 EP double rooms and a three-room suite. Many units have balconies and oversize beds, and all have cable color TV and private baths. It is also the home of one of the most distinguished restaurants in the vicinity, specializing in wild game and veal prepared in classic continental fashion. Recreational facilities include a large heated pool, whirlpool, sauna, game room, and a lounge with nightly entertainment and dancing. □ **Mod to Lux**

The Inn at Jackson Hole, *P.O. Box 328, Teton Village, WY 83025, (307) 733-2311* □ *Open all year* □ *12 mi NW of Jackson on SR 390* □ This modern, brown-shingled lodge is under the same management as Sojourner Inn and also lets you have children under 17 stay free in your room. It has 75 EP rooms, most of them spacious doubles, and suites. Some units have kitchenettes, fireplaces, balconies, or spiral staircases leading to sleeping lofts. The main Dining Room operates all year, and in the winter a second restaurant, La Fondue House, is open. Along with the Cocktail Lounge, which has a great view of the mountain, there is a comfortable guest lounge with handsome native-stone fireplace. The lodge has an outdoor heated swimming pool and tennis courts. ☐ **Mod to Lux**

Jackson Hole Racquet Club, *Star Route Box 362, Jackson, WY 83001, (307) 733-5740* ☐ *Open all year* ☐ *On Teton Village Road* ☐ Condominium units are available here in modern cedar-trimmed lodges nestled under trees. All units have fireplaces, kitchens, and balconies. ☐ **Lux** ☐ **T**

Ramada Snow King Inn, *400 Snow King Dr., Box SKI, Jackson, WY 83001, (307) 733-5200* ☐ *Open all year* ☐ *On US 26, 89, 187 in Jackson* ☐ Set at the base of the Snow King Mountain ski area, this modern, glass and timber, seven-story structure has multi-angled roofs. Inside, the four-story lobby has a stone fireplace that rises 30 feet to the ceiling. The decor throughout is contemporary, with a Scandinavian accent and lots of solid, unpainted wood. There are 207 rooms, studios, and suites. Some have private balconies, and all have picture windows with views of the surrounding mountains. The dining room is elegant but simple, with a long sloping roof and floor-to-ceiling windows looking out on the mountain. There's also a coffee shop, cocktail lounge with entertainment, a recreation room, health club, and child care center. ☐ **Mod to Lux**

Sojourner Inn, *P.O. Box 348, Teton Village, WY 83025, (307) 733-3657* ☐ *Open June–Sept and early Dec to mid-Apr* ☐ *12 mi NW of Jackson on SR 390* ☐ This alpine-style inn is a fraternal twin of the Inn at Jackson Hole, under the same management, and sharing the splendid mountain scenery. The 98 EP rooms (multi-room suites available) all have modern amenities and many have balconies. Children under 17 stay in your room free. There are a reasonably priced cocktail lounge with nightly entertainment, an outdoor heated pool, and a recreation room. ☐ **Mod to Lux**

LARAMIE: Vee Bar Ranch, *Rex Route Box 198, Laramie, WY 82070, (307) 742-5741* ☐ *Open all year* ☐ *21 mi W of Laramie on SR 130* ☐ The guest lodge here is an attractive two-story building of weathered clapboard with a white two-story porch looking out over the river and pastureland of the 800-acre working ranch. The Vee Bar was founded as a cattle ranch in the 1880s and the long one-story guest lounge dates from 1891. Guests stay in single and double rooms in the main lodge. Rates are EP, but three daily meals are available (bring your own booze), and the ranch has cookouts on an island in Little Laramie River. Saddle horses are available by the hour, by the day, or for trail rides. You can arrange trips into the Snowy Range Mountains either by horseback or on foot. The trout-filled Little Laramie River flows through the ranch for the benefit of fly-fisherman. You may play badminton, volleyball, and croquet on the ranch grounds. Winter activities include snowmobiling, cross-country skiing on the ranch, and downhill skiing at Medicine Bowl ski area 12 miles away. ☐ **Inexp** ☐ **R, S, F**

PINEDALE: Gateway to the magnificent Bridger Wilderness, and calling itself "Home of the Mountain Men," Pinedale is for those who like simple pleasures and the rugged outdoors. **Half-Moon Lodge,** (307) 367-2851, and **White Pine Lodge,** (307) 367-2434, offer lodging and pack trips. Swimming and boating is available in some of the 100 nearby lakes. Fishing is splendid in these lakes and in streams for rainbow, German brown, and other trout. Big-game hunting is available seasonally. For information, write the Chamber of Commerce, Pinedale, WY 82941.

INDEX

The Abbey, WI, 359
Abe Martin Lodge, IN, 140
Abominable Snowmansion Ski Hostel, NM, 230
Acorn Chalet Lodging, UT, 313
The Admiralty Resort, WA, 340
Aladdin, NV, 217
Alderbrook Inn, WA, 344
The Alisal, CA, 63
Allegany State Park, NY, 249
Allenberry, PA, 281
Alpenhof Lodge, CA, 54
Alpine Resort, WI, 355
Alpine Village, NY, 235
Alta Lodge, UT, 308
Alta Ski Area, UT, 308
Amberian Point Lodge, AZ, 34
Amelia Island Plantation, FL, 88
Anderson's Wonder View Motel and Cottages, CO, 73
Angler's Paradise Alaska, AK, 21
Antlers Resort, MO, 209
April Sound, TX, 301
Aqua Marine Lodge, OH, 262
Arapaho Valley Ranch, CO, 75
Arbor Vitae Motel, WI, 365
Arizona Biltmore, AZ, 27
Arizona Inn, AZ, 35
Arlington Hotel, AR, 41
Arrowhead Lodge, OK, 267
Asilomar Conference Center, CA, 55
Atwood Lake Lodge, OH, 262
Avalon Inn, OH, 265

Baca Grande Inn and Resort, CO, 70
Bahia Mar Hotel, FL, 92
Bahi'a Mar Resort, TX, 306
The Balsams, NH, 219
Barren River Lake State Resort Park, KY, 147
Basin Harbor Club, VT, 326
Bass Haven, MO, 210
Bay Shore Inn, WI, 355
Bear Mountain Inn, NY, 239
Beaver Creek Lodge, CA, 50
Beaver Village, CO, 83
Bedford Springs Hotel, PA, 281
Beech Mountain Resort, NC, 252
Belleview Biltmore, FL, 90
Best Western Alpenhof, WY, 376
Best Western Columbia Inn, OR, 279
Best Western Golden Tee, CA, 57
Best Western La Posada, TX, 303
Best Western Poco Diablo Resort, AZ, 34
Best Western Riverside, OR, 276
Big Canoe Resort, GA, 112
Big Moose Inn, NY, 235
Big Paw Resort, MI, 180
Big Sky, MT, 212
Big Sur Lodge, CA, 46
Bill Cody's Ranch Inn, WY, 371
Birchcliff Lodge, WI, 369
Birchmont Ruttger, MN, 197
Birkenhause, VT, 325
Bishop Pine Lodge, CA, 64
The Bishop's Lodge, NM, 229
Bittersweet Resort, MO, 208
Black Mountain Ranch, CO, 82
Black Mountain Ski Lodge and Motel, Inc., NY, 235
Black Point Inn, ME, 155
Blackwater Falls State Park, WV, 347
Blackwater Lodge Motel, WY, 372
Blue Church Lodge & Annex, UT, 313
Blue Paradise Resort, AR, 42
Blue Rock Inn & Golf Course, MA, 165
The Boar's Head Inn, VA, 329
The Boca Raton Hotel and Club, FL, 88
Bolton Valley Resort, VT, 315
Bonnie Oaks Resort, VT, 316
Bottle Hollow Resort, UT, 310
Boulders Inn, CT, 85
Boyne Highlands Inn, MI, 177
Boyne Mountain Lodge, MI, 177
The Breakers, FL, 103
The Breakers, SC, 292
The Breakwaters, MA, 165
Briarwood Mountain Lodge, ME, 154
The Broadmoor, CO, 69
The Broadwater Beach, MS, 206
Bromley Sun Lodge, VT, 315
Brooks River Lodge, AK, 22
Brookside Resort, MN, 201
Bryce Mountain Resort, VA, 329
The Buccaneer, GA, 114
Bullfrog Resort and Marina, UT, 311
Burntside Lodge, MN, 203

Cacapon State Park, WV, 348
Caesar's Palace, NV, 217
Callaway Gardens, GA, 115
Camanche Northshore Resort, CA, 49
Camp Denali, AK, 22
Campbell's Lodge, WA, 335
Canaan Valley State Resort Park, WV, 348
Cannon's, WA, 335
The Canterbury Inn, WA, 339
The Canyon Hotel, CA, 59
Canyon Inn, IN, 140
Cape Codder Hotel, MA, 166
Cape Coral Country Club Inn, FL, 89
Captain Gosnold Village, MA, 166
Cardinal's Manitowish Lodge, WI, 365
Carefree Inn, AZ, 27
Carolina Trace, NC, 261
Carson Inn/Nordic Hills, IL, 134
Carter Caves State Resort Park, KY, 147
The Cascades Inn, VA, 331

Castle Rock Ranch. WY. 372
Cataloochee Ranch. NC. 253
The Center at Foxhollow, MA. 163
Channel View. TX. 304
Chanticleer Inn. WI. 359
Charlton Motel. NC. 258
Charmey Chalet. CA. 51
Chatham Bars Inn and Cottages. MA. 166
Chautauqua Lodge. WA. 336
Cheeca Lodge. FL. 95
Chestnut Mountain Lodge. IL. 134
Chico Hot Springs. MT. 214
Childs Meadows Resort. CA. 54
The Chippewa Hotel. MI. 183
Chula Vista. WI. 370
The Chute Homestead Cottages. ME. 155
Circle Bar B Ranch. CA. 62
Circle R. Resort Ranch. TX. 299
Circus Circus Hotel/Casino. NV. 217
C Lazy U Ranch. CO. 75
Clark. Leland. and Clark Cottages. MA. 172
Cliff Dwellers. WI. 355
The Cliff House. ME. 155
Clifty Inn & Motor Lodge. IN. 138
The Cloister. GA. 117
Cobbly Nob Resort. TN. 295
Coco Palms. HI. 122
Coconut Grove. FL. 101
Coffee Creek Ranch. CA. 65
Coho Resort. WA. 344
Cohutta Lodge. GA. 112
The Colonial Inn. MI. 179
The Colony. ME. 154
The Colony Beach and Tennis Resort. FL. 99
Colorado Trails. CO. 71
Concord Hotel. NY. 239
Coolfont Recreation. WV. 346
Copper Mountain. CO. 70
Coral Cay. TX. 304
Coto de Casa and The Vic Braden Tennis College. CA. 64
The Cottages. ME. 153
Country House. WI. 356
Cove House. MA. 172
Covewood Lodge. NY. 235
Craigville Conference Center. MA. 167
Crescent Lake Lodge and Resort. OR. 273
Crystal Mountain. MI. 185
Cultus Lake Resort. OR. 271
Cumberland Falls State Resort Park. KY. 143

Daggett Houses, MA, 171
Daven Haven Lodge and Cottages, CO, 77
Dawn Hill Golf and Racquet Club, 42
Deer Park Lodge, WI, 365
Deer Valley Ranch, CO, 81
Deetjen's Big Sur Inn, CA, 46
DeGray State Park Lodge and Convention Center, AR, 40
Hotel del Coronado. CA. 48
Del Monte Hyatt House. CA. 56
Del Webb's La Posada. AZ. 28
Del Webb's Mountain Shadows. AZ. 28
Dells Mound Lake Resort. WI. 370
Dering Harbor Inn. NY. 248
Desert Inn and Country Club. NV. 217
Dexter's Inn and Tennis Club. NH. 223
Dillman's Sand Lake Lodge. WI. 361
Dixie Dude Ranch. TX. 299
Don Cesar Beach Resort Hotel, FL, 105
Donner Ski Ranch, CA, 51
Doral Country Club, FL, 102
Driftwood Lodge. SC. 292
Driftwood Lodge. WI. 363
Driftwood Motor Lodge. VA. 330
Drowsy Water Ranch. CO. 76
Ducey's Bass Lake Lodge and Pine Chalets. CA. 45
Dunes Hotel and Country Club. NV. 217
Hotel du Nord and Laura Maxwell Inn. WI. 357
Duval's Lakeside Resort. MI. 181

Eagle Knob Lodge. WI. 353
Earle Waters. WI. 359
East Bay Lodge. MA. 167
Hotel Edelweiss. NM. 232
Edgemont and Whiffletree Condominiums. VT. 318
Edgewater Motel. WI. 356
Edson Hill Manor. VT. 322
Elephant Butte Resort Inn. NM. 228
Elephant Head Lodge. WY. 373
Elkhorn Lodge. CO. 73
Elkhorn Village Inn. ID. 132
The Elm Arch Inn. MA. 168
Embarcadero Resort Hotel and Marina. OR. 278
Ephraim Shores. WI. 356
Erdmann's Eyrie. VT. 316
Esch's Serenity Bay. WI. 365
Eseeola Lodge. NC. 253

Fairway Villas. Diamondhead Yacht and Country Club. MS. 206
Fall Creek Falls State Resort Park. TN. 296
The Field's Buccaneer Inn of Sarasota. FL. 100
Finger Lakes State Parks. NY. 246
Flamingo Hilton. NV. 217
Flathead Lake Ranch. MT. 212
The Flying Cloud Inn. MA. 163
Forest Lodge. WI. 362
Fountainhead Lodge. OK. 267
Fox Hills Resort. WI. 364
Foxfire Golf and Country Club, NC, 259
Frances Langford's Outrigger Resort, FL, 98
Francisco Grande Resort. AZ. 24
French Lick-Sheraton Hotel and Country Club. IN. 137
Fripp Island Inn. SC. 288
Froelich's Sayner Lodge. WI. 366
Frontier. NV. 217
Frost Pond. ME. 153

Gaston's White River Resort. AR. 42
Gearhart-by-the-Sea. OR. 274
General Butler State Resort Park. KY. 142
Genesee Falls Inn & Motel. NY. 247
Gideon Putnam Hotel. NY. 250
The Glen. NH. 222
Glen Iris Inn. NY. 248
Glidden Lodge. WI. 356
Goff Creek Lodge and Motel. WY. 373
Golden Acres Farm & Ranch. NY. 240
Golden Eagle. VT. 323
Goose Cove Lodge and Cottages. ME. 158
The Gordon Folger Hotel and Cottages. MA. 173
Gordon Lodge. WI. 357
Grand Hotel. AL. 19
Grand Hotel. MI. 184
Grand Targhee Resort. WY. 375
Grand View Lodge. MN. 197
Grandview Lodge and Resort. ID. 130
Granit Hotel and Country Club. NY. 240
Great Smokies Hilton. NC. 253
Greenbo Lake State Resort Park. KY. 144
The Greenbrier. WV. 351
Grenelefe. FL. 91
Grey Bonnet Inn. VT. 318
The Grey Gull. WA. 339
Grizzly Ranch. WY. 373
Grossinger Hotel. NY. 241
Grosvenor Lake. AK. 22
Grove Park Inn and Country Club. NC. 254
Gulf State Park Resort. AL. 18
Gull Lake View Golf Club. MI. 185

Gunflint Lodge. MN. 203
Gurney's Inn. NY. 248

Hacienda del Sol. AZ. 35
Half-Moon Lodge. WY. 378
Harder Hall Golf and Tennis Resort. FL. 107
Hargood House. MA. 168
Harmel Ranch Resort. CO. 78
Harrah's. NV. 218
Harrison Inn. CT. 86
Hartness House. VT. 321
Hawk. VT. 320
Hawks Nest State Park. WV. 349
Hayloft Lodge. WI. 368
Heidel House Resort. WI. 360
Helm's Four Seasons Motel. WI. 357
Henry Horton State Resort Park. TN. 294
Hermosa Inn. AZ. 29
Hotel Hershey and Hershey Motor Lodge. PA. 282
Hidden Valley Guest Ranch. NY. 236
Hidden Valley Ranch. WA. 335
Hidden Valley Ranch. WY. 374
High Country Lodge Inc.. NM. 229
High Hampton Inn and Country Club. NC. 254
Hill's Resort. ID. 130
Hilton Sea Island. TX. 306
Hilton Shanty Creek. MI. 176
Holiday Acres. WI. 367
Holl's Inn, NY, 236
The Homestead, UT, 312
The Homestead, VA, 330
The Homestead at Glen Arbor. MI. 178
Homewood Inn. ME. 158
Hon-Dah Motel. AZ. 25
Hondo Lodge. NM. 232
Horseshoe Bend Guest Ranch. CO. 81
Host Farm. PA. 283
Hound Ears Club. NC. 255
Hueston Woods State Park Lodge. OH. 263
Hyatt Lake Tahoe. NV. 215

Indian Rock Resort. AR. 40
Indianhead. MI. 186
Indigo Inn. SC. 292
Inn of the Hills. TX. 303
The Inn at Indigo. FL. 91
The Inn at Jackson Hole. WY. 376
The Inn on Lake Waramaug. CT. 85
The Inn at McCormick Ranch. AZ. 29
Inn of the Mountain Gods. NM. 228
The Inn at the Mountain and the Town Houses. VT. 323
The Inn at Rancho Santa Fe. CA. 61
The Inn of the Seventh Mountain. OR. 271
Innisbrook. FL. 108
Innsbruck Lodge. NM. 232
Interlaken. WI. 361
Interlaken Inn. CT. 85
Interlaken Lodge. NY. 236
Interlochen Center for the Arts. MI. 180
I Pali Kai Cottages. HI. 123
Ireland's Rustic Cottages. OR. 276
Iron Springs Resort. WA. 336
Hotel Iroquois. MI. 184
Island House Apartments. TX. 301
Island Inn. FL. 106
The Islander Lopez. WA. 342
Itasca State Park. MN. 201
Iyanough Hills Motor Lodge and Golf Club. MA. 168

Jack O'Lantern. NH. 226
Jackson Hole Racquet Club. WY. 377
Jansen's Squirrel Lake Lodge. WI. 363
Jekyll Island Hilton Inn. GA. 114
Jenny Wiley State Resort Park. KY. 148
Joe Wheeler State Park Lodge. AL. 20
John Gardiner's Tennis Ranch. AZ. 30
John Newcombe's Vacation Resort. FL. 103
Johnny Seesaw's. VT. 315
Jot's Resort. OR. 276
June Lodge. CA. 50

Kaanapali Beach Hotel. HI. 126
Kah-Nee-Ta Vacation Resort. OR. 279
Kandahar Condominiums. NM. 233
Kapalua Bay Hotel. HI. 125
Kauai Surf Resort. HI. 124
Keauhou Beach Hotel. HI. 120
Kenlake State Resort Park. KY. 145
Kentucky Dam Village State Resort Park. KY. 146
Keweenaw Mountain Lodge. MI. 178
Key Biscayne Hotel and Villas. FL. 98
Keystone Resort. CO. 78
Kiawah Island. SC. 287
Killearn Golf and Country Club. FL. 108
Killington Village. VT. 318
King and Prince Beach Hotel. GA. 116
Kings Island Inn. OH. 263
Kingwood Inn and Country Club. GA. 113
Kohl's Last Resort. MN. 197
Kona Surf Resort. HI. 123
Kona Village Resort. HI. 120
Kulik Lodge. AK. 22
Kutsher's Country Club. NY. 241

La Costa Hotel & Spa. CA. 47
La Playa Beach & Racquet Inn. FL. 103
La Quinta. CA. 59
Lake Barkley State Resort Park. KY. 146
Lake Cumberland State Resort Park. KY. 144
Lake Guntersville State Park Lodge. AL. 18
Lake Lawn Lodge. WI. 354
Lake Mohave Resort. AZ. 24
Lake Morey Inn and Country Club. VT. 316
Lake Murray Lodge. OK. 268
Lake Placid Resort Hotel. NY. 236
Lake Powell Motel. AZ. 311
Lakepoint State Park Resort. AL. 18
Lake Shore Farm. NH. 222
Lakeside Inn. FL. 103
Lakeside Inn. MI. 187
Lake Tahoe Resort. NV. 216
Lake Texoma Lodge. OK. 268
Lake Toxaway Country Club. NC. 255
Lakeview Inn. WV. 346
Lakeway Inn and World of Tennis Resort. TX. 298
Land Between the Lakes. KY. 144
Lane Guest Ranch. CO. 74
Las Brisas. TX. 306
Las Casitas. CA. 44
Las Vegas Hilton. NV. 217
Leathem Smith Lodge. WI. 357
Lehigh Resort. FL. 99
Lemolo Lake Resort. OR. 277
Letchworth State Park. NY. 247
Liftline Lodge. VT. 325
The Lighthouse Inn. MA. 168
Linden Hill Hotel. MD. 160
The Lodge on the Desert. AZ. 36
Lodge of the Four Seasons. MO. 208
The Lodge at Pebble Beach. GA. 56
Lofty Pines Lodge. WI. 366
Longs Peak Inn & Guest Ranch. CO. 74
Lor-O Ranch. CA. 48
Lost River State Park. WV. 349
Lost Valley Ranch. CO. 81
Ludlow's Island. MN. 204
Lutsen Resort. MN. 204

Machins Cottages in the Pines. CO. 74
Madden's in Minnesota. MN. 198

Maggie Valley Country Club and Motor Lodges. NC. 255
Majestic Hotel. AR. 41
Malibu Beach Resort. MO. 209
Marco Beach Hotel & Villas. FL. 100
Marina Bay. FL. 93
The Mariner Motel and Cottages. WI. 358
Marriott's Camelback Inn. AZ. 30
Marriott's Casa Marina Inn. FL. 95
Marriott's Hotel Galvez. TX. 302
Marriott's Lincolnshire Resort. IL. 135
Marriott's Marco Beach Hotel & Villas. FL. 101
Marriott's Rancho Las Palmas Resort. CA. 59
Marriott's Santa Barbara Biltmore. CA. 62
Marriott's Tan-Tar-A Resort. MO. 210
Maui Surf Resorts. HI. 124
Mauna Kea Beach Hotel. HI. 121
Maxwelton Braes Resort. WI. 358
Mayan Dude Ranch. TX. 299
McGregor Mountain Lodge. CO. 75
MGM Grand. NV. 217
Mid Pines Club. NC. 259
Miller's Resort of the Woods. WI. 366
Miller's Shorewood Vista Resort. WI. 368
Millstone Lodge. MO. 208
Mirror Lake Inn. NY. 237
Mission Inn and Country Club. FL. 97
Mittersill. NH. 220
Mohawk Inn. NY. 237
Mohonk Mountain House. NY. 242
Hotel Molokai. HI. 127
Montauk Yacht Club & Inn. NY. 249
Montgomery Bell State Resort Park. TN. 294
The Moors Motel. MA. 169
Motel in the Woods. WI. 366
Mt. Summit Inn. PA. 286
Mt. Washington Hotel. Motor Inn. and Townhouses. NH. 219
Mountain Lake Hotel. VA. 332
Mountain View House. NH. 224

Naniloa Surf Hotel. HI. 119
Natural Bridge State Resort Park. KY. 148
Nautical Inn. AZ. 24
Nelson's Resort. MN. 204
Nevele. NY. 243
North Cascades National Park. WA. 338
Northern Resort. WI. 363
Northernaire Floating Lodges, MN, 200
Nor'Wester Lodge, MN, 205

Oak n' Spruce Resort. MA. 164
Oakland House. ME. 156
Ocean Crest. WA. 337
Ocotillo Lodge. CA. 60
Ohio State Parks. OH. 263
Ojai Valley Inn and Country Club. CA. 58
Oklahoma State Lodges. OK. 266
Oklahoma State Parks. OK. 266
Olympia Spa Golf and Country Club. AL. 17
On-the-Rocks Lodge. VT. 327
Opryland Hotel. TN. 295
Otesaga Hotel. NY. 245
Overniter. ID. 131

Padre Island Beach Hotel. TX. 302
Pala Mesa Golf & Tennis Resort. CA. 49
Palmetto Dunes Resort. SC. 290
Paradise Guest Ranch. WY. 371
Paradise Inn. WA. 337
Paris Landing State Resort Park. TN. 294
Park Place Motor Inn/Ranch Rudolph. MI. 186
Parry Lodge. UT. 310
Peaceful Valley Lodge and Guest Ranch. CO. 80
Pebble Beach (The Lodge at). GA. 56
Pennyrile Forest State Resort Park. KY. 143
Penrod's Au Sable River Resort. MI. 179
Perdido Bay Inn. FL. 104
Pick Point Lodge and Cottages. NH. 221
Pickwick Landing State Resort Park. TN. 296
Pier 66 Hotel and Marina. FL. 93
Pinehurst Hotel and Country Club. NC. 259
Pine Mountain Lodge. MI. 181
Pine Mountain State Resort Park. KY. 148
Pine Needles Lodge and Country Club. NC. 260
Pine Tree Point. NY. 251
The Pines. NY. 243
Pioneer Inn. ID. 129
Pipestem State Resort Park. WV. 350
Pisgah View Ranch. NC. 256
Placid Manor. NY. 237
Plantation Inn. FL. 90
Playboy Lake Geneva. WI. 362
Playboy Resort & Country Club at Great Gorge. NJ. 227
Pleasant Mountain Inn. ME. 151
Point Breeze Resort. NH. 226
Point West. MI. 182
The Pointe. AZ. 31
The Polynesian. WA. 340
Port Royal Plantation. SC. 290
Potowatomi Inn. IN. 137
Princeville at Hanalei, HI, 122
Prospector Square, UT, 313
Punderson State Park Manor House. OH. 264

Quadna. MN. 199
Quail Hollow Inn. OH. 264
Quail Lodge. CA. 47
Quartz Mountain Lodge. OK. 269
The Quechee Inn at Marshland Farm. VT. 321

Radisson Arrowwood. MN. 188
Radisson Ferncroft. MA. 170
Rain Forest Resort. WA. 341
Rainbow Cove Motel. NY. 246
Ramada Snow King Inn. WY. 377
Ramada Valley Ho. AZ. 31
Rams Horn Cottages. CO. 75
Rancho Bernardo Inn. CA. 61
Rancho de los Caballeros. AZ. 38
Rancho Encantado. NM. 230
Rancho del Rio. AZ. 36
Rancho Santa Cruz. AZ. 38
El Rancho Stevens. MI. 178
Rancho Viejo. TX. 300
Red Apple Inn, AR, 41
Red Canyon Lodge, UT, 309
Redfish Lake Lodge. ID. 131
Red Jacket Beach Motor Inn. MA. 169
Red Lion at Bowman's. OR. 277
Red River Motel. MA. 169
The Refuge Motor Inn. VA. 330
The Registry Resort. AZ. 32
Rio Bravo Tennis Ranch. CA. 44
Rio Hondo Condominiums. NM. 233
Rio Rico Resort. AZ. 26
Rio Verde Resort and Ranch. AZ. 32
Riveredge Resort. NY. 251
Riverside Villas Resort Motel. FL. 96
Roaring Brook Ranch & Tennis Resort. NY. 237
Roche Harbor Boatel. WA. 342
Rock Gardens Inn. ME. 157
Rock Harbor Lodge. MI. 182
Rock Springs Guest Ranch. OR. 272
Rocking Horse Ranch, NY, 244
Rockport Racquet and Yacht Club Resort, TX, 305
Rocky Point Inn. NY. 238
Roman Nose Lodge. OK. 269
Rondee Best Western. AZ. 34
Rosario Resort Hotel. WA. 343
Ross Lake Resort. WA. 339
Ross Teal Lake Lodge. WI. 360

Rough River Dam State Resort Park. KY. 143
Roundup Ranch. NY. 244
Royal Biscayne Beach Hotel and Racquet Club. FL. 98
Royal Lahaina Resort. HI. 126
Rustler Lodge. UT. 308
Rutledge Inn and Cottages. VT. 316

Sagamore. NY. 238
Saguaro Lake Ranch Resort. AZ. 25
Sahara. NV. 217
Hotel St. Bernard. NM. 232
Salishan Lodge. OR. 274
Salt Fork State Park Lodge. OH. 264
Salt Pond Motel. MA. 170
San Ysidro Ranch. CA. 62
Sand Castle. TX. 306
Sands. NV. 217
The Sands Beach Resort Motel. DE. 87
Sapphire Valley Resort and Fairfield Inn. NC. 256
Sawgrass. FL. 105
Schuss Mountain. MI. 185
Scottsdale Country Club. AZ. 33
Scott's Oquago Lake House. NY. 245
Seabrook Island. SC. 288
Sea-Gun. TX. 305
SeaMountain at Punalu'u. HI. 121
Sea Pines Plantation. SC. 290
Seascape Resort and Conference Center. FL. 102
Sebasco Lodge. ME. 157
7A Ranch Resort at Pioneer Town. TX. 307
Seven Springs Mountain Resort. PA. 281
The Shallows Resort Motel. WI. 358
Shangri-La. OK. 266
Shawnee Inn. PA. 285
Shawnee State Park Lodge. OH. 264
Shenvalee Lodge. VA. 332
Sheraton-By-The-Sea. GA. 114
Sheraton-Homosassa Springs Inn. FL. 96
Sheraton Marina Inn. TX. 301
Sheraton Molokai. HI. 127
The Sheraton at Sandestin. FL. 91
Sheraton Savannah Inn & Country Club. GA. 116
Sheraton Waiakea Village. HI. 119
Ship Harbor Inn. WA. 344
Shore Acres Resort Motel. VT. 320
Shore Lodge. ID. 129
Sierra del Sol Condominiums. NM. 233
Sierra Nevada Inn. CA. 54
Silverado. CA. 57
Sipapu Lodge Ski Area. NM. 233
Sitzmark Lodge/Sitzmark Guest Ranch. CO. 83
Skagit Motel. WA. 339
Skier's Lodge. UT. 314
Skytop Club. PA. 284
Smuggler's Notch Ski and Tennis Resort. VT. 317
Snow Lake Lodge. VT. 320
Snowbird Mountain Lodge. NC. 257
Snowbird Ski and Summer Resort. UT. 309
Snowblaze Condominiums. WA. 338
Snowshoe. WV. 346
Snowy Owl Inn. NH. 224
Sojourner Inn. WY. 377
Sonesta Beach Hotel and Tennis Club. FL. 98
South Seas Plantation. FL. 106
Spalding Inn Club. NH. 225
Split Rock Lodge. PA. 284
Sportsman's Lodge. MN. 188
Spring Mill Inn. IN. 139
Springside Inn. NY. 246
Squaw Valley Lodge. CA. 52
Starfire Lodge. NM. 230
Starved Rock Lodge. IL. 136
Still Waters. AL. 17
Stone Barn Inn. MA. 173
Stowe Flake/Nordic Inn. VT. 323
Stowehof Inn. VT. 324
Stratton Mountain Inn. VT. 326
Sugar Bowl. CA. 52
Sugar Hills Resort. MN. 200
Sugar Loaf Lodge. FL. 96
Sugar Mountain Resort. NC. 257
Sugarbush Inn. VT. 326
Sugarloaf Inn. ME. 152
Summit Lodge. VT. 319
Sun Mountain Lodge. WA. 345
Sun Valley Resorts. ID. 132
Sunrise Lodge. WI. 363
Sunriver Lodge. OR. 272
Surfside Inn. WA. 337
The Swampfox Motor Inn. SC. 293
Sylvan Dale Ranch. CO. 80

Tahoe Marina Inn. CA. 52
Tall Timber. CO. 71
Tamarron. CO. 72
Tamiment Resort and Country Club. PA. 285
Tanglewood Park. NC. 261
Tanque Verde Ranch. AZ. 37
Tarry-A-While Resort on Highland Lake. ME. 151
Taughannock Farms Inn. NY. 247
Tejas Village. TX. 302
Telemark Lodge. WI. 353
Tennanah Lakeshore Lodge. NY. 244
Tennis Ranch of Taos. NM. 231
Thousand Islands Inn. NY. 251
Thunderbird Lodge. MN. 201
Thunderbird Lodge and Chalets. NM. 233
The Tides Inn and Lodge. VA. 331
The Tiki. TX. 307
Toftrees Country Club and Lodge. PA. 286
Tolovana Inn. OR. 278
Topnotch at Stowe. VT. 324
Toro Hills Hotel. LA. 150
Trapp Family Lodge. VT. 325
Tropicana Hotel and Country Club. NV. 217
Tumbling River Ranch. CO. 77
Turkey Run Inn. IN. 139
Twin Elm Guest Ranch. TX. 300
Twin Falls State Park. WV. 349
Tygart Lake State Park. WV. 349

Val Chatel. MN. 202
Valley Ranch. WY. 374
Van Kirk's Lake Breeze Resort. WI. 369
Van Valin's Island Lodge. AK. 23
Vee Bar Ranch. WY. 377
Ventana. CA. 46
The Village Green. OR. 273
The Village of Loon Mountain. NH. 221
Virginia State Parks. VA. 333

Wagon Wheel Resort. IL. 135
Wahweap Lodge and Marina. AZ. 311
Walt Disney World. FL. 109
The Wanderer. GA. 115
Watoga State Park. WV. 350
Watts Bar Resort. TN. 297
Waunita Hot Springs Ranch. CO. 78
Wauwinet House. MA. 174
Wawbeek Inn. NY. 238
Waynesville Country Club Inn. NC. 257
The Waystation, CA, 53
Weatherby's, ME, 152
Wentworth-by-the-Sea, NH, 223
West Virginia State Parks. WV. 347
The West Wind Inn. FL. 107
Western Hills Lodge. OK. 269
El Western Resort. MT. 213
Westhampton Racquet Club. NY. 249

Westward Look. AZ. 37
White Eagle Motel. WI. 359
The White Elephant. MA. 174
Whiteface Chalet. NY. 238
White Pine Lodge. WY. 378
White Stallion Ranch. AZ. 37
White Sun Guest Ranch. CA. 60
Wickenburg Tennis and Guest Ranch. AZ. 39
The Wigwam. AZ. 33
Wildcat Lodge. WI. 367
Wilderness Lodge. MO. 209
Wilderness Trails Ranch. CO. 67
Williams Lake Hotel. NY. 244
Will O' the Wisp Resort. MD. 160
Willow Point Villas. AL. 17
Willow Valley Farms. PA. 283
Wilson Lodge. WV. 350
Winter Clove. NY. 245
Wintergreen Resort. VA. 333
Wolf Laurel Resort. NC. 257
The Woodbine Villa. MI. 180
Woodbound Inn. NH. 220
Woodcreek Resort. TX. 307
The Woodstock Inn. VT. 327

Yachtsman Resort Inn. SC. 293
Yankee Motor Lodge. MA. 164